THE INGLORIOUS ARTS OF PEACE

The Inglorious Arts of Peace

Exhibitions in Canadian Society during the Nineteenth Century

E.A. HEAMAN

UNIVERSITY OF TORONTO PRESS
Toronto Buffalo London

Toronto Buffalo London
Printed in Canada

ISBN 0-8020-4272-4 (cloth)

Printed on acid-free paper

Canadian Cataloguing in Publication Data

Heaman, Elsbeth, 1964–
The inglorious arts of peace : exhibitions in Canadian society during the nineteenth century

Includes bibliographical references and index.
ISBN 0-8020-4272-4

1. Exhibitions – Canada – History – 19th century. I. Title.

T395.5.C3H42 1999 907.4'71 C98-931769-2

This book has been published with the help of a grant from the Humanities and Social Sciences Federation of Canada, using funds provided by the Social Sciences and Humanities Research Council of Canada.

University of Toronto Press acknowledges the financial assistance to its publishing program of the Canada Council for the Arts and the Ontario Arts Council.

Contents

Acknowledgments

It is with considerable gratitude and relief that I acknowledge the many people who assisted me in writing this book. Above all, I thank Michael Bliss, who supervised my research as a doctoral student. His encouragement throughout the long years together, and his thoughtful criticisms of my thesis, made its completion possible, but it is from his own example as a historian and teacher that I have learned the most. Paul Rutherford and Allan Greer also served on the thesis committee and offered lively debate and penetrating criticisms. Examiners Carolyn Strange and H.V. Nelles made some important suggestions that influenced the revisions. I would also like to thank Clifford Orwin, who kindly allowed me to attend his remarkable seminars in political philosophy. Suzanne Zeller's book *Inventing Canada* originally inspired this project, and her seminar on science in Victorian Canada confirmed my interest. At McGill University, John Zucchi supervised the initial research paper and encouraged me to continue. For five years George Weisz showed me how to do rigorous intellectual history; I wish I had been a better pupil.

The Social Sciences and Humanities Research Council of Canada supported my research with a doctoral fellowship, for which I am deeply grateful. The history department at the University of Toronto provided funding in the way of scholarships and teaching assistantships that permitted me to complete the dissertation, and it provided a congenial atmosphere in which to work.

Gerry Hallowell of University of Toronto Press invited me to submit the manuscript, and he, Emily Andrew, Rosemary Shipton, and Darlene Zeleney helped me to make it publishable. I was particularly fortunate in the choice of readers for the Press and the Humanities and Social Sciences Federation: Their comments have greatly strengthened the manuscript.

Bridgit Schroeder-Gudehus took an interest in the project and made some important suggestions. Paige Raibmon and Ruth Philips made comments on reading chapter 10, and the latter showed me her unpublished work. Donald Fyson, Valerie Korinek, and Keith Walden also commented on chapter 9. That chapter appears as an article in the December 1997 issue of the *Canadian Historical Review*, and I am grateful to the *Review* for permission to publish it here.

Librarians and archivists provided crucial assistance. At the National Archives of Canada, Patricia Kennedy explained the bureaucratic mind to me, blazing a trail through the maze of pre-Confederation documentation, and Martin Tétreault guided me through Record Group 17. In Quebec City, Renauld Lessard helped me with the Quebec government archives; in Toronto, Leon Wormski did the same for the Ontario archives; and in London, England, Anne Barrett introduced me to the archives at Imperial College. The Atwater Library allowed me to consult the Montreal Mechanics' Institute Papers, and I am also grateful to the Marquess of Lansdowne and Madame Bonar for permission to consult family papers. At the University of Toronto, Marie Korey opened up to me the Ruari McLean Collection at Massey College, which contains many splendid and beautiful books relating to the Great Exhibition; the staff at Robarts Library, and especially the Fisher Rare Books Room, Inter-Library Loan service, Microfilm Reading Room, and Government Documents, were helpful, efficient, and friendly. I am also grateful to librarians at the Salle Gagnon in Montreal and the Baldwin Room in Toronto.

The personal debts are too numerous to mention, but I am particularly grateful to Raina, William, and Karen Fyson, who housed me on research trips and took a cordial interest in my work. Donald Fyson and Sovita Chander have been colleagues and friends throughout, and I owe Don in particular a great deal for his practical help and intellectual prodding. Valerie Korinek and Robin Brownlie propped me up during the PhD program and beyond. My new colleagues David Edgerton, Rob Iliffe, Lara Marks, and Andy Warwick have sharpened my understanding of history and make Imperial College an exciting place to work.

This book is gratefully dedicated to my parents. My mother proofread the entire manuscript and my father helped me with its preparation. My debt to them is, of course, much more profound than that.

THE INGLORIOUS ARTS OF PEACE

Introduction

Exhibitions are not the stuff of traditional historiography. Long dismissed as too ephemeral and frivolous for serious historians, they were left to local historians who unearthed details about the dates of the first county shows, the prizes offered, and who ran the refreshment booths. Cultural theory has changed all that. After all, Victorians attributed great economic, political, and cultural importance to exhibitions and pleaded to be judged by these curious displays. Ponderously they weighed exhibitions down with ideological and historical burdens that cultural theorists now vigorously debunk. Unfortunately, too many theorists content themselves with examining the exhibition as a cross-section of past culture frozen in time. There has been little serious historical inquiry into exhibitions, understanding historical to mean the examination of cause and effect over time. It is my contention that the exhibition was not just a reflection of culture but an important historical event in its own right. I have sought insight from the cultural theorists who write about this suddenly fashionable topic, but my ambition is the traditional one of constructing a narrative to discover what happened. An examination of the origin and development of the exhibition permits a more penetrating analysis of how culture worked in Victorian Canada both as discourse and as praxis.

The exhibition began as an idea, one that reflected the social thinking of the day, but it was always designed to provoke action. The exhibition foisted a particular theory of social cohesion on an unsuspecting Canadian population that was bribed to adopt it. I wish to analyse the relationships between theory and practice, intention and outcome, policies and politics. Their consequences, however unintentional, were profound, for exhibitions helped to shape relations among social classes, geographic regions, and governors and governed. They helped to unleash such powerful social forces as advertising and popular culture.

Exhibitions were a small but revealing episode in the history of the Enlightenment and its 'bastard' offspring.[1] They began as an idea, one reflecting all the ferment of new ideas about society and human nature. There were theories of progress through economic development based on self-interest, and theories about the empirical basis of knowledge. The content of the idea was new, and so was the determination to woo the public with it. Religious rhetoric had existed for centuries, but political and especially economic attempts at mass persuasion were fundamentally new. Historians have examined the development of political propaganda, and even economic propaganda in the form of advertising, during the eighteenth and nineteenth centuries, but have neglected the appeal by social and political elites to the productive powers of the labouring classes during this period. Exhibitions were one form of this appeal.[2] They constituted and appealed to an economic public that underlay and supported the political public. The liberal public sphere, Jürgen Habermas has argued, comprised reason, property, and publicity.[3] The exhibition bolstered all three of these Enlightenment and Victorian ideals: it educated and publicly advertised market activities. Chapter 1, on the idea of the exhibition, explores the theories of knowledge, the economic progress, and the growing recognition of public opinion from which the theory of the exhibition was derived. Although many of these ideas were developed in Britain and France, they gradually and imperfectly traversed the Atlantic Ocean.

Chapter 2 examines early agricultural exhibitions in central Canada, contrasting the very different structures in what is now Quebec and Ontario. These two provinces shared a common legislature at mid-century, when most of the legislation to encourage agriculture was put into place. For this reason, and to keep the project a manageable size, the eastern and western provinces have not been examined. The two central provinces remained distinctive entities. Quebec had been handed down by conquest in 1763 with a majority French-Canadian population, though it was increasingly intermingled with English Canadians, who came to exercise disproportionate political influence. Basic political rights were soon accorded Catholics, but an assimilation of sorts remained the long-term goal. English constitutional forms and criminal justice, as well as English economic forms and farming practices, were thought to be urgently required in the new British province of Lower Canada. Upper Canada was created subsequently as a home for Loyalists fleeing the American republic. Colonial leaders there prided themselves in their conservatism, but many early settlers brought with them Yankee democratic sentiments that occasioned much conflict in the Legislative Assembly. Canadian Toryism was complicated by the fact that the conditions that supported conservatism in England – including a hereditary aristocracy and a state church – had to be actively implanted in Canada. Whereas English Toryism defied innovation, Canadian Toryism invited it. Exhibitions, by celebrating the landed squierarchy, seemed to bolster a traditional English social structure in Canada, but the very appeal to public opinion to join in this celebration was innovative and, in the long run, politically unsettling. Moreover, because Canada lacked a wealthy landed class to stage rural exhibitions, the provincial government had to help organize agricultural societies and subsidize local subscriptions. This support served further to politicize what, in England, was a private initiative. Rather than uniting different classes and political groups in the pursuit of wealth, the exhibition became a focus for rising political tensions during the 1820s and 1830s, and played a part in events leading to the rebellions of 1837–8.

Chapters 3 and 4 examine the system of agricultural and industrial exhibitions in central Canada from the 1840s to the 1890s. After the failed rebellions, the Canadas were united in 1841 to form one province under one legislature, with representation divided equally between the French section, Canada East, and the English section, Canada West. A shared bureaucracy was gradually established, including a minister of agriculture and boards of agriculture and of arts and manufactures to oversee a burgeoning system of exhibitions. This system was consolidated after Confederation in 1867, when control of agricultural societies and mechanics' institutes was devolved to provincial governments. Chapter 3 examines local exhibitions in counties and townships as political institutions within the context of a liberalizing, increasingly democratic public sphere. Exhibitions were simultaneously elitist and democratic: Although the agricultural societies possessed elective executives responsible to members, the basic premise of exhibitions was that excellence trickled down from the best to the worst farmers. Even as they strengthened public opinion, exhibitions were designed to educate and undermine this opinion. This underlying tension was resolved differently in the two provinces.

Chapter 4 examines the large-scale provincial exhibitions that developed after mid-century. In both Ontario and Quebec, local jealousies decisively influenced the development of provincial exhibitions. However, in Ontario these regional jealousies fuelled the growth of ever-larger exhibitions, whereas in Quebec they fatally impeded the development of large-scale shows. This chapter also poses the question of who benefited economically from provincial exhibitions. The answer is provided in a survey of interest groups that formed around the exhibitions representing such industries as animal breeding and farm implements, as well as the developing business of improvement itself.

Chapter 5 gives closure to the section on agricultural exhibitions by illustrating the gradual devolution of historical agency. Exhibitions were founded by serious men of high purpose who determined to bring reason to bear on economic activity. Over the course of the century, exhibitions became increasingly frivolous, given over to commercialism and the carnivalesque. M.M. Bakhtin uses the term 'carnivalesque' to describe Rabelaisian wit, which, in its vulgarity and earthiness, subverted high-bourgeois seriousness.[4] Similarly, exhibitions came to symbolize the triumph of a vulgar popular culture that serious men of high purpose condemned. This chapter provides a detailed description of what visitors might see and hear on a tour of nineteenth-century Canadian exhibitions, from the rows of plump grains and cattle proudly displayed by organizers to the freak shows and drunken brawls that organizers failed to repress. The term 'exhibition culture' refers to the increasingly elaborate and eclectic forms of activity to be found on the fairground, but it also signifies an exhibition imperialism, whereby activities proper to the fair infiltrated the broader culture.

Part Two turns to the international exhibitions Canada attended between 1851 and 1900. Casting its splendid radiance over the second half of the century was the Great Exhibition of the Works of Industry of All Nations, held in London in 1851. Eric Hobsbawm wryly remarks: 'If Europe had still lived in the era of the baroque princes, it would have been filled with spectacular masques, processions and operas distributing allegorical representations of economic triumph and industrial progress at the feet of its rulers. In fact the triumphant world of capitalism had their equivalent. The era of its global victory was initiated and punctuated by giant new rituals of self-congratulation, the Great International Exhibitions.'[5] The glamour and success of the first of these spectacular gatherings does much to account for the growth of a Canadian exhibition-holding bureaucracy during the 1850s and the number of crystal palaces that suddenly sprouted in Canadian cities. However, narrative continuity has been sacrificed to thematic organization, for whereas the Canadian exhibitions vividly illuminate local political relationships, Canadian participation in the international exhibitions was more thoroughly mediated and controlled by government agencies that imposed meaning from above. However, as often as not they failed to achieve closure: Their message was lost on the audience or overshadowed by other meanings.

Canadian interest in the international exhibitions was largely restricted to self-advertising and profit-seeking. These gatherings provided an unequalled opportunity for a young colony to address the nations of the world in its quest for immigrants, markets, and status. Europeans rarely took notice of the British North American colonies, except to report on political crises and the trials of settlers. Imagine the excitement in British Columbia when, in 1861, it was invited to exhibit in London: 'An opportunity to represent British Columbia as she really is *before the assembled world* now presents itself to us, and it may be long before one so favorable again occurs. We have a good story to tell, and the ear of the world to listen to ... The old report of the cunning and designing Hudson's Bay Company that it is a bleak and inhospitable region, infested with wild beasts, mosquitoes, and savage Indians, and that it is totally unsuitable for the reception of a civilized population, remains in a great measure uncontradicted. Let us proclaim to those who in May next will be assembled from all nations that the report is a wanton slander cast with foul design upon one of Britannia's fairest daughters.'[6] The discourse was one of self-advertisement, but it was more than that. British Columbia was anxious to insert itself into the meta-narrative of progress and history, concepts that were deeply informed by the international exhibitions. Moreover, it was not simply a matter of packaging a pre-existing commodity, for little systematic information about British Columbia existed at all. Exhibitions were designed to generate knowledge that was useful to administrators and businessmen both at home and abroad. Canada emerged as a nation during what Henry James called 'the age of advertising,' and the international exhibitions were one of its earliest means of self-expression.

Part Two is divided into three chapters shaped roughly by chronology and geography. Chapter 6, a survey of exhibitions until 1867, addresses the obvious and inevitable question of Canadian national identity. Did the early exhibitions, where British North American provinces gazed at one another's displays with varying degrees of admiration, envy, and scorn, have any impact on the decision to form a confederation?

Chapters 7 and 8 further develop the relationship between exhibitions and Canadian development after Confederation. The question shifts from how exhibitions helped to shape Canada as a set of geographical relations among provinces to how they influenced or reflected Canada as a community of interests. Whereas exhibitions in Canada were organized by local communities that lobbied for government support, the government itself tended to seize the initiative in exhibiting overseas and had to prod a reluctant public to show some interest. Most people had little to gain from exhibiting overseas, whereas the government had a great deal to lose if

Canada showed badly, for it would bear the blame. The government worked to make Canada more exhibitable and to make it a showcase. The extent of its intervention and its effectiveness is considered in these two chapters, which address European and American international exhibitions.

Part Two is concerned with both the content and the form of the international displays. Here I raise questions about the epistemological status of the exhibit, in terms of its relationship to the thing exhibited, a relationship that was seen to deteriorate over the century. Like the agricultural fairs, international exhibitions changed considerably over the course of the century, becoming increasingly popular and frivolous, until the carnivalesque aspects drowned out the statements of authoritative truth that the exhibition tried to promulgate. At the same time, the very ideal of promulgating authoritative statements of truth about the world gradually proved unsound and impracticable. In sum, these three chapters address the effect of international exhibitions on Canadian unity, on Canadian institutions, and on exhibitions themselves.

Part Three studies exhibitions and subaltern identities. Exhibitions were devised by white men to consolidate and enhance the economic world of white men, be they farmers, mechanics, manufacturers, or merchants. Chapters 9 and 10 examine the ways in which exhibitions represented women and native peoples. Representation is considered in terms of material depiction and of providing exhibitors with a public voice. This part of the inquiry has been placed in a separate section because it requires a reading of exhibitions 'against the grain.' Postmodern theory suggests that categories of race and gender cannot be seamlessly inserted into mainstream historical narratives because they destabilize the concepts and categories that prop up this mainstream history. Certainly, reference to gender and postcolonial theory requires a rethinking of exhibitions. The system of exhibitions that developed in the nineteenth century was designed to illustrate relationships of superiority and inferiority, and this system was nowhere more applicable than to relationships of race and gender. Women and aboriginals were not considered full historical agents: white men were considered more intelligent, more energetic, and the prototype of the rational economic agent who propelled historical development. As a theoretical system, exhibitions could not accommodate these 'excluded others.' However, as chapters 9 and 10 suggest, women and aboriginals began to make use of exhibitions in ways that permitted them to act as historical agents in their own right. Women, long prohibited from public address, used exhibitions to advertise their economic agency. Aboriginals, by holding exhibitions on reserves, formed themselves into an economic public and claimed an autonomy that the federal government was not quite ready to accord them. Finally, the very concept of native culture destabilized the category of culture that underlay the system of exhibitions. It subverted the messages about historical progress and the hierarchy of civilizations that exhibitions were supposed to embody.

1

The Theory of the Exhibition: An Overview

Before tracing the history of exhibitions in Canada, it may be useful to identify what an exhibition was. Reduced to its distinctive abstract qualities, the exhibition was a competitive means of circulating knowledge and artifacts simultaneously. It resembled traditional mechanisms of economic circulation – the market and the fair – but was intended to be more explicitly educational. The exhibition was an idea that the heirs of the Enlightenment imposed on market activity. This introductory chapter discusses the ideological origins of the exhibition as an Enlightenment project to improve humanity, as well as some of its implications. It identifies the European origins of Canadian exhibitions as a set of ideas and practices developed by such theorists as Adam Smith and John Stuart Mill. Illustrations are drawn from the Canadian context to show that these ideas did, in fact, circulate among, and were adapted by, Canadians. This chapter does not address the historical development of exhibitions in their economic and social context, but limits itself to a discussion of their intellectual pedigree. It describes the way exhibitions were intended to function, while the chapters that follow examine how they actually did function. The discussion is divided into two sections, reflecting the two concepts that shaped discourse about the exhibition: competition and competitive emulation, and knowledge and education.

For the sake of clarity, exhibitions should first be distinguished from markets and fairs. Both are as old as society itself, and they developed in colonial Canada wherever a growing population could sustain them. Markets are gathering places for buyers and sellers which occur weekly or daily and serve a local population. Fairs are 'low-frequency commercial gatherings held at regularly spaced intervals and involving the distribution of merchandise not destined for consumption on the spot.'[1] During the Middle Ages, fairs sustained the trade in leather, wines, woollens, livestock, and other staple goods, but they declined about 1300 as more regular economic channels developed. Some remained important centres for trade into the nineteenth century, but many came to resemble circuses, dominated by sideshows, thieves, and rowdiness. In 1685 Sir Robert Southwell described London's Bartholomew Fair: 'The main importance of this fair is not so much to merchandize, and the supplying of what people really want, – but as a sort of Bacchanalia, to gratifie the multitude in their wandering and irregular thoughts.'[2] The state began to repress these gatherings and closed Bartholomew Fair in the mid-nineteenth century. Markets also declined as village stores emerged and the transport and packaging of foods improved.[3]

Markets and fairs soon developed in Canada. Montreal had only a small weekly farmers' market in the eighteenth century, but boasted several by the early nineteenth.[4] Newark got a weekly market in 1792 as one of the acts of the first lieutenant governor of Upper Canada, John Graves Simcoe; Kingston obtained one in 1801; and York, in 1814; by 1881 the province accommodated seventy-one markets.[5] In 1836 the Kingston *Chronicle* reported approvingly that the 'valuable European custom of holding Fairs semi-annually is fast gaining ground in this Province.'[6] That year, 300 head of cattle and over 1200 bushels of grain were sold at the Napanee fair. At the Perth fair in 1835, about twenty yoke of oxen were sold for £50 to £70 each, and a few cows for smaller prices. Many communities aspired to hold fairs, which were supposed to stimulate market growth, although in 1838 Inspector General John Macaulay remarked that they had nowhere 'given the expected stimulus to agriculture.'[7] To hold a fair, one had first to petition the Governor in Council, as Portage du Fort did in 1847: 'That a numerous and fast increasing Population have within a very few years sprung up in this and the neighbouring Townships on both sides of the Ottawa, subduing the Forest and Producing a large market.'[8] In 1856 the minister of agriculture reported that 'the applications for holding periodic Fairs are becoming more numerous, and several were authorized during the past year.' These numbers increased with the rise of the livestock trade in the 1860s.[9]

Exhibitions were also intended to encourage market activity. They differed from fairs only by the introduction of prizes awarded to participants who could show the finest cattle, grains, or homespun. Prizes were occasionally awarded at early fairs in Canada, but they became common only in the nineteenth century, after the government began to contribute to their cost.[10] Local historian Robert Sellar describes a successful exhibition-fair (the terms were used interchangeably) founded at Beauharnois in 1828:

> It was a meagre affair, of course, a few horses and more oxen; a group of cows and an odd sheep or two. Seventy two dollars was paid out by Mr. Norval in prizes and then the day was celebrated by a carouse. With the exception of 1830, when the funds would not permit, shows were held yearly thereafter, generally during the last week in September, and were of increasing interest ... The attendance grew more quickly than the exhibits, show-day coming to be regarded as the harvest holiday, and, acting thus, one report declares that the shows 'promote cordiality and goodwill, amalgamate persons of different races, and spread information ...' After a while, the shows were held on Prince Arthur Square, and began to take of the nature of fairs, the settlers coming in to buy and sell [and some buyers coming from a distance] with rolls of bills in the pockets of their long coats and leather breeches. The evening was given up to drinking, and the taverns and stores were scenes of disorder.[11]

Prize-giving, the innovation of the exhibition, had two lineages: the turf and the learned society, where savants and sporty aristocrats passed around rewards. Having long bred racehorses, the landed gentry turned to cattle and sheep, and began to dispense prizes for the finest specimens; at Smithfield cattle market in London, for example, a Smithfield club began to award prizes in 1798. Societies for cultivating knowledge flourished during the Age of Enlightenment, proffering prizes for essays and practical innovations in all fields, including agriculture.[12] The Royal Dublin Society began to distribute prizes in the mid-eighteenth century. In the 1790s the Bath and West England Society began to give prizes for new crops like roots and for planting trees, and it held a few cattle shows. The Highland and Agricultural Society, which replaced an earlier Edinburgh Society of Improvers, began to reward high yields and root crops in 1785. Most of the early prize-giving associations subsisted on private subscription. The English government finally developed an interest in improvement, and from 1793 until 1822 it paid £3000 annually to a Board of Agriculture in London, which Rosalind Mitchison describes as 'an interesting and unsuccessful muddle' of bureaucracy and voluntarism.[13] Its primary task was to collect and publish information, but, in 1821 and 1822, in a vain attempt to stave off collapse, it held exhibitions. Agricultural societies also appeared in France in the 1760s and in the United States in the 1780s.[14]

There were also industrial competitions. In the 1750s the London Society of Arts staged exhibitions to encourage English commerce and design.[15] This was an isolated series, but France began to hold regular exhibitions, beginning in 1792, when the revolutionary government inherited warehouses filled by state enterprises like Gobelins and decided that an exhibition might attract buyers. This exhibition became a biennial event, open to private exhibitors across France. These aggrandized fairs were supposedly didactic, as indicated by the term 'exposition,' which, like 'exhibition,' referred to the public examination of schoolchildren. Inventor Charles Babbage organized industrial exhibitions at meetings of the British Association for the Advancement of Sciences in the 1830s. At mid-century the London Society of Arts decided to hold a national exhibition, but, on the advice of the prince consort, opened it to foreign nations, and the first Great Exhibition was held.[16]

The exhibition was one of many Enlightenment ideas for rational community festivals. Jean-Jacques Rousseau devised competitions for fleetness of foot or maidenly modesty, and French revolutionaries came up with one festival after another devoted to abstractions like 'Reason.'[17] The exhibition was a festival devoted to economic development. As such, it had a particular appeal to Canadians. It celebrated the farmers' labours and their importance to the economic basis of the community. One Canadian described an exhibition in 1871 as 'a rational harvest home, when we, the proprietary tillers of the soil, owning our own fields, which most of us have reclaimed from a state of nature, meet together with our wives, our children, and shall I say our sweethearts, to compare the progress we have made in the sovereign art of agriculture, and learning lessons from the experience of one another; gathering hope from the successes of one another, take a new departure for future advancement.'[18] J.E. Hodgetts, surveying the exhibitions, along with the museums, schools, lectures, and libraries aimed at producing a rural utopia, mused: 'Here, surely, on the harsh Canadian soil, we see the last full-flowering of the Age of Enlightenment.'[19]

Canadians drew on European ideas as well as their own experiences in Canada in developing a system of agricultural and industrial exhibitions. The result was a hybrid system that underwent considerable alteration over the course of a century. Morris Berman's study of the Royal Institution, established in 1799 to promote agricultural improvement, has shown how its activities altered from one decade to the next, as membership and political ambitions changed between 1799 and 1844.[20] Exhibitions also evolved over the course of the nineteenth century, reflecting changes in the epistemology and political ambitions of their organizers as well as a long process of adaption to the social and economic conditions of the

Canadian provinces. Much of this book concerns this process of adaption, but this opening chapter identifies the Enlightenment tenets that originally gave it such a powerful impetus: progress through commercial competition, and social engineering through education.

Competition (Passion)

The South Brant Agricultural Society was repeating a truism when, in its annual report for 1867, it hailed 'commerce, the chief medium for diffusing knowledge and the arts – the great civilizer of the human race.'[21] The idea of progress through commerce involved a rehabilitation of human competitiveness. Life, said the seventeenth-century philosopher Thomas Hobbes, was a race without a finish line, in which 'continually to out-go the next before is *felicity*. And to forsake the course, is to *die*.'[22] Hobbes thought that people were essentially selfish, and that the natural state was one of war, of 'every man, against every man.'[23] The civil state differed from the state of nature in channelling naturally hostile impulses towards common self-preservation. Private self-interest and the public good were not in opposition but on a continuum, he wrote, in an attempt to ease the bloody civil strife convulsing England. In the next century, David Hume and Montesquieu argued that the pursuit of self-interest benefited society by restraining political passions. In the words of Samuel Johnson, 'There are few ways in which a man can be more innocently employed than in getting money.'[24] Bernard Mandeville and Adam Smith agreed, adding that a healthy self-interest lay at the basis of national prosperity. The invisible hand of the market reconciled competitive self-interest and, by the division of labour, transformed it into cooperation. By the nineteenth century, ambition was no longer something to be restrained or redirected, but something to excite and provoke. The report on the French Exposition of 1878, held after France's humiliation by Prussia, was uncompromising: 'Le XIXe siècle est par excellence le siècle de mouvement. Il n'est plus possible de s'arrêter. Le repos n'est plus permis, ni aux peuples, ni aux personnes. Il faut courir ou mourir.'[25] Canadians gravely intoned the same rhetoric throughout the century: 'Will Canada content herself with slowly jogging on, while the rest of the world is progressing with daily increasing speed? We trow not.'[26]

People had to be persuaded to compete in the race of life, for they seemed all too prone to jog slowly on. They subsisted, content to reproduce themselves when they might be growing wealthy. Exhibitions were designed to stimulate market competitiveness, as a Canadian agricultural journal explained:

> We take our beef, as we do the rest of our surplus produce, to the English market; which is, in fact, the world's market, where we meet the world's competition. To be able to carry off the best prizes we must produce the best articles. When a farmer competes for a prize at our Provincial Exhibition, or at a County or even a Township Cattle Show, the field of competition is comparatively narrow; but still he prepares for the friendly contest. He perhaps intends to exhibit an ox, the breeding and symmetry of which he regards as coming up very close to the line of perfection: but he knows a neighbour who intends to exhibit a similar animal: he therefore does all he can to add to the weight and increase the quality of his own; conscious, that without great care on his own part, his more active neighbour will carry off the prize. He is sure of being second, if not first, of ten competitors: his breed has been selected with care; he has paid the most scrupulous attention to the mode of feeding; and when the anxiously looked for day of exhibition arrives, he carries off the prize. If he had exercised less care, judgment, and industry, his neighbour would have left him behind in the race of competition. This is an illustration, on a small scale, of the preparation to be made, and the competition to be encountered, at the shambles of London and Liverpool. There he has to compete with beef fed in every conceivable way [upon high-protein oil cakes, oats, turnips, rather than upon grass, as in Canada]. He finds, in short, that he gets the very lowest market price for his beef: the high prices, which may be regarded in the light of premiums, are all carried off by others. But still he has no self-reproaches; he has sold his beef in Amherstburg, or London, or Hamilton, or Toronto, or Cobourg, or Kingston, or Montreal, and he gives himself no further trouble about it: he knows that he has *sold* it; but he does not know, and apparently he does not care, whether he has been able to obtain the highest market price. This general apathy must be overcome.[27]

To obtain a reasonable return on one's labour was 'apathy'; only the best price would do. Improved farmers were intolerant of the unimproved because low-grade products lowered the price for everyone. If much Canadian flour was lightweight or full of weeds, or Canadian beef was tough, then all Canadian farmers suffered from lower prices and reduced demand.

Exhibitions, in short, encouraged the creation of *homo economicus*, the rationally self-interested producer who sought to maximize economic gain.

The very attempt to construct this economic being amounted to an admission that he was unnatural and did not really exist. If the market had offered sufficient reward for effort, there would have been no need to resort to the artifice of exhibitions. Exhibitions aroused material self-interest by hitching it to a deeper human passion – the love of distinction.

It was a commonplace of the day that the desires of the imagination were insatiable, far outstripping those of the body. Mandeville developed, and Smith and Hume domesticated, the idea that love of renown was the deepest human motive. 'To what purpose,' Smith had inquired, 'is all the toil and bustle of this world? What is the end of avarice and ambition, of the pursuit of wealth, of power, and preheminence? Is it to supply the necessities of nature? The wages of the meanest labourer can supply them.' No, he concluded, 'it is not ease or pleasure, but always honour, of one kind or another, though frequently an honour very ill understood, that the ambitious man really pursues.'[28] In 1830 the radical Canadian Reform leader William Lyon Mackenzie expressed similar ideas in a treatise on education:

> By what motives are we influenced in the society in which we move? – By the intense desire which we feel, of the *favourable regards* of mankind; Whatever are the trains of thought, whatever is the course of action, which most strongly recommends us to the favourable regards of those among whom we live, these we feel the strongest motive to cultivate and display ...
>
> What are the ordinary pursuits of wealth and of power, which kindle to such a height the ardour of mankind? – Not the mere love of eating and of drinking, nor all the physical objects put together, which wealth can purchase or power command. With these every man is in the long run satisfied. It is the easy command, which those advantages procure over the favourable regards of society, – *it is this which renders the desire of wealth unbounded, and gives it that irresistible influence which it possesses in directing the human mind.*[29]

Glory was a kind of power, and if the love of glory had hindered the peaceful and progressive arts in the past, it would now bolster those arts.

Exhibitions promised both gain and glory to competitors. There was the money prize (a handsome sum for the best bull, a more modest reward for the best pillowslip), but there was also a public ceremony, replete with patriotic music, speeches, and a priest or mayor or even a governor to distribute prizes before a cheering crowd, and then the honour of seeing one's name published. In 1852 William Hutton, secretary to the Canadian Bureau of Agriculture, argued that agricultural societies promoted progress not with vulgar money prizes, but by 'the very notoriety which the association of numbers does not fail to secure.'[30] Glory was crucial to the whole idea of the exhibition, and it rested on public approbation.

This attitude was a fairly radical reformulation of traditional ideas. It did away with the old-fashioned notions that gain and glory were mutually incompatible and that self-interest was a low human motive. Where Andrew Marvell's 'Ode to Cromwell' had rained contempt on 'the inglorious arts of peace,' the Victorians attached honour to work done well in any field, from breeding cattle to laying down manure. Thomas Shaw, a stockbreeder, professor at the Guelph Agricultural College, and editor of a livestock journal at the end of the nineteenth century, saw no difference between heroes of antiquity and those of Victorian Canada except in scale of reward: 'The Peloponnesians of a forgotten age treated their heroes better.'[31] The Canadian state distributed prizes to the best farmers and manufacturers, just as medieval monarchs had bestowed trophies and conquered lands on their greatest warriors. The love of glory was equated with a weaker love of gain, and this passion was declared an eminently social virtue.

This was a hybrid response to the age-old problem of competition among citizens for resources. Centuries earlier, Francis Bacon had attempted to transform this political problem into a technical one, resolved by wringing more wealth from nature to make resources less scarce. But when wealth, in the form of prizes and the higher prices these trophies brought, was equated with glory, it became, by definition, scarce. If glory is not restricted to the few, it loses all its value, so it must always be unequally distributed.[32] This limitation creates two problems. First, a society that encourages its pursuit runs the danger of encouraging domination, for glory, as Hobbes pointed out, is a power one exercises over others.[33] Second, those who are deprived of this power will be discontent. It was for this reason that Rousseau declared all those passions that lead us to compare ourselves with others a threat to order and happiness: 'Self-love, which regards only ourselves, is contented when our true needs are satisfied. But *amour-propre*, which makes comparisons, is never content, and never could be, because this sentiment, preferring ourselves to others, also demands others to prefer us to themselves, which is impossible ... what makes man essentially good is to have few needs and to compare himself little to others; what makes him essentially wicked is to have many needs and to depend very much on opinion.'[34] Rousseau wrote a book to show how man could be educated to be good and escape the tyranny of opinion and *amour-propre*. The Victorians eagerly studied Rousseau and then applied his theories backwards, trying to excite competition in all things.

Rousseau's solution was rejected because comparison and reliance on public opinion came to seem unavoidable. Hume argued that reason and morality were not intrinsic, but developed through experience. Smith developed this idea into a 'spectator' theory of morality. He advanced two arguments that ultimately provided a theoretical grounding for the exhibition as a form of social engineering (and Smith, like Hume, subscribed to the early Edinburgh improvement society).[35] First, he argued that man can only know 'of the propriety or demerit of his own sentiments and conduct, of the beauty and deformity of his own mind,' by looking into the 'mirror' of society and comparing his own ideas with those of others.[36] Rousseau, it followed, was wrong to think a person's own desires could be understood except through this process of comparison. Second, he argued that acceptable sentiments and conduct were those likely to impress an imaginary impartial spectator: for example, gratitude rather than ingratitude. In other words, citizens should consider themselves as displays at an exhibition, and their failure to do so was one of the great ills of society: 'If we saw ourselves in the light in which others see us, or in which they would see us if they knew all, a reformation would generally be unavoidable.'[37] Or, as his famous countryman apostrophized, 'O wad some power the giftie gie us / To see oursels as ithers see us!' The exhibition was designed to accomplish this goal: each farmer would step forward with his productions to be truly judged by his neighbours.

Perhaps most striking about Smith's spectator morality was the burden it placed on public opinion as an objective arbiter. Citizens performed their actions before the general public, whose reaction – whether applause or catcalls – should serve as a guide to the propriety of those actions. In fact, he recognized that there was a vast gap between the impartial spectator of the imagination and actual public opinion. Smith argued that few people possessed judgment enough to decide propriety for themselves and that most would rely on public opinion, but, he admitted, the public admired the wrong sorts of qualities. The 'most universal cause of the corruption of our moral sentiments' was, he argued, the 'disposition to admire and almost to worship the rich and the powerful, and to despise, or, at least, to neglect the persons of poor and mean condition.' The rich and powerful tended to squander wealth rather than investing it sensibly, and their idleness was no suitable example for the industrious classes.[38] The solution was, of course, to educate and direct public opinion.[39]

The education of public opinion was one of the great goals of the Victorians, and it reflected a feeling that the political and epistemological power of public opinion was manifestly growing. In 1820 British prime minister

Robert Peel asked a friend, 'Do you not think that the tone of England – of that great compound of folly, weakness, prejudice, wrong feeling, right feeling, obstinacy, and newspaper paragraphs, which is called public opinion – is more liberal – to use an odious but intelligible phrase, than the policy of the Government?' A scholar remarks that this was 'a factor of paramount importance, one which changed radically the character of British culture and political life.'[40] In the 1840s John Stuart Mill and Alexis de Tocqueville grappled with the 'tyranny of the majority' in politics and in culture. This tyranny was particularly oppressive in Canada, where a high level of property ownership created an unusually broad franchise. The Toronto *Leader* explained the value of mechanics' institutes in 1866: 'Class education was a very excusable thing when the work of government and the guidance of public opinion were supposed to be the privilege of a class; but the inevitable tendency of social and political power to the masses, the confusion and intermixture of ranks ... warns us that if we would preserve the State in its integrity, we must as liberally and as fast as we can educate to the highest point every member of the state.'[41] Earlier reformers like William Lyon Mackenzie complained that public opinion was too weak in Canada: 'The road to honour, power and preferment in the United States is public opinion, not the courted favours of "a mushroom aristocracy," and their captive governor.'[42] Conservatives believed that social elites should set the tone for public opinion, while the radicals believed that the mass of humanity should prevail. The nineteenth-century struggle for political power was also a struggle for the control of public opinion and culture.

Hume and Voltaire had been the first to argue that something broad and inchoate called 'culture' exercised as much influence over the course of history as the reasoned decisions of statesmen.[43] A century later, writing in the decade before the Great Exhibition, the prominent mid-Victorian thinker John Stuart Mill argued that the first cause of historical change was knowledge and opinion: 'Every considerable advance in material civilization has been preceded by an advance in knowledge: and when any great social change has come to pass, either in the way of gradual development or of sudden conflict it has had for its precursor a great change in the opinions and modes of thinking of society.'[44] Mill thought even philosophers must remain slaves to public opinion because they used words whose meaning was determined by common expression. The solution was to 'educate up' opinion to the standard of knowledge. Public opinion began to matter in a way it had never done before, because morality, politics, and prosperity rested on it. Exhibitions developed because elites realized that public opinion did not simply flow from the top down but also from the bottom up; that history was not shaped only by the reading public but also by the illiterate masses. They tried to reverse this flow by providing direction for it. Social elites, Tory and Reformer, in Canada as in Europe, created agricultural societies and mechanics' institutes where they could lead popular opinion, directing its attention to the scientific principles underlying breeding or boilers. Much of the information conveyed was practical and useful. Much of it was also saturated with class values.

There were two models of prize-giving: subscribers could either compete among themselves or distribute the premiums to worthy inferiors. Prizes might be a reward for unparalleled excellence or an encouragement for those who could not compete with the best – a Duke of Bedford or a Lady Pigott – but who had made some improvement. When the upper classes met in a learned society and lauded one another's scholarly contribution, they were largely amusing themselves. When they rewarded agricultural labourers for having planted forage crops or for having raised the most legitimate children without parish help, they were instructing the popular classes how to behave; in other words, they were practising cultural hegemony.

Exhibitions seemed objective in so far as they invited spectators to verify the truth with their own eyes, in true Baconian spirit, but they were not ideologically neutral. Canadian mechanics' institutes were run by social elites[45] who lectured on political and economic issues, but were chary of letting labourers do the same.[46] In the countryside, it was lamented that 'there is scarcely a Society established for the promoting of Agricultural Improvement, but what mainly owes its existence and support to the exertions and influence of merchants, gentry, and others who are not directly connected with Agricultural pursuits.'[47] In Toronto, it was candidly admitted that mechanics' institutes did the most good by teaching values, not facts: 'Whatever practical advantage may attend any special study, whatever necessity there may be for its pursuit, the great end, the highest object is the *moral issue won by engaging the working classes in the pursuit of knowledge.*'[48] Montrealers welcomed an institute because, irrespective of what the mechanics actually learned, 'the habit of study will produce the most beneficial results upon the moral principles of the mechanical community.'[49]

At the Canadian agricultural exhibition, advanced farmers who had adopted British crop rotations and had imported British livestock presented themselves as the standard to which all should aspire. Improved farming was a set of techniques, but also a set of values that amounted to a whole way of life, one dedicated to orderliness and profits. Exhibitions were intended to bolster hegemony, which Gramsci defined as 'the "spontaneous" consent given by the great masses of the population to the general direction imposed on social life by the dominant fundamental group.'[50] Exhibitions could approximate this goal by publicly bestowing honour on tokens or fragments of the model life, but agricultural reformers dreamed of 'model farm' competitions where the mass of farmers would study the model life as it was lived. 'Could our slovenly, untidy farmers (they are a tremendous host) be made to march in long array around about the buildings of Hillcrest and through them, they would immediately move away asking themselves what manner of men they were.'[51]

Improvers were always surprised by how much resistance they met from farmers. Some farmers were just obtuse, but they could be excited to curiosity about improvement at an exhibition. 'We can scarcely conceive of the dullest fellow who ever trod in "how-low" or sported "smock" walking through the alleys of an agricultural show without having some new thought awakened, or some slight desire to know what it was all about.'[52] In Nova Scotia, John Young's Agricola letters, which began to appear in the *Acadian Recorder* in 1818, advocated cattle shows as having 'a wonderful effect in stirring up the most generous and powerful passions' towards improvement.[53]

Some farmers resisted improvement because they were too proud and jealous. One writer commented in 1842: 'Men are generally prone to doubt every thing that is not presented to their own observation, and this is especially the case with regard to improvements which imply the mental superiority of others over ourselves.'[54] The president of the Upper Canada Agricultural Association lectured at the Provincial Exhibition of 1849:

> The naturally intelligent, who is necessarily also a reading and thinking man, will adopt of his own accord every feasible suggestion, whilst the other will feel indignant at the idea of receiving instruction from any source ... Such men should be cautiously dealt with; any attempt to force upon them an improved system is almost certain to fail. 'A horse loose in a pasture,' says an experienced agricultural writer in reference to this matter, 'can rarely be caught if you approach him swinging the bridle – the emblem of his subjugation – before his eyes; but if you go to him, shaking only the measure of oats before him, and concealing the bridle under your coat, you can generally expect to take him without difficulty.'[55]

French-Canadian farmers were also 'too wise to be taught': 'few read at all, and still fewer understand what they read; any expense imposed upon them for agricultural information, in which they believe themselves all sufficient, is at once rejected ... "men must be taught as if you taught them not."'[56] Exhibitions were the oats that hid the bridle, the school that did not seem to teach. This amusing festival lured the farmer towards improvement despite himself. It showed him magnificent cattle and horses that it was impossible not to covet and stirred his pride by showing how others prospered. 'L'homme ne voit jamais d'un oeil indifférent les signes réels d'une heureuse et croissante prospérité dans l'état de son compétiteur. Il en recherche et sonde les causes il en poursuit les moyens ... l'amour-propre survient et active puissamment ses efforts.'[57] Even the insouciant French-Canadian peasant, it seemed, could be converted from economically and socially irrational behaviour to proper market values. The Quebec Agricultural Society in 1820 congratulated itself that, by exhibitions held throughout the district, 'Agriculture and Agricultural Improvements have, at all those Assemblies of People, become the subject of conversation: excellence has become a title, not only to profit, but also of distinction, and consequently of ambition ... Even the objections, the jealousies and ridiculous fears which follow, will, in the end operate favorably.'[58]

Exhibition organizers did not use the term 'hegemony.' Instead they spoke, relentlessly, of the 'noble passion' of emulation. In his *Dictionary* of 1755, Samuel Johnson defined 'to emulate' as 'to imitate with hope of equality or superior excellence.'[59] Emulation, according to Adam Smith, 'the anxious desire that we ourselves should excel, is founded in our admiration of the excellence of others.'[60] The improvers who gave copious speeches at exhibitions in Canada prefaced the term with the adjectives 'healthy,' 'noble,' or 'laudable.' They interpreted class and ethnic hostility as a sign of admiration and eventual imitation. This attitude presumed a commonality of interest and of experience that was probably more convincing in Ontario than in Quebec, for lack of a traditional distinction between land-owning seigneurs and tenant-farmer habitants (which, however true in fact, did provide a vocabulary for making qualitative distinctions between these two groups). It was also more convincing in agriculture than in industry, because rich and poor farmers had more in common than did capitalists and workers. The presumed commonality on which hegemony rests was new, for there had been few previous attempts to open channels of communication among the estates, other than in organized religion. As Gramsci remarks, earlier ruling classes conceived of themselves as a closed caste; only the bourgeoisie encourages emulation, thinking itself 'capable of absorbing the entire society, assimilating it to its own cultural and economic level.'[61] A measure of this change is provided by the Canadian Tory educator and Anglican bishop John Strachan, who, in 1824, preached a sermon against envy and advised members of his congregation to compare themselves with the poor rather than the rich; thirty-four years later, on rereading the sermon, Strachan found it 'very inferior.'[62] The argument against envying one's betters had lost the ring of conviction. Exhibitions reveal that the elites consented to the principle of hegemony, but it remains to be seen if they were effective instruments of hegemony, bearing in mind that social elites spoke through exhibitions from a position of weakness rather than strength. Exhibitions were designed to woo a broadly based public opinion, using appealing stratagems. As Karl Marx remarked: 'Whomever one seeks to persuade, one acknowledges as master of the situation.'[63]

Knowledge (Reason)

Exhibitions, by acting on the passions, ploughed the field of the mind to prepare it for the seeds – knowledge itself. They offered a new kind of knowledge, appropriate for an unlettered, prejudiced audience that resisted formal instruction. Exhibitions were in the vanguard of popular education: State support for them preceded the development of a common school system in Canada. The kind of knowledge transmitted at an exhibition was easily grasped even by an illiterate *habitant* because it was made identical with reason itself, as reason was understood at the time.

The Enlightenment overturned old ideas of human reason as innate and of divine origin. In 1690 John Locke wrote in his account of human understanding that it was wholly empirical. He argued that simple sense impressions were the building blocks of all complex ideas. A swan was composed of the simple ideas of whiteness, long neck, red beak, etc. Locke believed that one obtained accurate knowledge of ideas 'by *comparing them one with another*, finding their Agreement, and Disagreement, and their several Relations and Habitudes.'[64] Here lies the idea behind exhibitions: that by directly comparing one object with another, laying them side by side, one could discover their relationships and their true nature. A cow is neither fat nor thin until it is placed by another cow and seen to be fatter, and, thus, preferred breeding stock. The possibilities did not end with comparing like objects, cows and cows, ploughs and ploughs. This sort of comparison was the goal of agricultural exhibitions, but industrial exhibitions had higher ambitions. Mill wrote in his 1843 *System of Logic* that 'the order of nature, as perceived at a first glance, presents at every instant a chaos followed by another chaos. We must decompose each chaos into single facts.'[65] At the Great Exhibition just eight years later, material artifacts from all around the world were exhibited as discrete facts, made intelligible by physical proximity to other objects of the same class. Another prominent philosopher of science, William Whewell, who described the Great Exhibition as a royal road to knowledge, declared its overarching classification to be a first step towards a 'permanent and generally accepted classification of all the material, instruments, and productions of human art and industry' by which the world could be reduced to 'a permanent order.'[66] The principle of association informed educational discourse during the nineteenth century and was applied to design, to make commodities a 'visible language for the internalization of a universally accepted social code.'[67]

Learning had been associated with scholarly treatises mastered after years of study by brilliant individuals, but now learning was what people did in everyday life, taking in information from the objects around them: 'All our knowledge is ultimately reduced to sensations that are approximately the same in all men.'[68] Two examples from the 1840s illustrate this change. William Evans, an Irishman who farmed at Lachine and thought exhibitions frivolous, insisted that 'the wisdom of Nature has confined the gift of intellectual ability to that proportion among mankind whom the public interest requires to be employed in intellectual pursuits.'[69] By contrast, Jesse Beaufort Hurlbert, a Methodist teacher and exhibition commissioner, told the Toronto Mechanics' Institute: 'All science, indeed, may be reduced to facts, and, therefore, every man whose organs of sensation are in a sound state, is capable of observing the elements of science. That one man excels another in the discovery of truth, is chiefly owing to his mind being more particularly directed to the contemplation of certain objects and relations.'[70] The exhibition was an attempt to make objects more intelligible, so that spectators would arrive at the right conclusions, and to accustom them to study objects more closely. The actual process of thought could be made to take place outside the brain, in the physical arrangement of objects, as in a mechanical computer.[71] Error was not possible, nor was mental resistance. A popular guide to the Great Exhibition of 1851 remarked: 'The most listless lounger in the Exhibition was there at school. Consciously or unconsciously, he received at every sense, lessons which cannot be altogether without effect in after life.' Geologist Henry Youle Hind described what Canada tried to do at the Great Exhibition: 'It is one thing to show specimens of inexhaustible supplies of minerals, forest, or agricultural products, but it is another to teach the eye and understanding at a glance the wide application and general usefulness of the raw materials.'[72]

Exhibitions had a direct path to the brain through the eyes. To the Victorians, vision was 'not only the most useful but in every way the most powerful of the bodily senses.'[73] In 1801 J.H. Pestalozzi published an influential pedagogical monograph in which he developed the concept of visual education, using 'object lessons' (pedagogical objects) to teach students through activity rather than with dry lectures.[74] He catapulted into education the model of knowing-by-doing that Bacon had imposed on scientists.[75] In a report to the Upper Canada House of Assembly in 1836, reformer Charles Duncombe vaunted the 'inductive system,' whereby 'children are taught facts from observation and the examination of natural substances, which are presented to as many of the student's senses as are accessible ... and natural science is taught by observation, not by the recollection of arbitrary names and almost incomprehensible descriptions of ideas obtained only by long and constant application from books, but by the easy and perfect natural channels of the senses.' The Legislative Council rejected his program of education, but in the 1840s the new superintendent of education, Egerton Ryerson, introduced many of its provisions.[76] Ryerson's *Journal of Education for Upper Canada* advocated visual education on the grounds that 'the impressions made upon the mind through the eye are more vivid and distinct, than those made through hearing, tasting, or smelling.'[77] John Langton, a member of the provincial parliament and vice-chancellor of the University of Toronto, was taught by Pestalozzi, and in 1854 lectured to the Peterborough Library Association on the need to give mechanics a scientific and practical education 'not from books alone,' but with lectures and a museum of philosophical apparatus.[78] David Boyle, an artisan who became a teacher and archaeologist in the late nineteenth century, was a Pestalozzian who advocated museums for the 'education of the masses' and exhibited minerals and archaeological specimens.[79]

Enlightenment ideas of knowledge and vision, which provided the theoretical basis for exhibitions, were common currency in Canada, the sort of topic a professor or clergyman might discuss at a learned society or a mechanics' institute.[80] For example, in 1846 Brown Chamberlin, a university student, defended Locke's theory that knowledge was acquired.[81] Chamberlin became a prominent journalist and member of the provincial assembly who did much to promote exhibitions in Canada: in 1856 he wrote legislation affecting mechanics' institutes; in 1860 he organized an industrial exhibition in Montreal; and in 1862 he accompanied the Canadian exhibit to London. Whewell's remarks on the Great Exhibition appeared in the *Canadian Journal of Industry, Science, and Art* in 1854, as did an article on the Crystal Palace, recently moved to Sydenham to serve as a permanent museum: 'The whole of the scheme now working to completion, known as the Crystal Palace, might be properly described as one vast and combined experiment of visual education; and I think it would be easy to show that its educational powers and design constitute its legitimate claims to the support of all civilized Europe.'[82] At the International Exhibition of 1876, held at Philadelphia, the Ontario school system exhibited the best collection of object lessons in the world.

The newfound emphasis on practical knowing reflected a distrust of words and an impulse to liberate knowledge from the vagaries of rhetoric. Since the creation of the Royal Society in the seventeenth century, science has sought authority in linguistic transparency.[83] The *Encyclopédie*, with more than 3000 lavish illustrations, conveyed direct knowledge of the arts, while 'placing the philosopher at a vantage point, so to speak, high above this vast labyrinth, whence he can perceive the principal sciences and the arts simultaneously.'[84] The exhibition was a three-dimensional encyclopedia that recreated the world, rather than verbally representing it. It replaced words with things, for, as Johnson remarked, 'words are the daughters of earth, but things are the sons of heaven.'[85]

Another explanation for the distrust of words is suggested by the work of M.M. Bakhtin, who argues that novels parody conventional texts and undermine their authority.[86] Charles Dickens was, to Bakhtin, the archetypal parodying novelist. In his *Household Words* appeared 'The Catalogue's Account of Itself,' in which the Exhibition Catalogue chatted gaily and irreverently to the public: 'I mean to flow out now into a tide of gossip; to pour into your ear, confidentially, a stream of information on the subject of my early life, and to unbend – if I may say so, to un-catalogue myself; to loosen myself from the accustomed bondage by which I am compelled to travel only on a certain path ... I shall begin by quoting from a high authority, namely myself.'[87] But another article compared the English and Chinese exhibits, juxtaposing machines, clocks, model ships, instruments, even art, to draw the conventional conclusion that here lay 'the comparison between Stoppage and Progress.'[88] Words could be debunked, but objects could not – at least until the advent of an André Breton or an Andy Warhol.

Modern scholars have become more distrustful of the 'empire of the eye.' A recent book by Martin Jay surveys French intellectuals who have denounced this most 'mystifiable' of the senses,[89] and new books appear almost monthly deconstructing 'visual culture.' Michel Foucault was foremost among these critics, recasting in a sinister light the 'Ring of Gyges' principle. Plato suggested that if a man were given a ring conferring invisibility, he would immediately perform the most vile actions he could imagine. Foucault argued that over the past two hundred years, society has been redesigned to increase the visibility and thus the orderliness of citizens' actions. The process began in institutions like the asylum, the hospital, and the prison, but was extended into everyday life, creating a society of 'surveillance.'[90] Exhibitions clearly worked on the principle that if men's actions were exposed to public scrutiny, those men would pursue activities that would bring glory and serve the community. At the same time, exhibitions permitted a certain democratization of visual power, which was distributed among the people rather than horded by the bureaucrats and experts Foucault studied.[91]

This democratization of visual power also throws into question the work of Guy DeBord. The Marxist DeBord argued that individuals become passive spectators before the spectacle of industrial capitalism, left to feel impotent at the exercise of vast impersonal forces. The society of the spectacle is, he argued, a society of 'separation, estrangement and nonparticipation.' Exhibitions provide evidence both for and against this theory, for they invited citizens to be both spectators and participants. Questions related to the politics of seeing will be examined later through particular examples.[92]

Suffice it here to remark that exhibitions were about power, knowledge, influence, and money. Their history sheds light on how these forces interacted. Enlightenment theorists discovered that every activity was, in some sense, a transaction involving, at least potentially, knowledge and money. These qualities were almost as ubiquitous as modern scholars now find relations of domination to be.[93] Simply staring at one's surroundings could be made an intellectual act and a profitable one: George Wilson, director of the Edinburgh Industrial Museum and Regius Professor of Technology, asked at mid-century: 'How many of the young men who visit foreign countries or the colonies, bent on commercial enterprise could tell gold from mica or pyrites, or diamonds from rock crystal, or platina ore from iron sand? ... How many of the youths in question could tell whether the exudation from a tree was gum, a sugar, a manna, a resin, a gum-resin, a camphor, a caoutchouc, or gutta-percha?'[94] Scientific exhibits, he asserted, would dispel this costly ignorance. Similarly, everything a farmer did could affect his yields, according to improvers who drew up extensive programs to occupy every minute of his every day. The farmer was to be trained to rational and profitable habits from which only sleep and death could relieve him. If every act was a reasoned choice with economic consequences, then all human operations could be understood as intellectual operations, the building blocks of a brave new world and the stuff of improving manuals, lectures, and exhibitions.

Exhibitions brought together the best and the worst of Enlightenment theories of human nature and history. On the one hand, their exponents encouraged rational and critical activity, to get people thinking for themselves about their own best interests and the means to realize them. They provided a material apparatus for improvement, by creating markets where farmers could sell their goods more easily, even in the backwoods, and they gave small amounts of public money first to farmers and, later, to mechanics. On the other hand, they believed all too naively in the benevolent effects of capitalism and publicity. Exhibitions did not merely cultivate reason and convey fact, but also disseminated the upper-class values of their organizers. They appealed to covetousness, jealousy, and materialism. They were designed to overcome class rivalries by pitting people of the same class, or the same trade, in competition against one another.

Above all, exhibitions were designed to make the invisible hand clearly visible. Participants symbolically re-enacted those two rituals of the market: selection and purchase. The marketplace supposedly rewarded people according to their deserts, distributing good prices to good farmers and bad prices to bad ones, but this process was simply not efficient enough. Exhibitions spelled out the rules of the game, denoted winners and losers, and demanded of the public that it enter wholeheartedly into the game and promise to obey the rules. In real life, people did not pay enough attention to the objects of study, did not think they had anything to learn from their betters, and were content to subsist rather than pursue the greatest possible rewards. Exhibitions promised to overcome these problems because they fought indolence with envy and made it an instrument of knowledge and virtue, glory and gain. At least, that was the theory of exhibitions. The practical result was somewhat different.

PART ONE

Exhibitions in Central Canada

2

The Early History of Exhibitions in Canada, 1789–1837

The period from the late eighteenth to the early nineteenth century was one of economic and social upheaval in Europe and in Canada. In Western Europe and especially in England, economic conditions changed significantly: 'For good or evil, Britain's countryside was being kneaded like dough.'[1] As population growth increased demand, yields rose: between 1750 and 1770, mean gross yield per acre was 18 bushels of wheat; by the years 1795 to 1800 the yield had risen to 21½ bushels; by 1810, to 23 bushels; and by 1850, to 28 bushels.[2] B.A. Holderness attributed this increase to open-field cultivation, new fertilizers, under-drainage, and the fact that weaker soils had gone out of wheat production; as he remarked, 'The achievement of the age was to bring average husbandry up to a standard approaching that of the best practitioners.'[2]

The Canadian economy was based on resource extraction – fur, timber, fish, and wheat. Montreal and Quebec were commercial entrepôts and accommodated some industrial activity. Upper Canada was still a backwoods in 1790, but towns gradually sprouted across the province to service the farmers, as a wheat staple slowly developed.[3] In Lower Canada, wheat exports actually declined in the early nineteenth century, to the distress of the Canadian commercial elite. The men who made fortunes in commerce and land sales occupied the highest ranks of society. They accrued official positions and privileges, and formed in each province a Tory compact that dominated the Executive Council and battled reformers in the Legislative Assembly.

Canadian society did not much resemble Britain, but the conservative elites tried to foster any resemblance there was. The governor of Upper Canada, Sir John Graves Simcoe, promised to make Upper Canada's constitution the image of Britain's. This goal required that he coax into existence the two classes considered to be the bulwark of British liberties: a landed gentry and a sturdy yeomanry. As much for political as for economic reasons, both backwoods farmers and French-Canadian peasants had to be remade on the image of the improving British farmer. Simultaneously, the wealthy elites of Canada tried to turn themselves into a gentry, with all the regalia and privileges of the British aristocracy. These ambitions dovetailed at the agricultural exhibition, where the gentry ostentatiously displayed its own fine livestock or root crops and rewarded tenant farmers who had improved their condition. Exhibitions could provide a would-be aristocracy with an unparalleled means to display its wealth and leadership, while, simultaneously, it enhanced its future prosperity and reassured itself of a stable social and economic order.[4]

Exhibitions were a useful tool for a class whose power was rooted in spectacle. For the traditional nobility, the conspicuous display of grandeur and wealth was an act of power rather than a private choice of lifestyle. Exhibitions provided an opportunity for self-display and pageantry, the natural forms of expression for nobility and wealth. Whether the elites who ran the early agricultural exhibitions personally paraded their horses and cattle up and down the fields, or merely orchestrated and judged competitions among their inferiors, they established their superiority through these acts. English author Edward Lytton Bulwer remarked in 1833 that members of the aristocracy added 'to the weight of property, and the glitter of station, the influence of a personal popularity when they mingled with the people at agricultural and county meetings.'[5] Paul Langford has argued that by promoting agricultural improvement, the nobility could be seen to exercise condescension and leadership.

Although spectacle was a traditional form of communication for the nobility, agricultural improvement and agricultural exhibitions were, according to British evidence, undertaken not by traditional aristocrats, but by men serving the interests of commerce. Langford mentions a few high Whigs who held agricultural fairs, such as the Duke of Bedford and Coke of Norfolk, but suggests that 'gentlemen farmers' – 'a new breed of tenants whose wealth had made them genteel, but who lacked the proprietorial standing of the squierarchy' – were the most natural practitioners of this form of display.[6] Peter Mandler also distinguishes between the Grand Whiggery – the very wealthy aristocrats at the pinnacle of society – and the younger and lesser gentry who made common cause with the rising mercantile elite. The grandees espoused political rather than economic reforms, and were amused when the gentry took up improvement: 'An odd note of earnest rusticity was introduced into the salons accustomed to worldliness, flash, and flair.'[7]

As spectacle passed into the hands of businessmen, its function changed. Jürgen Habermas and Michel Foucault have argued that the 'society of the spectacle' was on the decline during the nineteenth century.[8] Habermas described the traditional royal court as a theatrical spectacle performed by a few choice actors before the nation, which was essentially a passive audience. It was replaced, in the nineteenth century, with a more participatory public sphere, to which every citizen, at least in theory, enjoyed access. Similarly, Foucault described *ancien régime* justice as something performed before an audience, as, for instance, the spectacular execution of the regicide Damiens. He contrasted it with nineteenth-century justice, based on principles of institutional organization – principles of surveillance – which infused society as a whole. Spectacular justice had to be abandoned because audiences refused to imbibe passively the lesson that was meted out and, instead, turned public executions into riotous events. Both Foucault and Habermas chart a growing resistance among the popular and middle classes to passive observation of political and other forms of spectacle. For the same reason, the attempt to use the traditional medium of aristocratic display to promulgate the new message of economic rationality to a popular audience was unlikely to succeed.

Canada's would-be aristocracy certainly believed that exhibitions could help to secure recognition for their social pretensions. In 1841 one Upper Canadian editor wrote enviously of aristocratic condescension at agricultural shows in Britain: 'There you will witness immense crowds of very respectable people exhibiting the choicest productions of the land, the best cattle and the finest horses; and mingled see noble Lords, Dukes and Earls, striving with the entry of the land to be foremost in the exhibition of the choicest stock, or vying with their rival neighbors in rewarding the labors of industry, and in bestowing prizes on the fortunate winner.'[9] But three major problems occurred in trying to adapt this scenario to Canada. First, the Canadian elite was not a landed gentry but a mercantile group that enjoyed little contact with or interest in rural pursuits; moreover, it generally delegated the organization of improvement to intermediaries, or self-styled experts, who were not themselves very spectacular. Second, the medium of the spectacle was ill suited to this purpose. The founders of exhibitions in Canada wanted to use the spectacle to convey a message that the social order was rooted in industry. They wanted the spectacle to be about something – themselves – but, instead, it turned out to be about the spectacle itself. As exhibitions took on a life of their own, growing increasingly spectacular and vulgar, the note of high seriousness was banished and, with it, the social pretensions of the founders. The spectacle was an aesthetic experience to which the improvers introduced instrumentalism.

This is not to say that the self-display of the aristocracy was neutral; on the contrary, it was, itself, an act of power because the act of observation was one of deference. But improvers sought to introduce technical information, of both facts and values, into this relationship of identity which existed between power and display. This instructional goal inserted a gap between the medium and the message which had not existed before. The third and final problem concerned the recalcitrance of the audience. The elite's social pretensions were contested by the farmers, who, when they showed any interest at all, fought to control the agricultural societies and to exhibit their own accomplishments. They drew strength from a wide range of popular and spectacular amusements that increased during the nineteenth century.[10] These contradictions worked themselves out gradually and unevenly over the course of the century. This chapter surveys the conditions under which agricultural exhibitions were organized in central Canada from the 1790s to the 1830s, and assesses them as a medium of social communication.

By the end of the eighteenth century it was clear that one of the things progressive-minded elites did was sponsor agricultural societies. '"Improvement" had proved some of its points, had even become fashionable and patriotic.'[11] The governors of Upper and Lower Canada, Simcoe and Lord Dorchester, had their paths cut out for them. In both provinces, agriculture seemed to need encouragement. Rough-and-ready pioneer agriculture in Upper Canada could, it was thought, benefit from advanced British techniques, as could the traditional methods practised in Lower Canada. French farmers were thought to be slovenly ever since agronomist Arthur Young branded them so on the eve of the revolution in France.[12] North American farmers practised extensive farming suited to a country where labour was expensive, land cheap, and only wheat found a ready market, but it was frowned upon by those who were accustomed to British standards of intensive farming, applying manure and rotating grain and forage crops.

Both Dorchester and Simcoe sponsored agricultural societies, but they did so, not through the apparatus of the state, but through private patronage and their own personal influence. In Quebec City on 6 January 1789, at Dorchester's instigation, 'the rank and fashion, nobility and clergy of all denominations, as well as commoners, crowded at the Chateau St. Louis to enter their names as subscribers to the Quebec Agricultural Society.'[13] The sixteen directors (divided equally between English and French) ordered oats, barley, peas, and hemp seeds from England, distributed fruit saplings and sheep, and published papers, which were mainly directions for steeping grains to prevent rust.[14] The society also sought laws to prevent owners from allowing their animals to roam and tried to convince the bishop of Quebec to cancel the religious holidays, which were, it calculated, costing the province £100,000 annually. It planned to award premiums when finances permitted, but finances never permitted. Subscriptions of one guinea dropped from fifty-nine in 1791 to twelve in 1792. Secretary Hugh Finlay, himself a merchant, politician, and seigneur, complained that the gentry lost interest after Dorchester left and they turned to electioneering. Moreover, the *habitants* remained aloof. The curé of Yamaska found no takers for hemp seed, and Curé Fortin of Lotbinière explained why. If the *habitants* sowed wheat and raised more than they needed, they were assured of selling it; moreover, they needed the straw as fodder. If they reduced their grain production to grow hemp and linen, and there was a shortage, as frequently happened, then wheat would be an exorbitant price.[15] Later offers of hemp bounties by the London Society of Arts and the Canadian government had no more effect, for production was labour intensive. In 1795, amid the worst harvests of the decade, the society folded. Not least to blame was a new hostility between English and French, after the Terror in France made Britons suspicious of all things French. Finlay himself wanted to 'make the people entirely English' by legislating the use of English in schools and courts.[16]

The Upper Canada society was hardly more successful. Before he left England, Simcoe had told Joseph Banks, the president of the Royal Society of London and a patron of agricultural and industrial improvement, of his plans to encourage agriculture and the arts. By 1793 an Agricultural Society had been formed at Newark (Niagara), to which he subscribed ten guineas and donated Young's agricultural works.[17] The society was a social club, and members dined together each month and discussed the Hessian fly, smut, fertilizers, and breeding. It also imported varieties of fruit and held a fair in Queenston in 1797, offering prizes for the greatest quantity of maple sugar produced in one year and the largest amount of wheat raised on one acre. In 1805 it disbanded and donated fifty books to the Niagara Library.[18] An Upper Canada and York Agricultural Society existed in York in 1806–7.[19] Fairs continued to be held at York and Niagara, along with informal cattle shows, but agricultural societies disappeared for several years. They had failed to reach the common farmers, and the country could not support social clubs of this sort, either because there were too few farmer-intellectuals or because these societies were too far apart in a pioneer country where communication was difficult.

Not until the end of the 1810s did agricultural societies reappear. In 1818 the provincial governments of Nova Scotia and Lower Canada began to provide funds for their support; in 1820 New Brunswick did the same, as did Prince Edward Island in 1827 and, finally, Upper Canada in 1830. Graeme Wynn describes the early Nova Scotia efforts, which followed publication of the Agricola letters, but he finds them largely a failure, unable to attract more than a few members.[20]

Lower Canada

In Lower Canada, support for agricultural societies developed out of a mounting concern about the condition of agriculture. It was widely believed that French-Canadian farming practices were antiquated and inefficient; that, by sowing wheat crop after wheat crop, *habitants* depleted the soil of its minerals, with the result that yields, and export sales, dropped. Scholars dispute the accuracy of this perception: The decline in exports may be explained because the land was exhausted by outdated methods,[21] because the *habitants* chose to grow other crops,[22] or because home consumption increased.[23] Marvin McInnis has argued that French-Canadian farming practices were not so far behind the English as is generally thought, but his work has been seen to rest on some questionable assumptions.[24] There is also some suggestion that 'crisis' is the wrong way to describe a long-term restructuring of agriculture.[25] Still, merchants showed an understandable concern at the shrinking exports. If industry is limited by demand, agriculture is limited by supply,[26] and it was to the unit of production, the farmer, that the province's leading men turned their attention.

In 1815 frosts and the Hessian fly ravaged the crops in several regions of Lower Canada.[27] When it met in January 1816, the Legislative Assembly was bombarded with petitions of distress; it named a committee to inquire into agriculture, 'to examine its progress or decline in this Province, and the means of encouraging the same.' The committee reported that agriculture was in a bad way, and it made two major recommendations: a Board of Agriculture should be established to perform experiments and distribute prizes; and laws should be made to prevent owners from allowing their animals to stray, eradicate thistles, and stop people from being so litigious, for time spent in courts was time lost to agriculture.[28] That year, harvests failed south of Quebec City. When the assembly met, debates raged as to whether the government should aid the starving by charity, loans, or an embargo on exports. The 'science' of political economy dictated that charity would only drive up prices, injure other regions by taxation, and hinder recovery.[29] Some relief was provided, but the new governor,

Sir John Sherbrooke, recommended in his Throne Speech that the legislature eradicate the first causes of distress: 'The failure of last year's harvest is, I find, generally attributed to an unpropitious season – It may however be a subject worthy of investigation whether other circumstances may not have had a share in producing the present scarcity, and if any such causes exist, to consider how far it may be in the power of the Legislature to prevent their future influence.'[30] Montreal writer Charles Grece, who had come to Canada in 1805 as an expert on hemp, similarly advocated that improvement be subsidized: 'It is very probable that premiums distributed in the parishes to promote improvement, for the finest field of wheat under particular management, for root crops to feed and fatten stock, and also for the rearing and fattening of every kind of animals; might create emulation of the country people.'[31] In 1817 agricultural societies formed in Quebec and Montreal, and they petitioned the assembly for aid.[32] The result was a law, passed in 1818, which granted £2000 in total: £800 each to the districts of Montreal and Quebec, and £400 to Three Rivers; later, Gaspé was added.[33] The societies were 'to offer rewards or premiums' up to £20 each for useful grains and vegetables, improved implements, and for 'good and strong' cattle, genuine Canadian horses, and sheep and hogs. Later acts extended the provisions, regularized the funding and accounting, approved purchases of books, seed grain, and models of implements (1821), named clergymen and legislators as honorary members (1829), and required that militia captains supervise meetings (1834).

The Lower Canada societies bustled with activity. In 1820, for example, the Quebec Society spent £450 to distribute seed grain and implements to one thousand persons. The societies reported on the state of agriculture and demanded better roads, education, agricultural journals, and model farms. They tried to sell the *Letters of Agricola*, but there were few purchasers.[34] Above all, they held exhibitions. There were local shows in each county, grand district exhibitions every fall, and fat stock shows for butcher meat in winter and spring. British Canadians excelled in the district shows, but in the counties, French Canadians won most of the prizes: three-quarters of all awarded in the Montreal District in 1819, and all but thirteen of the 450 prizes awarded in fifteen county shows in the Quebec District in 1820.[35] The Quebec shows drew 426 competitors and 2050 spectators in 1819, and 732 competitors and 4400 spectators in 1820. Most prizes went for animals: in 1820 there were 253 horses, 238 cattle, 39 sheep pens, and 31 swine, but there were few implements or grains, and the ground was too hard for ploughing matches.[36] By 1824 there were seven auxiliary societies in the Quebec District, dropping to four in 1827 but rising to eleven in 1831. According to the 1831 report, the local societies had died out when Quebec officers stopped attending the county shows, 'and it became evident that to derive assistance from these auxiliaries, it would be necessary to assume over them a parental control.'[37] By 1827 Montreal had two satellite societies. The Three Rivers Society supported an auxiliary in Sherbrooke, but by 1824 it complained that it lacked the funds to continue, though it recommended that exhibitions should be local: 'There being more of pageantry than real Advantage in District Exhibitions, where a few wealthy individuals are sure to carry off every premium, to the discouragement of the humbler practical farmer.'[38] In their first five years, the Montreal Society paid £2100 in prizes; Quebec, £3023; and Three Rivers, £700.

The Lower Canada legislature had determined that agriculture should be encouraged with public money by means of a system of exhibitions. In England, this goal took the form of direct interaction between two social classes, whereas, in Canada, it was mediated by the state. There were obvious practical reasons for this difference in approach: reformers argued that the absence of a hereditary, landed aristocracy in Canada made the government's intervention necessary. Only the government commanded sufficient wealth and authority to reach the public. There was also philosophical support, for Jeremy Bentham declared that 'the business of government is to promote the happiness of the society by punishing and rewarding.'[39] Punishment would not work, because the state could only regulate relations between individuals so as to prevent harm and was powerless against irrational acts performed quietly on the farm. It could only offer rewards to combat economically vicious behaviour. Inspection and grading, demanded by the boards of trades, penalized those who sold second-rate products, but exhibitions promised actual improvement. If the Canadian state took a greater interest in agricultural improvement than the British, this attention can be attributed to greater reliance on agriculture as a source of revenue and as a basis for general prosperity. Only by increasing yields could surplus value and, thus, taxable revenue be increased. The colony's perpetual shortage of cash made this point an important consideration. Writing in the 1830s, William Evans deplored the national loss incurred by subsistence farming: 'A farmer who systematically consumes annually the produce of his farm, without accumulating some part of the production in stock, in useful improvements, or in money, contributes nothing towards individual or national riches. If our inclinations and endeavors to produce are only equal to our disposition to consume, we can make no saving – no accumulation for settling our families –

no capital – no fund for the maintenance of labour.'[40] Evans feared that his adopted country might grow more like his native Ireland: unable to support a diversified economy and peopled by a subclass of impoverished Catholic peasants vulnerable to famine. State intervention, in the form of education, agricultural improvement, and public works, would prevent this fate, but intervention required a healthy basis of taxation.

The government sanctioned patterns of behaviour as norms to which everybody should aspire. It told the mass of farmers to behave like their social betters and paraded examples before their eyes. Simultaneously, it installed a quasi-administrative structure to make the activities of the population more visible to those who were governing. In short, with one stroke, the Canadian government installed not only Gramsci's hegemony but also Foucault's governmentality, based on panoptical principles of visibility. Foucault developed this concept to describe the managerial tendencies of modern government, understood as disposing of things and men, as opposed to the earlier paradigm of sovereignty, which consisted of simple obedience to law.[41] Foucault identified economic causes for this transformation in government, arguing that economic and political discipline were intertwined. He interpreted the drive to measure and control population as an exercise in wealth and power in eighteenth-century France.[42] In Canada, it was not the population per se, but the agricultural population, which needed to be managed more efficiently. In France, early epidemiologists began to count populations and to generate bureaucratic institutions, whereas, in Canada, the agricultural reformers, physiocrats all, led the effort to examine and to tabulate society. By these means, they can be said to have initiated governmentality.

Foucault used the term 'hegemony' from time to time, but his concept of communication differed from that of Gramsci. For Foucault, as Axel Honneth has argued, communication was wholly coercive,[43] whereas Gramsci's understanding of hegemony straddled the realms of consent and coercion. Gramsci elaborated a dichotomy between coercion and consent, instead aligning hegemony with consent, but he also called it an unreasoning deference and used it to explain why the Italian proletariat failed to respond to Marxism's powerful appeals to reason and interest. When it is contrasted with violence, hegemony stands for communication; but as a form of unreasoning deference, as Jackson Lears argues (in a Foucauldian vein), hegemony stands for the silences in discourse, the principles that are beyond discourse.[44]

However one defines hegemony, exhibitions were hegemonic because they were designed to appeal to reason and deference alike. Both, to the

Victorians, were exercises in judgment, and judgment must be wooed.[45] The improvers thought they were appealing to the free use of reason by opening channels of communication between French and English, bourgeois and *habitant*. By this standard, the state's attempt to popularize improvement was an attempt to generate rather than to impose shared standards of judgment. Of course communication could be coerced: Antoine Plamondon wrote to Evans's agricultural journal demanding laws requiring all farmers to subscribe to the journal and develop a taste for it.[46] It was not, he insisted, tyranny to force someone to make a profit. This suggestion was not considered seriously, but it does convey the urgency that improvers attached to communication.

However objective they may have seemed to their organizers, exhibitions were riddled with class biases. The standard remained that of the English farmer, and attempts to generate local standards were overruled. For example, the government specified that unpedigreed French-Canadian livestock, descended from the animals of the first settlers, should be awarded prizes, but agricultural societies objected to this category as perpetuating scrubby stock or encouraging the *habitants* to lie about their mixed-blood animals.[47] Moreover, the exhibitions instituted a monologue rather than a dialogue because the governing classes controlled the apparatus of communication by which the ideology of improvement was promulgated – the popular press and platforms at exhibitions. Canadian newspapers disagreed incessantly with one another about political policies, but they agreed, almost without exception, on the need to popularize knowledge and values appropriate to market behaviour. This uniformity made it difficult for the targeted audience to air any disagreement publicly. However, other forms of resistance were possible, including non-participation or appropriation of the agricultural society for other ends. Indeed, although exhibitions may have been intended to overcome class and ethnic hostilities and to secure economic assimilation, the practical effect was just the opposite.

For two decades, from the first vote to the rebellions of 1887–8, the agricultural societies waxed and waned. They grew increasingly dependent on government funds: Montrealers subscribed only £3 in 1828. But public funding was unreliable, as the governor and the Legislative Assembly fought over who was to pay the expenses of government. The assembly refused to vote funds for the civil list unless it had greater say over how the money was spent, and the governor, refusing to concede, chose, instead, to scrape by on other sources of income he controlled, such as military funds.[48] In 1819, when the Duke of Richmond advanced funds to the societies, the Colonial Office reprimanded him for having missed a rare opportunity to act in concert with the assembly: 'It would have been far more satisfactory to have learnt that the Legislature had made provision for a service so strictly Provincial as the improvement of the Colonial Agriculture, and upon the expediency of encouraging which it was to be applauded that no difference of opinion could exist.'[49] Richmond died of rabies before this message arrived, but his successors were more circumspect. In 1820 the Montreal Society cancelled its fall show for lack of funds, after a *coup* by grain merchant and politician Thomas Porteous, who had packed a meeting with family members, weakened the society's feeble popularity. The *Gazette* complained: 'What a miserable picture of the Agricultural interest in the District of Montreal, when the nomination of its officers is left to a dozen of members, and the half of these form a *family compact.*'[50] In 1824 the society collapsed and did not recover for two years, while an informal Farmers' Club held small shows.[51] Lord Dalhousie, governor of Lower Canada between 1820 and 1828, was an enthusiast of agricultural societies, having encouraged one in Halifax, but his hands were tied by political opposition, as he told a querulous Quebec Society in 1827: 'It is with great regret that I must candidly declare to you, that so long as those causes continue to paralise the Executive Government, so long will the Agricultural Societies, & all such public Institutions, feel the unhappy consequences.'[52] The Three Rivers Society, folded the following year. There was also, very briefly, a Gaspé Society, formed in 1823, which sought £100 that year to purchase seed grain; after receiving the money, its president, James Shearer, decided that 'all prospect of a Society of that nature was hopeless, in fact, the Country was too much in a State of Nature, to anticipate any thing of the kind for several years to come' – and he spent the money on roads. Agriculture in the Gaspé at that time was 'embryonic,' subservient to the fisheries.[53]

Bad harvests and deteriorating political relations injured the agricultural societies. The Quebec District was particularly hard hit by the failure of crops in the early 1830s,[54] and the exhibition became a 'beggarly affair of empty boxes.' The Agricultural Society president was John Neilson, a publisher who had supported the Patriote party in the past but who later rejected its increasing radicalism. As a result, the society dwindled to a handful of English members and, by 1835, was 'at best in a state of abeyance.' Neilson donated its library to the Literary and Historical Society.[55] In Montreal District, the lack of fodder for livestock resulted in spring fat stock shows where the animals could almost be counted on one hand.[56] Much importance was attached to these shows because farmers, it was said, starved their animals in winter so that, come spring, the animals could be lifted by their tails. But high politics among the directors of the Montreal Society also discouraged entries. At first, both the Quebec and the Montreal societies excluded directors from competition: They might receive honourable mention, but the cash went to poorer farmers. The law encouraged this outcome by specifying that prizes should be awarded to native breeds. The disparity between rich (usually English) farmers breeding animals imported from England and the French-Canadian farmers with their native animals was particularly striking around Montreal. Canadian horses were prized as racing and working animals across North America, but the cattle, although hardy and good milkers, looked scrubby compared with imported Durhams or Herefords. The temptation to reward British standards of excellence was too great, and the Montreal Society began to award prizes to directors during the 1820s. From 1826 (when the society was re-established) to 1828, only 25 of 279 prizes awarded went to French Canadians.[57] The society then organized separate competitions for French and English breeders, and other societies soon followed suit, as Lord Durham noted with disgust in his report, seeing it as yet another lost opportunity for contact between the 'races.'[58]

In 1832, another attempt was made to reform the Montreal Society. The French press attacked it as subsidizing English farmers, and secretary William Evans reported that the directors had lost the public's confidence and should forgo prizes. The society's president, Louis-Joseph Papineau, leader of the Patriote party and Speaker of the Legislative Assembly, adopted this proposal. But at the exhibition that fall the top breeders stayed away, so although the money was disseminated, the display was poor.[59] Papineau and Evans were voted out of office, and the rule was overturned. Evans wrote angry letters to Patriote newspapers, while Ludger Duvernay, editor of *La Minerve*, rained invective on English-Canadian 'machiavélisme' for the backwardness of habitant farming.[60] By 1834 separation was complete: Papineau and Evans controlled the Montreal County Society, leaving the District Society to the British.

By the 1830s Britons dominated most of the county societies in the Montreal District. Those in Stanstead, Shefford, Missisquoi, and Rouville were almost wholly English, though, in 1836, Captain Luc Fortin had the best farm in Rouville.[61] In Two Mountains, in 1834, French Canadians won one-tenth of the prize money.[62] In Terrebonne, Beauharnois, and L'Acadie, Britons formed a smaller majority. French Canadians made up one-quarter of the Beauharnois members and took home one-quarter of the prizes. At L'Acadie, French-Canadian membership rose during the 1830s from 41 per cent to 49 per cent, but dropped to 42 per cent after the rebellions. In Chambly and Laprairie, French Canadians ran the society and, in Laprairie in 1831, won three-fifths of the prizes.

Most of these societies survived the troubles and received a modicum of funding from the Governor in Special Council.[63] Membership at the L'Acadie Society, which had been 80 in 1834, rose to 119 in 1837 and fell to 104 in 1838; Missisquoi rose from 86 to 89, while Rouville dropped from 92 to 61 over the same two years.[64] (These membership figures pale before total population figures in 1831: L'Acadie, 11,419; Missisquoi, 8801; Rouville, 18,115. Shefford had 133 members between 1834 and 1836,[65] of a county population of 5087.) At L'Acadie, French and some English left the society after 1837, but other French Canadians joined. In Beauharnois the proportion of French members shrank, for, in 1845,[66] 47 of 343 members were French Canadian. The society held exhibitions in seven parishes, six of them English Canadian, and also sponsored a county show: English farmers won 203 prizes to the French Canadians' 42. But it is hard to believe the Beauharnois Society ever provided a congenial atmosphere for French-Canadian members. One of its organizers, Lawrence G. Brown, an agent for the Seigneur Edward Ellice, was rabidly anti-French. In 1830 he described *habitants* as degraded specimens of humanity, 'an almost reduction to mere brute capacity.'[67] The Missisquoi Society was also a bastion of British loyalism, according to one Tory paper's account of the county show held on the eve of the Rebellion:

> O ye city gentlemen why did ye not leave your Day Books and your Ledgers, and your traffic and come out to breathe the fresh air of the country, last Thursday? You would have seen a large meeting of men well fed, well clothed, well mounted, well pleased with one another, without a word to assail your ears about politics, or grievances or Papineau, or any such stuff, but ready to sing every one
>
> 'My Wheat, my Potatoes, and Oats
> My Horse that I ride and my Ox,
> My Beef, Butter, Mutton and Pork
> Are all from the ground which I work.'
>
> In every direction were waggon loads of rich cake and pies and sweet apples; and before it was dark, our brave and loyal yeomanry contentedly and proudly measured their way home on their prancing steeds, or easy gigs, or in their swift-going waggons, over the smooth, hard roads.[68]

Upper Canada

In Upper Canada, exhibitions developed more slowly. In 1818 the Upper Canada Agricultural Society was formed at York 'to promote competition and emulation to excel in the various branches of Agricultural pursuits, improvements & productions' by awarding 'premiums, prize medals, or other pecuniary or honourary marks of distinction.'[69] Membership was ten shillings, but one guinea for directors. Rev. John Strachan, a prominent member of the Tory Family Compact, helped found the society. He had earlier opposed Robert Gourlay, a Scottish immigrant and would-be improver, who had been thrown out of the Bath and Wiltshire Society in England for trying to organize the tenant farmers and sue his landlord, the Duke of Somerset, and who now denounced the 'empirical pretensions and coxcombical exhibitions' of agricultural societies.[70] Gourlay was impressed by the rural tours of English radical William Cobbett and the work of the Board of Agriculture, and he decided to produce information about Upper Canada that would be useful to farmers and immigrants. He had the temerity to organize meetings of farmers around the countryside and to circulate a series of questions, most of which were harmless except for the last: 'What, in your opinion, retards the Improvement of your Township in particular, or the Province in general; and what would most contribute to the Same?' Many respondents blamed land speculation among the province's governing class, and, as criticism mounted, Gourlay was first thrown into prison, then banished. Strachan led the fight against him, expressing horror at the 'impudence of a stranger calling public meetings.'[71] Thus, while the lower provinces funded societies with public money, Upper Canada did not.

The Upper Canada Society held its first exhibition in 1820, when 35 guineas ($147) were distributed among eleven competitors. The Midland District Society had its first show the previous autumn, distributing $147 among a score of competitors, and another $112 in 1820 to an entirely different slate of competitors. By 1822, however, the Kingston *Chronicle* was inquiring what had become of it. Branch societies picked up the slack, holding shows in Addington, Hastings, and Frontenac counties in 1821.[72] The Addington reporter rejoiced at the 'large concourse of people who assembled,' and that 'the number and appearance of the articles exhibited were favourable indications of an emulative spirit among the farmers, and were reputable to the Country.'[73] Only the Frontenac Society survived the initial enthusiasm: In 1825 it awarded $175 in prizes for livestock, ploughing, crops, and domestic work.[74] Shows were held in York, Cobourg, and

Port Hope during this period. A few years later, societies were formed in Lanark County and Northumberland County; the latter, with 164 members, held its first exhibition in 1829.[75] That same year a society in the Western District, with the 'liberal' patronage of Colonel Talbot, awarded $100 in prizes.[76]

These farmers' groups finally drew the notice of the House of Assembly. In January 1830 Charles Fothergill, a moderate reformer and enthusiast of scientific projects, proposed funding agricultural societies to a total of £100 for each district that raised £50 among subscribers. The resolution passed almost unopposed, although Reform leaders W.W. Baldwin and Jesse Ketchum worried that the country was still too primitive to maintain the societies.[77] Fothergill's biographer sees the motion as proof of his 'inability to understand the rank-and-file farmers he lived among in Durham County,' since the societies 'merely acted as meeting-places for gentlemen-farmers who already knew about modern practices.'[78] This assessment seems overly harsh. First, it was thought that if the best farmers – the vanguard – traded experiences, their knowledge would trickle down to the poor ones among them. Second, patronage, not efficiency, motivated the vote.[79] Here was an opportunity to buy the good will of the farming population, at a time when most voters were farmers. The society was sufficiently physiocratic to believe that wealth came from the land and that merchants and politicians were parasitic upon farmers. The Reform press decried the Upper Canadian leadership as 'a motley generation of upstart would-be gentlemen to set us the fashions – a race alike despised by and despising the great body of the Canadian people; rioting in luxuries this infant province can ill afford to spare them and possessing as it were the soil and its serfs on a rack-rent lease seemingly neither knowing nor caring who shall follow, in their train after them.'[80] The bill would send money flowing the other way. As James Wilson said during the debate: 'Let the Legislature encourage the farmers, who have to support the many thousands that do not labour.'

Tories and Reformers could agree on the resolution, though the Tories voted down William Lyon Mackenzie's call for payment by population rather than by district. But a suggestion that the assembly should set the rules was rejected because 'the prosperity of all societies and institutions depended on their being established on liberal principles, and if we wished the agricultural societies to be successful, they must be untrammelled.' The law did not save Fothergill's career – he lost his seat in 1830 – but it was welcomed by the ten societies, most of them hastily formed, that applied for the grant before the end of the year in the counties of Leeds, Northumberland, Durham, Grenville, in Talbot for the Western District, and in the Ottawa, Niagara, Home, and Midland districts.[81] Some quickly disappeared again, but others took their place: In 1831 Prince Edward County and Gore District applied. A county society claiming to represent the whole district could seize the entire £100, so there was much underhanded jockeying for the funds. In 1830 Fothergill almost lost the Durham money to Northumberland, which applied for the Newcastle District grant.[82] In 1832 the Carleton County Society claimed the funds for the Bathurst District: miller-politician Hamnett Pinhey, told treasurer A.J. Christie, a journalist and phoney doctor, to 'keep the affair snug – for if Morris or Lewis get scent of it – they will certainly catch the Bonus that we have our Eyes on – this would be *too bad.*'[83] This society had nine justices of the peace among its directors and a bevy of military men. A few years later it was overrun by Bytown raftsmen, the Shiners, who packed the meeting and elected their own president. Michael Cross notes that the genteel character of the society made it 'a natural target' for the Shiners. Farmers were a pawn between gentry and lumbermen.[84] Across the river at Hull, Philomen Wright, a lumberer and large-scale farmer, had tried and failed twice, in 1821 and 1824, to establish an agricultural society.[85]

The York Society also had a stormy beginning. The Reformers called a public meeting at a Yonge Street tavern to organize the society, but, because the roads were bad, only about forty attended. The resolutions were laid over, including the suggestion, reminiscent of Gourlay, 'that it is expedient that the communications of the members of the Society shall not be confined merely to agricultural details, but may be extended to geographical and statistical accounts of the district.'[86] This was too much for the government party, which took control of the next meeting when most of the prospective audience went off to watch a libel case involving Mackenzie. A leading Compact politician, D'Arcy Boulton Jr, called another meeting for late May at the court-house. When this meeting began, only a few Compact members were present, so one of them was sent to the market to rustle up some farmers while the others gave speeches and 'looked wistfully towards the door, in the fond and anxious expectation of some little countenance from public opinion, some few practical agriculturists to carry an appearance of a farmers' meeting.' The £50 was easily raised, which Squire Gapper called 'a gift from the gentlemen of the town of York to the farmers in the country. – We give and they receive.' At a later meeting, attended by about twenty farmers and thirty others, rules were drawn up, including a provision for some sixty officers. When Mackenzie mocked the idea of an agricultural society having more directors than the East India Company, he was punched and thrown out the door.[87] This incident proved to be a dress rehearsal for the day when, in December 1831, Mackenzie was bodily removed from the assembly. Rather than providing an opportunity to cooperate on shared goals, agricultural societies were sparks to the tinder of partisanship. The Family Compact won that round, though the Reformers had more success in organizing a mechanics' institute; the rival Compact association died within the year.[88]

The Home District Society's first agricultural exhibition was held that fall at York. The plough match attracted a crowd, but the exhibition was a failure. Seven prizes were awarded for livestock, and two of the winners returned the prize money. The spring fair in 1831 was 'tolerably well attended,' but apart from a few good stud horses, the animals were few and poor-looking. At the fall show of 1831, competition was, Mackenzie observed, 'not very spirited': forty-nine animals on the grounds (twenty-one horses, thirty-one cattle, eight sheep, and six pigs) shared £29 in prizes. This paltry showing in a district of over 20,000 inhabitants probably resulted from the society's decision to restrict competition to members. In 1832 there were complaints that the wealthy won the prizes.[89] The exhibitions continued and, in 1834, were joined by art and horticultural exhibitions. Many winners had 'Honourable' before their names, though the Horticultural Society also offered prizes to gardeners 'for the advancement of the art and the benefit of their employers.'[90]

The societies continued to grow during the 1830s. Some preferred to invest in livestock. The Talbot Society imported Durhams, Devons, and horses, selling them at a profit that it invested in a steam mill.[91] The Midland Society bought cattle and clover and had regular exhibitions at Waterloo; in addition, it encouraged branch societies in Frontenac, Lennox and Addington, and Prince Edward counties. At the Frontenac County show in October 1833, $129 went to livestock and $11 to wheat, oats, and barley.[92] The Ottawa District held shows annually, awarding £64 in 1830 and £292 in 1837. When the Bathurst Society was reclaimed by the upright citizenry (four justices of the peace, one squire, one merchant, and twenty-four 'yeomen') in 1836, it dispersed £91 in prizes and imported grain and livestock. The Eastern District, with its large proportion of Highlanders, purchased a horse, a bull, and 200 bushels of seed oats from Scotland.[93] In 1838, prizes worth £875 were awarded in Upper Canada. The societies were hardly disturbed by the rebellion. The Talbot Society collapsed, which permitted a newly formed Western District Agricultural and Horticultural

Society, based in Sandwich, Essex County, to claim the district grant from under the noses of the Talbot and the Kent societies, but by 1841 the Sandwich Society was 'disgracefully dormant' again.[94]

The agricultural societies were run by economic and social elites, who paid higher subscription rates than ordinary members. In 1838 the Dumfries Society counted farmers among its directors but not its officers, two of whom were merchants, one a lawyer, and one a land developer.[95] Some officers such as Allan Napier MacNab, later a Conservative leader, were figureheads, and others, such as John Macaulay and Hamnett Pinhey, were zealous promoters.[96] A survey of the founding officers in 1830 reveals many legislative members of both parties, as in the Eastern District, where radicals Alexander Chisholm and Alexander McMartin held office alongside Tory incumbent Alexander Fraser, a prosperous farmer and magistrate.[97] Sheriff Treadwell in the Ottawa District was a seigneur, a miller, and office holder, and his co-directors included George Hamilton, one of the great Ottawa Valley timber barons, and David Patee, also a miller and longtime enemy of Hamilton. The Midland Society was dominated by Compact members such as Isaac Fraser and John Macaulay, but it also included Hugh Thomson, a moderate reformer. The Northumberland Society had several land speculators interested in developing the country. Perhaps this concentration of wealth and power accounts for one director's recollection of there being much prejudice against the early agricultural societies.[98] Members of the elite often gave prizes: in 1838 Thomas Mairs put up £35 for livestock bred in the Midland District. In Beauharnois the seigneur, Edward Ellice, made annual donations to the society's coffers and gave away Ayrshire calves. This paternalism did not signal cordial relations between seigneur and *censitaire*. Ellice had some of the highest rents in the land and, in 1826, he took twenty-six *censitaires* to court.[99]

The question of elitism takes a different character in Upper and Lower Canada. The Lower Canadian officers condescended to the *habitant*, who, bewilderingly, remained impervious to their blandishments. The Quebec District exercised 'paternal' supervision of local societies, and it sent seed grain to farmers in the Gaspé District, an act of benevolence that contrasts markedly with the sneaky attempts in Upper Canada to seize moneys earmarked for neighbouring counties. Montreal directors were greedier, but adopted the same paternalistic attitude. The Quebec Society assumed that the public grant and the subscriptions raised by commercial elites were to be spent on the poorer farmers. This *noblesse oblige* approach existed in Upper Canada only in the wishful fantasies of a few Compact members. In that province, there was almost no hint of philanthropy: the funds came from the farmers and were returned to the farmers – as a right, not a gift. By contrast, Quebec elites recognized no debt to agriculture. In 1835 John Neilson remarked: 'In truth, the inhabitants of Lower Canada nine tenths of whom are chiefly employed in agriculture, have always depended for a supply of everything but food and some coarse clothing of family manufactures, on exports produced by some other employments. In early times, it was hunting & fishing, latterly cutting, getting out, preparing & shipping Timber & Lumber.'[100]

In Upper Canada, agricultural societies were 'societies' of citizens, to be founded on liberal principles of self-government. In Lower Canada, they were philanthropic organizations to benefit a population treated as minors, not quite capable of recognizing their own best interest. Agricultural societies were to the economy as municipal organizations were to the polity: local training schools. According to Tocqueville, and to his disciple Lord Durham, who popularized his theories in Canada, municipal institutions provided a field for training in self-government, preparing individuals to participate in national representative government.[101] Agricultural societies brought farmers and local elites together to discuss how to spend money in the best interests of the community: whether to have an exhibition or to import grains or animals. In Upper Canada and later in Ontario, societies were left largely to themselves to make this decision. So long as no outright fraud or theft occurred, decisions made communally by self-interested individuals could not but reflect their best interests; otherwise, liberalism itself was flawed. In Lower Canada, English elites distrusted the workings of public opinion. Habermas has argued that the liberal public sphere developed from the reading public; but in Lower Canada, many farmers who met property qualifications for suffrage were illiterate and were, some argued, so ignorant as to be unworthy of the vote.[102] The first superintendent of education in Quebec, J.B. Meilleur, complained that *habitants* could neither read nor think for themselves: 'It is of little use, in an Agricultural Education, to talk of systems and precepts to men who do not know how to reason, and to offer them books and written rules when they are unable to read.'[103] Distrust of the farmers' judgment persisted long after Confederation, at which point control of the agricultural bureaucracy, vested in the agricultural societies during the Union, was placed under direct control of the provincial executive.

A contrast with Prince Edward Island reinforces this point. In that colony, where social relationships were all the more fraught with tension because of the absenteeism of many landlords, the Executive Council retained the power to choose the office holders of the Royal Agricultural Society. In

1853 the government defied the stated wishes of the society and turned out a secretary from the opposing political party. The opposition press, outraged, spelled out the principles on which both society and the Agricultural Society should rest: 'Self Government, properly speaking, consists in the people managing their own affairs by municipal election; but that result the Snatcher Government oppose by every means in their power ... The hostility of the present Government and Legislative majority to elective institutions is telling heavily against the Agricultural Society, as it must in every case where the hand of the Executive is unnecessarily, ignorantly, or corruptly thrust.' This issue precipitated the failure of the society: 'The attempt to constitute a Society so thoroughly a creature of the Government of the day, was enough to account for its unpopularity, and was, under pretence of inviting farmers to join an Agricultural Society, just treating them as children, considered as unfit to manage anything, an attempt to join the voluntary on the establishment principle; the voluntary having in all cases to succumb.'[104] Overreliance on state initiative and subsidy was seen to enervate the population, robbing it of the spirit of enterprise that exhibitions were supposed to generate. But this economic objection, although widely voiced, was less serious than that of political enervation. Again, in Manitoba into the 1880s, the liberal press complained that the Board of Agriculture lacked public confidence because it was nominated, not elected.[105] The agricultural society was a microcosm of the larger community, considered from an economic and a political point of view, and it was supposed to operate on similar principles. Upper Canada was not yet a liberal society in 1830, but the liberal principles called for were more easily instituted there in small matters like the organization of agricultural societies, and this attention or inattention to small matters presaged and hastened the transition to a liberal society. Upper Canadian elites were more disposed than their counterparts in Lower Canada to believe that a society could be organized along lines both liberal and hegemonic.

The rebellions, which brought the independent history of Upper and Lower Canada to a close, prove the limited success of the early agricultural societies. These societies were supposed to bolster economic improvement and political stability. Both the economic crises and the consequent political confrontations reveal that these ends remained elusive. The societies were strongholds of privilege and wealth, and they stirred up political hostilities that they were specifically supposed to erase. Popular participation was limited in both provinces. What little consent that was secured was not spontaneous but adversarial, after the fashion of political loyalty. Oddly enough, the lofty speeches praising exhibitions as occasions for the whole community to cooperate and overcome the jealousies of class, ethnicity, and interest grew more frequent and fervent in the decades after the rebellions, although the foundation for such claims remained as flimsy as ever.

3

Exhibitions as Politics in Central Canada, 1841–1891

After the rebellions, Canada was reconstituted with one legislature to serve both provinces, its representation equally divided between Canada West, as Upper Canada was now called, and Canada East, the new name for Lower Canada. French Canadians were outnumbered in this legislature by the combination of representatives from Canada West and English-speaking members from Canada East, a situation that was meant to hasten the political and economic assimilation of the French. This absorption did not happen: instead, reformers from Canada East and Canada West collaborated to achieve responsible government and a host of 'modern' educational, legal, and political institutions.[1] Agricultural and industrial exhibitions were one aspect of this program of modernization, and they were established on a sound financial and administrative basis during the 1840s. By the 1850s, reformers were in the minority and John A. Macdonald and George-Étienne Cartier oversaw a conservative alliance that continued to invest in economic development. By the 1860s, however, this *modus vivendi* had collapsed and Macdonald could no longer command a majority in Canada West. In both parts of the country, there was widespread demand for the repeal of the Union, which was seen as a hindrance to more efficient local political institutions. The English-Canadian Reform party, led by the editor of the Toronto *Globe*, George Brown, complained that Canada East impeded the stronger Canada West economy, while French-Canadians demanded local self-government to regain control of their cultural institutions. The solution finally arrived at was a federal union, with a local government in control of education, cultural matters, and some power over economic matters (including the agricultural societies), and a federal government to determine matters of national concern, including relations with Britain and foreign countries. Canada West became Ontario, Canada East became Quebec, and these two provinces were joined by Nova Scotia and New Brunswick in 1867, with the addition of western Canada and Prince Edward Island by 1873.

The business of exhibitions expanded tremendously from the 1840s on. Canada West and Canada East were endowed by the Union legislature with a similar agricultural apparatus, but they developed in very different ways. The gap was already apparent by 1854, when Canada East had 53 agricultural societies and Canada West had 190. In 1871 about 20 per cent of Ontario farmers and 6 per cent of Quebec farmers belonged to agricultural societies. Two decades later, Ontario had more than three hundred exhibitions, with gate receipts above $100,000, while Quebec had less than one hundred exhibitions, with receipts totalling about $5000.[2] As well, Ontario had provincial exhibitions every year from 1846 to 1889, while Quebec could manage one only every two or three years, and then they were much smaller affairs. From the 1860s, Ontario also supported large, profitable, private shows like the Western Fair in London, but they appeared in Quebec only in the last decade of the century. While Ontario exhibitions flourished, Quebec exhibitions floundered. The explanation for this difference sheds light on more than the exhibitions themselves.

The exhibition was a tool designed to improve the farming interest, on both an economic and a political front. From a purely economic perspective, exhibitions were designed to make farmers more proficient and prosperous, and were, thus, an economic lever towards a wealthier and more stable society. Exhibitions were also a political lever, in so far as they organized farmers as an interest group and unified them as a farming public. This public existed as a mass of people gathered together on fairgrounds and in a series of agricultural societies formed on democratic principles. George Brown repeatedly described the Upper Canada provincial exhibition as a 'grand national exhibition' and as 'the only occasion on which the whole people of U. Canada are called together with a common object, at which friends meet who are separated by great distances and political opponents mingle on common grounds.'[3] The system of agricultural societies formalized this arrangement by representing farmers to themselves and to the broader public sphere. Henry Ruttan, sheriff of the Newcastle District and president of the Upper Canada Agricultural Association during the 1840s, argued that the tasks of agricultural societies should be to foster knowledge and friendly rivalry among farmers and 'by a combination of talent and knowledge, and an union of purpose, to adopt such measures as shall bring before the public, and place in its true position, the paramount importance of Agriculture, and its claim to the support of every class of the community.'[4] Agricultural societies were meant to enhance the political presence of farmers by providing them with a forum. This political aspect of exhibitions is examined in this chapter, and chapter 4 considers their economic significance.

Exhibitions were a contested political terrain in the mid-nineteenth-century liberal public sphere. Although they predated the development of this public sphere, they had been conceived along liberal principles and were made to adapt. The history of the Canadian public sphere is still being written, but traditional historiography locates the introduction of responsible government – self-government through institutions responsible to the electorate rather than to England – in the 1840s.[5] Discussions of self-rule for individuals and for societies through representative institutions proliferated during this period, and even the Conservative Party finally rallied to these principles in the 1850s.

The exhibition was conceived of as something like an economic version of parliament, an assembly of citizens in their economic capacity as farmers and, from mid-century, as industrialists and mechanics. Members of the producing classes came together by township, county, province, and even country, to take the measure of their labours and learn from one another at this commercial leviathan. Exhibitions permitted citizens to affirm common ideals, and they seemed visibly to transform self-interested private activity into public activity that served the whole community. Yet this use of political forms within economic organization was as much a repudiation as a glorification of politics. Exhibition promoters distinguished their activity from politics, which for them was a term of contempt referring to a degraded activity characterized by bribery and greed. Richard Lewis told the Toronto Mechanics' Institute in 1862 that 'it is not in the political arena alone that the power and hopes of nations lie. A nobler field is opening to them, free from the factious agitation, the unhealthy excitement and the false expectations of politics, and offering them far richer and more lasting rewards. Political power is not the ruling power in a country ... The true ruling powers of a country are its wealth and intelligence.'[6] He denounced Canadian politicians who had hindered progress with their squabbles and corruption, but he also denounced politics itself as an activity. Politicians met to argue, but farmers met to trade and to show what they did as farmers. Political speeches were so much specious eloquence; farmers were like scientists who exchanged knowledge, proudly displaying for empirical verification the cattle and crops that proved them knowledgeable. Politicians defended special interests against the common good; exhibitions, by bringing the work of all classes together, showed how the interests of farmers, manufacturers, mechanics, and merchants coalesced.

The rise of exhibitions coincided with a period of agricultural ferment when farmers turned against politicians. Farmers became very vocal during the 'Hungry Forties' after the repeal of the Corn Laws left them to compete in an unprotected English market.[7] In the past, farmers' needs had not been distinguished from general needs – for roads, canals, railways, and so forth.[8] Now farmers began to form a distinct interest as they demanded either protection or tariffs, for they saw no reason why others should be protected when they were not. But instead of granting more power to farmers, politicians granted them more representation in the form of exhibitions and an agricultural bureaucracy set up during the Union. This development probably weakened farmers as a political force; for the next two decades they argued incessantly about the election procedures for the quasi-official boards of agriculture, which had no political power and resolutely ignored politics. These debates concerned form rather than substance, as happens, Harvey Mansfield Jr suggests, in representative institutions, where the concern becomes how to make government 'truly' representative.[9] Because nothing short of full presence can achieve closure, this sort of debate tends to be inconclusive.

The attempt to harmonize political and economic interests by means of public and private collaboration in boards and societies had the effect of diverting farmers from political action. Exhibitions did not replace political dispute with economic consensus and, in fact, themselves provoked ugly political squabbles. As Vernon Fowke remarks, politicians remained politely disinterested in the farmers' concerns or bought them off with puny grants.[10] One prescient letter in the *British American Cultivator* in 1843 objected to exhibitions for that reason: 'Why not attempt at least to make them masters instead of slaves? – why not make them understand their own importance? – why amuse them with District Councils, District Agricultural Societies, with Shows of fine cattle, large roots, spans of matched horses, social conversation, flaming reports with a Provincial Society to head or eclipse all the others, manual labour schools and every other invention of the wise to impose upon and transfer the farmers' hard earned dollars into their pockets?'[11] Not until the Grange movement emerged much later in the century did farmers mount a serious political defence of the agricultural sector.

Exhibitions accommodated three interests: organizers, exhibitors, and spectators. Exhibitors and organizers often overlapped, but some of the key organizers were professional improvers with their own goals to pursue.

The government sometimes took the part of the improvers, sometimes of the exhibitors. Spectators developed as a separate interest group over the course of the century: In 1830 the spectators were all supposed to be farmers and potential participants, but, by 1890, exhibitions had become mass entertainment. This chapter outlines the shifting balance of interests around exhibitions, by tracing the laws regulating them, the political controversies surrounding them, and the system that emerged as a result, by focusing on the frontlines of the struggle – the county and township shows. Chapter 4 examines the system of provincial exhibitions that developed from mid-century.

Exhibitions were placed on a sound institutional basis during the 1840s and 1850s by the United legislature. In 1845 Canada West agricultural societies were granted three times the amount they subscribed, up to £250 per district. In 1846–7, during an intentionally uncontroversial and 'crushingly dull' session,[12] the legislature granted Canada East societies up to £150 each and incorporated a provincial society to promote education, publish a journal, collect statistics, and hold exhibitions. In 1850 another law created a Board of Agriculture in Canada West to circulate information about agriculture, manage an experimental farm in Toronto, and hold exhibitions.[13] Statistics were not available, and the experimental farm attracted few students, even fewer of whom ever returned to the countryside. The board published a journal of record from 1855, as did the English and New York societies.

Finally, in 1852, legislation was passed creating a Bureau of Agriculture, complete with a minister who doubled as president of the Executive Council. There was much opposition, albeit in vain. The young Conservative leader John A. Macdonald and the aging radical Louis-Joseph Papineau objected on the grounds that it was 'a job, an unblushing, shameless and corrupt job.'[14] This was true. Francis Hincks, the Liberal co-premier, created the position to woo Malcolm Cameron to his government. Cameron had described the presidency of council as a sinecure, so this duty had to be added before he could accept the post.[15] Some complained that a clause empowering the minister to collect statistics encroached on civil liberties. Canada East members objected that inefficient agricultural societies should be replaced by model farms, though a few defended the societies.

Canada West members thought their system worked well, but objected that the law did not represent the will of the farmers, who would have a politician they had not elected imposed on them. George Brown, who published *Canada Farmer* as well as the *Globe* and thought that public grants were demoralizing, resisted the idea that agriculture was a special interest group and insisted it was adequately represented by ordinary members of the legislature: 'Agriculture, sir, is the great interest of this country – all other interests depend on it – and every member of this House is too happy when he can find an opportunity to advance it. (Hear, hear.) What is the member for Oxford but a Minister of Agriculture? And what else is the Chief Commissioner of Public Works? Are not the great ends of the departments of these officers, to regulate our commercial policy, and cheapen the prices of freight, so as to produce the welfare of the farmer? What suggestion is ever made for the promotion of agriculture that does not receive prompt and earnest attention?'[16] Everyone represented farmers except, it seemed, the Board of Agriculture, for Brown flatly contradicted arguments by David Christie, a breeder, board member, and member of the assembly, that farmers wanted a minister. Brown anticipated what was to become the great problem of the agricultural bureaucracy: Organized to represent the farmer, it represented itself first, as a discrete interest group. It didn't help that the early ministers, ranging from Allan Napier MacNab to Thomas D'Arcy McGee, rarely knew the first thing about agriculture.

Lower Canada also got a Board of Agriculture. In creating it, the bureau canvassed leading agriculturists, only to set aside their recommendations in favour of a political choice.[17] This board was packed with members from Montreal, so they could meet easily; as a result, most exhibitions were held in Montreal. In both Canadas, agricultural societies each voted on four new members to the Board of Agriculture each year. Because societies could not coordinate their efforts, the four retiring members were always re-elected, stirring up great resentment against the 'parcel of old fogies at the Provincial Board, who, by the well devised scheme of getting themselves nominated by circular, year after year, seem to have acquired a vested right to manage the association and work it to the exclusive benefit of three or four places.'[18] When the cranky populist William Weld began to publish the *Farmer's Advocate* in 1866, he built a readership by board-bashing.

These arrangements continued largely unchanged, even after Confederation, when the societies were transferred to provincial jurisdiction. Both provinces spent about 3 per cent of their budget on agricultural improvement.[19] The boards were reformed: In Ontario they were elected directly by regions, and, in Quebec, elections were cancelled and members were named by the executive. There were scandals when, after a change of government, noted agriculturalists – Philippe Landry, in 1878, and Louis Beaubien, in 1887 – were dropped from the board, now named the Council of Agriculture. This Quebec council introduced a panoply of rules governing the behaviour of the agricultural societies, specifying how much they could spend each year on exhibitions, farm competitions, and seed grain. In Ontario the government left the societies to their own devices, intervening only to ban gambling.

Exhibitions flourished in Ontario from mid-century, a boom period for agriculture. Ontario farmers prospered under the Reciprocity Agreement with the United States from 1854 to 1866 and, although its abrogation coincided with the decline of grain production (due to soil exhaustion, parasites, and western competition), this setback was mitigated by a rise in dairy production and the development of a meat trade with Britain. After the slump of the mid-1870s, first grain and then meat prices declined, but increasing use of machinery and the growth of local urban markets offset these losses.[20]

Quebec agriculture continued to lag behind Ontario after mid-century, though there is some dispute about how long the recovery from the depression of the 1830s took. J.I. Little's study of Winslow Township shows that French Canadians had fewer animals and lower yields than the Scottish farmers around them. A measure of prosperity returned with the development of dairying after Confederation, and the consequent increase in manure did much to replenish soils. But most Quebec farmers continued to experience hardship, and many left for the cities and the New England factories. The migration began before mid-century, but increased so much in the last three decades of the century that Quebec lost one-tenth of its people to the United States.[21] Laziness, ignorance, and new desires among country folk for luxury goods and alcohol were all blamed for the hemorrhage, and the remedy was said to be threefold: education, association, and greater pride in the occupation of farming. Exhibitions ministered to these goals, but not with obvious efficacy.

At every exhibition in Ontario, some local dignitary was certain to mount a platform and tell the gathered farmers: 'Look around you, witness our prosperity, the sturdy cattle, plump grains, ripe cheeses, and the large crowds, all of it proof that the system is working.' Quebec reformers complained that nothing seemed to be working at all. Such was the finding of a parliamentary commission in 1850 presided over by Dr Joseph-Charles Taché, member for Rimouski. After interviewing large-scale farmers, deputies, and curés, he concluded that ignorance and want of capital hampered development. He recommended that exhibitions be expanded to reward crop rotations and manuring rather than just cereals and animals, and also that inspectors be sent around farms.[22] Another committee, in 1868, concluded that the agricultural societies 'are far from having effected all the good that may be desired.'[23] Within months, the government moved to reform the agricultural societies. Ontario agricultural improvers demanded only that progress, already occurring, be sped up, and they approved of exhibitions as reflecting and intensifying processes already at work. In Quebec, reformers demanded that agriculture be given an entirely new direction and complained that exhibitions seemed to perpetuate what already existed. The rich, winning all, got richer, while the poor, not winning, stayed home and languished. This was, in fact, more true in Quebec than in Ontario, because the two systems developed differently.

In Ontario, a system of township societies penetrated every nook and cranny of the countryside, bringing a modicum of competition to every farmer. Township societies received three-fifths of the county society's money, and most Ontario farmers who joined agricultural societies belonged to a township society. Exact figures were not tallied, but an estimate, based on reported subscriptions, suggests that, in 1868, township societies had 26,525 subscribers and county societies another 14,132, for a total of 41,132.[24] This number accords with contemporary estimates that one-fifth of Ontario farmers belonged to the societies.[25] The average in the townships was about 24 per cent, but in some townships over 50 per cent of farmers subscribed to the society. In Quebec, many counties had two societies, and a few large, empty counties like Gaspé and Ottawa had four or five small and struggling societies, but a system of township societies never developed. Membership in the county societies in 1866 was 12,586, not far behind Ontario county societies, but far below the total membership there. Membership rose to 15,712 by 1892, but the numbers failed to keep pace with population growth, and total subscriptions declined from $21,088 to $20,689.[26] In Ontario, subscriptions rose from $18,109 in 1853 and $40,412 in 1867 to $66,013 in 1879 and $90,248 in 1890.

Most Ontario societies spent their funds on exhibitions. There were 200 township and 73 county shows in 1869, and 237 and 73, respectively, a decade later, including some union shows. Grey, Hastings, Huron, Middlesex, and Wellington counties sustained more than a dozen fairs every year by the 1890s (see map 1); these were rich farming lands where the cattle trade flourished (see table 1 in the appendix). By the early 1890s, large township fairs at Markham, Woodbridge, and Newmarket would draw 2500–3500 entries and 10,000 visitors per day. Ontario exhibitions earned tremendous amounts in gate receipts: $59,000 in 1877 and $107,387 in 1890, for county and township shows alone. Quebec held fewer than one hundred exhibitions, held only in alternate years, and paid admissions came to $1008 in 1874 and $5883 in 1890.

MAP 1: AMOUNT OF PRIZE MONEY PAID ANNUALLY BY ONTARIO AGRICULTURAL SOCIETIES, 1889-1890

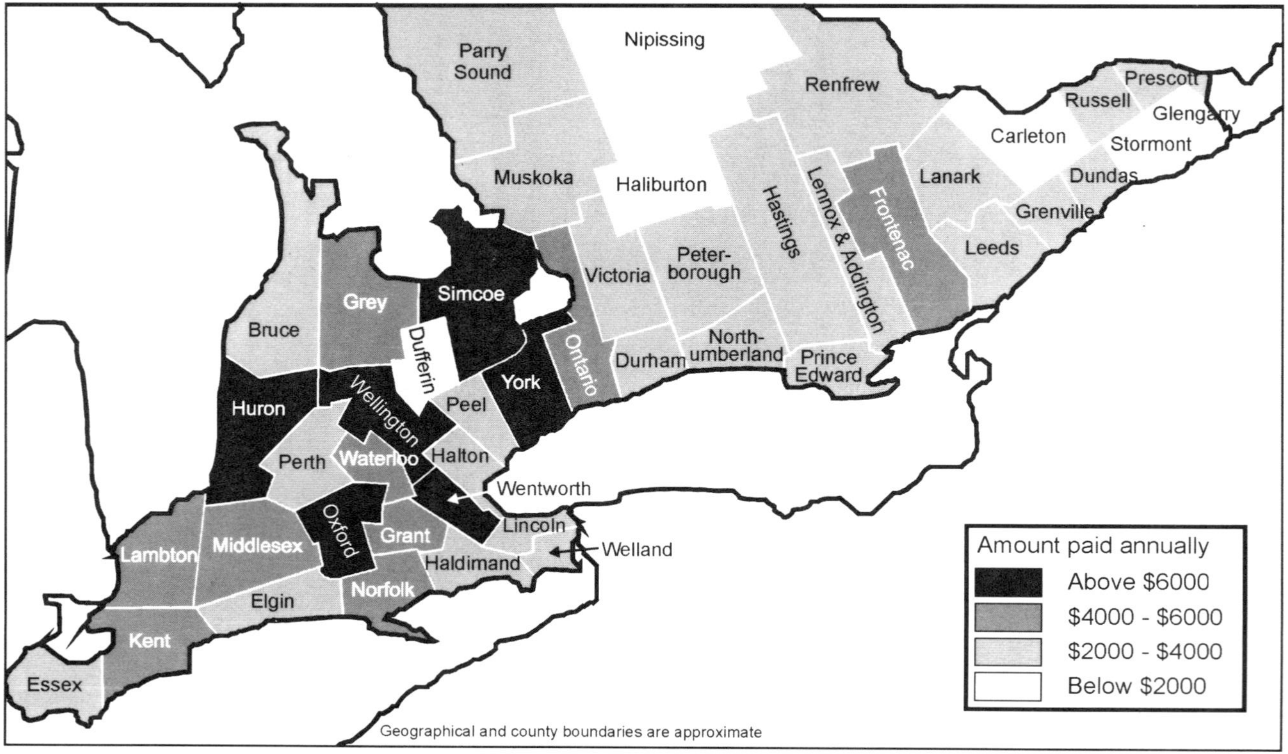

The popular success of exhibitions in Ontario can be attributed to the township societies. The North Hastings County Society reported in 1870 that 'township shows seem to be the choice of the people in general; in rear Townships roughly situated, and thinly populated, to go a distance of ten miles to an exhibition, driving stock, drawing grain, and other articles, is as far as any one wants to go and return back home, in one day; and to spend two days does not pay.' Another observer thought that local shows were the most equitable: 'It is not probable, nay it is not possible, that the farmers of a part of the country recently settled, such as this is, with undrained fields, thickly studded with undecayed stumps, which are an effectual preventive to good cultivation and the use of improved implements, with insufficient and incomplete buildings, with many miles of bad roads, and generally with inferior stock, can compete successfully with farmers who live in older settlements, who have few or more of those evils to contend with, and who are in possession of all those things which early settlers stand in need of.'[27]

But most leading Ontario agriculturists would have abolished the township shows if they could. County societies, coveting the township grants, asked the government to abolish these small shows, which were 'a frittering away of strength and resources that united make a respectable array, but divided, only show the nakedness of the land.' Township societies retorted that the small shows belonged to farmers, not fancy breeders, and attracted practical farm stock rather than plump and pampered beasts.[28] The agricultural press was divided on the matter: George Brown's *Canada Farmer* sided with the counties; William Weld's *Farmer's Advocate* with the townships.[29] The townships defended themselves successfully, although the struggle never abated. In their distribution of moneys, the county and township exhibitions did not differ markedly, although the smaller the show, the more likely it was to invest in livestock and field produce (see table 2).

The two levels of exhibitions corresponded to the two models of prize-giving noted earlier: emulative and distributive. In 1890 the minister of agriculture, John Dryden, a wealthy breeder, explained the emulative model: 'I think what we ought to aim at in this country is to so educate our farmers that the great mass of them will be able to come up somewhat to what we see at exhibitions. I am of those who believe in a man in any pursuit having an ideal – some high purpose and object. A man may achieve a little temporary success by accident, but to continue in a successful course he must have some ideal before him. Where are the mass of our people to get their ideals? I say largely they must get them at the exhibitions. The man who goes to the exhibition cattle shed or the horse ring or among the grain or vegetables will be able to see what his ideal ought to be, and will be able to copy it.[30] Or, as Viscount Monck remarked, the point was to create a 'sound standard of taste.'[31] But this meant that taxes and subscriptions paid by the mass of farmers were showered upon men with education, ambition, and means: professional breeders and rich farmers who had invested in high-grade animals, chemical fertilizers, or drainage and had better crops than their poorer neighbours. The poorer farmers subsidized the better ones. The *Canada Farmer* explained that the point of agricultural societies was 'to offer prizes of sufficient magnitude for improved farms, stock and crops to induce men of capital and large ideas to set a good example of what the country is capable of. Even Weld insisted that 'if the indolent, ignorant, or inactive drones do not exert themselves to improve their minds, their stock or their crops it is right that they should be made to pay something to aid those who are doing good for them.' Or, as one Dundas farmer argued, 'If any individual in a community has enterprise enough to introduce better stock than his neighbors, the community must be benefited, and if the prize money *does* come out of our own pockets, the argument is in no wise impaired, the principle holds good, that they who share in the benefit, should in some measure pay for it.'[32]

Improvers discouraged a fixation on the money and insisted that the knowledge which even the losers won was infinitely more valuable. Weld often quoted Bacon: 'Knowledge is power; and no sane man can attend such an exhibition without acquiring knowledge,' or 'Knowledge is power, and power is equivalent to money.' The president of the Hamilton Farmers' Club declared in 1853 that 'knowledge is capital in the most compact and available form in which it can exist.' Money was beside the point, as South Victoria's directors complained: 'We observe a petty selfishness prevailing amongst many, who, when you approach to solicit their annual subscription, look upon you as a beggar asking for charity, and before actually subscribing have first to calculate what they are to get in return.'[33] Ideally, farmers would compete for honour, love of learning, and the welfare of the community.

Farmers were less philosophical. Because public money was involved, they felt they had a right to it. One improver complained of the 'taunts and insinuations of his unprogressive neighbours who tell him that they are taxed to help pay the premiums awarded to him by Agricultural Societies.'[34] 'N.F.' of Nassageweya objected that at a central show, 'a few individuals would monopolize the whole thing and gobble up all the money.' He added:

> It is becoming quite customary to open County Shows to other Counties, and here we see a few men of capital driving round all the Fall with a sort of menagerie of domestic animals, fed up on purpose for showing, and in a majority of cases, these animals are not a fair representation of the stock kept by these men; they merely show what can be done by lavish feeding, such as is beyond the power of the small farmer to accomplish. If these men wish to compete with each other, the Provincial Exhibition is a very proper place to do it, but I will ever use what little influence I possess on behalf of those Shows where poor men may come in for a share of the spoils, believing that the man who can increase the produce of a garden of potatoes is as much entitled to consideration as the man who pretends to greater things.[35]

County shows did not attract '*that class* which most need a spirit of emulation infused amongst them.' Some provision had to be made for ordinary farmers.[36] But a distributive exhibition was not very competitive. In defiance of all improving logic, prizes went to scrub stock. One critic remarked that at a township fair, spoils were distributed but no improvement effected, and there was 'very little honour and satisfaction in obtaining prizes where there is little or no competition.'[37] A few scrawny beasts and weedy roots hardly achieved the effect desired and could be a public embarrassment. Some counties shared this problem: North Bruce, tucked up between Georgian Bay and Lake Huron, cancelled exhibitions because 'competition in most classes was always slight, and very meagre indeed were the benefits to be derived from a view of the animals or articles placed on exhibition.' Grey County also languished: In 1860 'there was no great attendance of either man, beast, or any article,' and, in 1862, the display was 'decidedly of the most meagre description.'[38]

Ontario accommodated both kinds of exhibition: Small prizes were distributed at small shows, while valuable prizes at larger exhibitions went to the leading farmers and breeders. There were always complaints that the wrong sort of people were winning prizes, but, because everybody had a chance to win somewhere, there was little impetus for reform. The state regretted the surfeit of local exhibitions, but refused to intervene. John Carling, minister of agriculture in 1870, thought that the small shows were 'a serious drawback to agricultural progress,' but prefaced his comments with the remark: 'I have always been favorable to the allowing of the members of Societies to decide matters of local interest for themselves, as they ought to be the best judges of what they require.'[39]

After listening to the reformers of Canada West, those of Canada East were loath to introduce local shows. The two most influential improvers in the Union period were William Evans and Joseph Perrault, the one an Irish immigrant and the other a graduate of agricultural schools in England and France. Both, in turn, served as secretary to the Board of Agriculture, edited its journal, and had *rouge* sympathies. Both vehemently opposed small-scale exhibitions, insisting that only large shows could stoke ambition and inspire emulation. However, county shows reached so few farmers that they were clearly inadequate to the task of improving agriculture. Improvers denounced the county shows for being simultaneously elite institutions that concentrated wealth among the best agriculturalists and democratic, liberal organizations designed to reflect popular will and encourage popular agency.

Exhibitions did concentrate wealth. At mid-century, the average amount each successful competitor won in Quebec was $6.92 ($6.27 for French Canadians), and in Ontario, $5.09 ($4.44 at township, $5.78 at county, and $7.08 at district exhibitions, respectively).[40] Individual societies are a better measure, but only scattered records survive. In Missisquoi County over four years in the 1880s, 42 per cent of competitors lost money, 47 per cent won up to $10, and only 11 per cent exceeded $10. But the top 11 per cent made off with 82 per cent of the prize money.[41] In a French county, Montmagny, in 1891–2, 6 competitors out of 167 won 40 per cent of the prize money.[42] Ontario examples are comparable: in Gore District, in 1847, five men won from 41 per cent to 55 per cent of the prize money in township shows, but only 28 per cent at the district show; five men also took home 35 per cent of all prize money in Addington County from 1853 to 1856, and 50 per cent in Darlington Township from 1858 to 1860.[43] In short, concentration occurred in both provinces. John Kenneth Galbraith's recollections of twentieth-century Elgin County fairs probably applied to earlier shows. His family shared the Shorthorn cattle prizes with another family each year: 'Neither we nor the Browns were in the slightest measure stimulated by the exercise. It was simply that there was no way one could so easily pick up thirty or even forty dollars in the course of one day. No ordinary husbandman would have dreamed of anything so impractical as to try improving his herd in order to compete with us. We had no need to improve to get the money.' They left the garden and fancy-work prizes to the wives of poorer farmers.[44]

Although concentration of wealth was the norm, the problem in Quebec was exacerbated by the scarcity of prizes and by ethnic divisions. Most of the leading breeders for most of the century were British. At provincial exhibitions they won nearly all the prizes for purebred stock, though French

Canadians often did well in grade cattle and in horses.[45] At the same time, most of the county agricultural societies were run by French Canadians: forty-two of the societies in fifty-nine counties spoke mostly French (see table 3). Usually the ethnic make-up of the society resembled the ethnic make-up of the county, but English farmers were more than 10 per cent overrepresented in Argenteuil, Compton, Drummond, Hochelaga, Megantic, Missisquoi, Montreal, Ottawa, Pontiac, Richmond, Sherbrooke, and Shefford. In Laprairie, however, English-speakers made up 15 per cent of the county population, but not one from this group belonged to the society. Until 1869, when it was forbidden, most societies with a mixed membership held separate competitions. French Canadians probably won most of the prize money distributed at county shows, although the English minority was disproportionately subsidized by public money raised from taxes on the majority.

The most successful French-Canadian agricultural societies did not hold shows, but found other ways of dispensing the government's bounty. In 1872 seventeen societies counted as members 10 per cent or more of the farmers living in the county. Five of these were English and spent their money on exhibitions. Three had mixed membership (Jacques-Cartier, Hochelaga, and Beauharnois) and bought livestock as well as holding exhibitions. The rest were all more than 90 per cent French, and all save Laprairie spent every penny they could on grains. Eight years later, thirty-four societies had 10 per cent or more of the farmers in the county as members (see map 2). Large empty counties like Ottawa and Rimouski had as few as 2–3 per cent, while at the other end of the scale, Huntingdon had 34 per cent and Verchères had 28.5 per cent. Eighteen societies were 90 per cent or more French, and only one of these, Laval, did not make large purchases of grains. Three others were predominantly English and spent their money on prizes; the rest had mixed membership and mixed spending patterns. On the whole, English societies had more members when they held exhibitions, while French societies, unless situated near the Montreal market, flourished when they purchased grains, and those with English and French members compromised. French societies preferred non-competitive and egalitarian methods of distributing the government grant. However, their methods were repudiated by the improvers.

Quebec improvers suggested four alternatives to exhibitions: newspapers, schools, model farms, and *cercles* (or farmers' clubs). Evans monopolized all the provincial society funds from 1847 to 1852 for journals

MAP 2: PERCENTAGE OF FARMERS RESIDING IN QUEBEC COUNTIES SUBSCRIBING TO AGRICULTURAL SOCIETIES, 1880

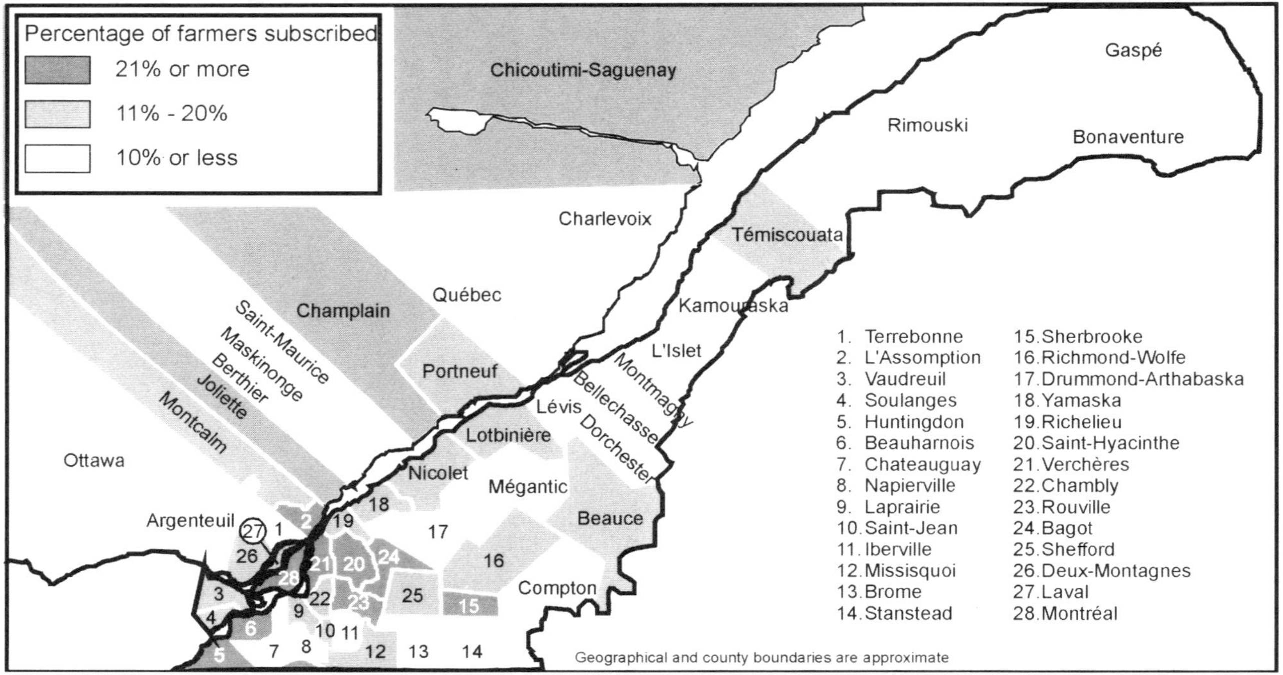

(published in both languages), and Édouard Barnard, Quebec's 'director of agriculture' from the 1870s to the 1890s, also edited the Council of Agriculture's official paper. These costly journals attracted few subscribers and relied on government subsidies. Their editors argued that journals conveyed information better than exhibitions and should replace them. By contrast, Ontario papers like *Canada Farmer* and the *Farmer's Advocate* relied on subscription and advertising, not on subsidies, so they wrote to please their show-loving constituents. The one independent agricultural paper in Quebec was an *ultramontane* organ, the *Gazette des campagnes*, published by Firmin H. Proulx in Ste-Anne-de-la-Pocatière from 1861 to 1895, which advocated better butter-making along with papal infallibility. Proulx twice printed jeremiads against exhibitions as diverting the young from serious pursuits as well as inculcating pride and materialism: 'Une prime excite parfois trop notre orgeuil, sentiment qui tient dans notre pauvre nature une assez large place pour n'avoir pas besoin d'être developpé; l'argent passe assez souvent en dépenses inutiles, et puis, est-il bon, est-il moral de faire toujours de l'argent le couronnement de toutes choses?'[46] But most of the time, Proulx approved of exhibitions as providing education to people who 'ne lisent pas ou lisent peu' and were easily led by example: 'L'exemple est le moyen infallible de propager le goût des améliorations. Le peuple canadien peut-être plus que tout autre est porté à l'imitation.'[47] Exhibitions stirred up a 'noble et louable émulation,' and did honour to agriculture and to French Canada. When the Chicago World's Fair was announced in 1890, Proulx predicted glory for French-Canadian Catholics and counselled, 'à l'oeuvre donc et pas d'hésitations.'[48] Independent agricultural journalists supported exhibitions, but state-paid ones did not.

The argument for schools as an alternative to exhibitions was best stated by Abbé Félix Buteau, director of the agricultural school at Ste-Anne-de-la-Pocatière, who remarked in 1874 that, after twenty-five years and $1.5 million, agricultural societies could only report a 'commencement d'amélioration graduelle et générale dans la masse des agriculteurs.'[49] What farmers lacked was not motivation to improve themselves but the means to accomplish this goal. Buteau likened them to patients in a hospital: 'Or, comment fait-on sortir les infirmes et les malades de l'hôpital? Est-ce en leur offrant des prix d'agilité, de force, d'appétit? Que répondraient ces pauvres infirmes, à une telle proposition? – Messieurs, commencez par nous donner des rémèdes propres à guérir nos infirmités, pour nous faire sortir de l'hôpital, après celà nous acceptrons volontiers le défi.' Schools were more obviously educational than shows, but less popular. The first agricultural school in Quebec failed during the 1830s;[50] another attempt in 1851 also failed. In 1859 Abbé François Pilote attached a farm school to the classical college at Ste-Anne-de-la-Pocatière, and another was established under clerical control at L'Assomption. They attracted only a handful of students and eked out a miserable existence with paltry government funding.[51] Agricultural societies, though more frivolous, had a wider scope.

Proponents of schools, like Barnard, who yearned for one of his own, insisted that the money paid over to agricultural societies would be one hundred times more effective if it were spent on schools.[52] Graduates might be few, but each one would be a model farmer, an example to his neighbours, who would then see how profitable the improved methods were and adopt them. This idea might be good in theory, but it proved to be unsuccessful in practice. Buteau's successor, Abbé L.O. Tremblay, complained that graduates were drowned out by the mass of unimproved farmers 'qui méprisent leurs enseignements et se moquent de leurs bons exemples.'[53] Philippe Landry, a wealthy graduate of the college, practised scientific agriculture in Montmagny at vast expense and to little effect. Farmers in the parish habitually referred to unproductive soil by saying, 'C'est comme la terre à Landry.'[54] The French-Canadian farmer, it was said in 1886, would only learn from 'un de ses égaux.'[55]

Agricultural reform had to balance two issues: setting a good example and catering to popular opinion. If the model was too far ahead of what farmers knew and what they could afford, they ignored or ridiculed it, but if agricultural societies went too far towards catering to popular prejudices, improvement would come to a standstill. Perrault complained that directors were slaves to the will of ignorant members and flattered their errors and caprices. He distrusted their judgment, and demanded that county exhibitions be replaced by larger regional shows to ensure a better quality of exhibits.[56] Evans also complained that the agricultural societies gave in too easily to the demands of the rank-and-file members.[57] This contradiction was the dilemma of agricultural societies: they were democratic bodies, responsible to public opinion, but their promoters distrusted that public opinion and sought to transform it.

Another solution was a model farmer who was elevated from within the community, rather than arriving, diploma in hand, as an expert accredited elsewhere. Then his neighbours would say, 'C'est un des nôtres,' and imitate him. Reformers could not convince the government to invest in model farms, so they advocated prizes to distinguish the best farms in each county. Evans had first advocated farm inspections to ensure that strict justice would be meted out, not for one hand-picked bag of grains or a bull fatted at the expense of the other animals, but for the entire system of farm management. The free-thinker Perrault and *ultramontane* Proulx convinced the Board of Agriculture to introduce this system shortly after Confederation. All agricultural societies had to follow the new rules, on pain of forfeiting their grant. The board dictated exactly what criteria farms should be judged by: so much land in pasture; all weeds, stones, and stumps removed; good roads and fences; a rotation with green crops to replenish the soil after cereal crops; and at least one-quarter acre of roots. The societies were ordered to hold an inspection every other year and to award five prizes of $50, $40, $30, $20, and $10.

The societies greeted this fiat with dismay and anger. Many farmers grew no roots and were ineligible.[58] Roots served as fodder and encouraged winter feeding of livestock, but they required much labour, did not suit all soils, and proved less profitable than corn. The rule was overturned in 1875. Some societies already had farm competitions, but even they might object to the standards imposed. The Missisquoi Society, for example, complained: 'Our successful farmers are all dairymen, our land is broken, the farms are of irregular shape and often composed of fields scattered at a considerable distance one from another, the pastures are frequently rugged, rocky and from necessity permanent grazing grounds, as our experience teaches us the wisdom of breaking up the land only as a means of keeping up the growth of grass and this only when it is imperatively necessary.'[59] Some objected to the enormous prizes as taxing poor farmers to subsidize the rich. A society in Drummond County complained that 'les petits cultivateurs ne pouvaient songer à concourir avec les ancien cultivateurs plus riches qu'eux. Il leur semblait que c'était jeter l'argent de la Société entre les mains de ceux qui en avaient le moins besoin. Le fait est que le mécontentement devient à peu près générale, et, si nous continuions ce système, il s'en suivrait, d'une manière presque certaine, une dissolution de la Société.' Another in Ottawa complained that the system 'jeterait tous les fonds de la société entre les mains de cinq ou dix membres,' and then the others would refuse to subscribe.[60] These objections were well founded, as the example of Montmagny shows (see table 4). The Beauce Society lost sixty members whenever it had no exhibition. Temiscouata Society disbanded rather than hold the inspection, and overall membership in the societies dropped by nearly 10 per cent, from 12,586 five years earlier to 11,452. In 1875 many societies objected so stridently that their reports were suppressed.[61]

Requests for exemption were numerous and some were granted, especially in the newer and poorer counties. Charlevoix was exempted when it complained that any competition or exhibition would be 'tout-à-fait inutile, et j'oserai même dire que les résultats en serait ridicule.'[62] Until 1890 the model farm competition was a failure, as even its leading proponent Barnard agreed.[63] Then the 'Mérite Agricole' that Quebec premier Honoré Mercier imported from France made it more popular. Farmers who had won county prizes entered a larger district competition, where their farms were inspected by some of Quebec's leading agronomists. These inspectors published a report detailing good and bad points on every farm, something county judges had never done, and prizes were distributed with great ceremony.[64] However, this system did not replace the exhibitions.

Exhibitions were held despite interdictions. In 1879, when the law required farm inspections, two-thirds of the seventy-five societies that reported had an exhibition as well or as a substitute. Many societies also bought huge quantities of grain or paid back the grants directly to members, often in the form of 'consolation' prizes. The council and Barnard could do little because, if they docked the grants, outraged farmers complained to their political representatives, who then pressured the department to pay in full. Farmers considered the grant a bribe for their vote, so the members were eager to support them. In 1881, when the grants were late, the deputies for Bagot and Ottawa warned that 'un tel état de choses ne peut continuer longtemps sans amener de graves résultats contre le gouvernement.'[65] When the council banned prizes to male animals without pedigree, a restriction that would have prevented ordinary farmers from winning prizes, the government exercised its veto. Farmers also had allies in the department: Georges Leclère, the council's secretary, was too lazy and disorganized to exercise the surveillance he should have, and Siméon LeSage, deputy minister of agriculture, thought that the system of exhibitions worked reasonably well.[66]

A final offensive was launched by priests and the press, led by Barnard, in favour of *cercles agricoles*. At a *cercle*, or club, a handful of parish farmers would meet to discuss new ideas. Members subscribed to an agricultural journal, where their discussions would be published, thus creating content and a market for Barnard's *Journal d'agriculture illustré*, which the government required agricultural societies and clubs to purchase. *Cercles* resembled scientific societies with free debate among members, but they also bolstered conventional authority. The Catholic episcopacy told parish priests to encourage these gatherings, and many did: one-third of clubs (and only a few agricultural societies) in 1894 had a priest as president or secretary.[67] The most prominent priest in the movement, Abbé T. Montminy, described them in these words to the Dairyman's Association, or the Société d'Industrie Laitière: 'Le cultivateur canadien catholique aimera ces réunions présidées par le curé qu'il aime comme un ami, comme un bienfaiteur. Là, le prêtre sera comme le père au milieu de ses enfants, il sera comme le centre de cette union qui pourra faire des merveilles, il sera le gage du succès. Avec le prêtre point de discussion, point de rivalité funeste, point de jalousie, point de politique, point de spéculation.'[68] This was not the Enlightenment notion of intellectual exchange. Barnard and the bishops drew up a model constitution for clubs dedicated to St Isidore. Members promised to forgo blasphemy, lawsuits, luxury, and intemperance. The curé decided who could join. When Colonel William Rhodes, the commissioner of agriculture, heard that the department was circulating this document, he had it withdrawn.[69] Clubs helped to bolster lay as well as clerical authority because their organizers demanded lecturers from schools and from the department. During the 1890s the clubs began to receive government funding, so long as they held at least two lectures per year, and the government began to pay professional lecturers. Self-instruction at the exhibition gave way to formal instruction.

Reformers touted the clubs for being different from agricultural societies, and they even asked the government to transfer all funding to the clubs. The Dairyman's Association took this position in 1892, after hearing G.A. Gigault, secretary to the Department of Agriculture, argue that 'les expositions ont pour but de faire admirer l'effect, le produit du savoir et de l'habilité et non d'établir la cause. Et la première chose que nous devons chercher à établir, c'est la cause.' In fact, many clubs held exhibitions, beginning with Abbé Montminy's at St-Agapit in 1882. In 1887 two clubs applied for grants to hold shows: The government disguised these allocations as honoraria for lecturers. The club at Ste-Rose praised the exhibition as a 'magnifique fête de famille en même temps qu'un grand exemple d'union et de progrès.'[70] As it became clear that clubs also wanted to hold exhibitions, Barnard and his minions began to argue that this was a good idea, for it would bring all farmers into the rivalry. The Council of Agriculture, however, tried to prevent an explosion of exhibitions by forbidding clubs to hold them.[71]

Clubs did not attract many members until a grant was allocated in 1893, but then they grew rapidly. Membership in agricultural societies in 1892 was 15,712; by 1894 membership in clubs and societies together had almost tripled, to 39,000 French and 5000 English.[72] Quebec had 750 parishes: in 1892, 222 parishes were entirely unrepresented in agricultural societies, and another 200 had only a few members.[73] By 1894, 515 parishes had a club with at least twenty-five members. Reformers claimed this increase showed dissatisfaction with the societies. In fact, clubs enjoyed some privileges. Societies could spend only half the $2 subscription on forage seeds, whereas clubs could spend the whole amount on seeds, and most did.[74] Membership in many societies had been higher thirty years earlier when they, too, could spend freely on grains: Bagot, in 1866, had 845 members who subscribed $1311, but only 409 members in 1890 and 325 in 1892.[75] In 1882 a society in Montmorency set an upper limit of $40 per subscription, and 233 members subscribed $2975.[76] From 1890 to 1892 its membership hovered between 60 and 83. When the council limited grain purchases, Rouville's membership dropped from 400 to 200.[77] In Lotbinière, former Quebec premier Henri Joly predicted that 'nous allons perdre prèsque tous nos souscripteurs; nous ne sommes pas encore assez avancés pour nous passer de cet encouragement.' The fact that members now grew forage crops proved the usefulness of the society, he argued.[78] When one council member toured Charlevoix in 1893 to find out why farmers had stopped subscribing to the Agricultural Society, he was told that, ordered to buy pedigreed animals and forbidden to buy grain, 'ils ont préféré ne pas entrer dans la société et faire leurs achats eux-mêmes,' and that 'si l'on veut que les Sociétés se maintiennent, il faudrait les donner autant d'avantages qu'aux cercles.'[79]

Farmers liked to hold exhibitions and/or to buy grains, but reformers fought these initiatives. In 1868, 465 St-Hyacinthe farmers subscribed $500, spending it on parish farm inspection, a county exhibition, and forage grains, but not on Perrault's journal, where they were denounced for being so anti-progressive as to shun 'la lumière avec l'obstination du hibou.'[80] Membership in this society was 243 in 1892. Perrault remarked sarcastically, referring to the 269 members of the Portneuf Society who received $2.50 in grains for every dollar subscribed in 1866: 'O Progrès! voilà bien de tes inspirations.'[81] Portneuf membership in 1892 was 189. Barnard also complained of excessive self-interest: 'Malheureusement un trop grand nombre de souscripteurs aux sociétés d'agriculture souscrivent dans l'unique but de soutirer la plus grand part possible de l'octroi fait par le gouvernement à ces sociétés. Ainsi, si l'on a des animaux à vendre dans l'automne, on commence par souscrire une piastre à la société d'agriculture, souvent on retire de suite pour au delà d'une piastre en graines de mil et de trefle; puis on cherche à obtenir le plus de prix possibles, à l'exposition; après cela on vend ses animaux, le plus souvent à la boucherie, sans s'inquieter d'avantage au progrès de l'agriculture.'[82] Early agricultural societies would have boasted of such success. Barnard's ideal *habitant* was not economic man, but something between bureaucratic man and the noble *colon* who farmed in poverty for the greater good of his race.

Exhibitions were designed to get people thinking about how best to serve their own interests, since liberal economic theory held that people could best serve the public by pursuing their self-interest. The council subverted this goal by overruling agricultural societies' deliberations. Every year, the Charlevoix Society at Baie-Saint-Paul met and discussed at length how to spend its money; every year, the council rejected its proposals and told the society how to spend it. To descend to details: in 1895 the society decided to buy Leicester sheep, Ayrshire cattle, Yorkshire pigs, implements, $50 worth of fertilizers, some Dakota potatoes, and a boiler, and to spend $50 for prizes on beets, fertilizers, pastures, and the like. The council wrote back that the society must buy Shropshire, not Leicester sheep; that it must spend more on the prizes (the list to be sent in for approval) and less on the fertilizers; and that it could not purchase the boiler.[83] This interference happened year after year. In the suppressed reports, Temiscouata voiced its objections forcefully: 'Les Directeurs de chaque Société sont plus en état de comprendre ce qui leur convient que des personnes ne connaissant aucunement le besoin et les ressources de leur Comté. Nous demandons encore une fois à être libre de conduire les affairs de nos Sociétés comme nous l'entendons et nous pouvons assurer au Conseil d'agriculture que nous sommes disposés à suivre la loi à la lettre, et à travailler dans l'intérêt des contribuables de ce Comté et à l'avancement de l'agriculture en général.[84]

Most of the criticism of agricultural societies came from Barnard, the director of agriculture. He visited the societies, inspected their books, and was appalled to find that accounts did not balance, reports were late, and rules were broken. He exaggerated his case, claiming that, in several cases, thousands of dollars had been embezzled.[85] Embezzlement occurred twice in a quarter century, and most books were out by small amounts, sometimes a few pennies. Barnard accused the societies of bad faith, but most acted in good faith. The problem was that neither Barnard nor the Council of Agriculture trusted the *habitants'* judgment. Barnard doubted their good intentions, and the council doubted their wisdom. However, Barnard's criticism of the societies was really an attack on its supervisory body, the council, whose functions he wanted to usurp, as he admitted frankly. At the Provincial Exhibition of 1881, Barnard and council president L.-H. Massue came to blows. Massue complained when Barnard described the council as a body with no head or soul, but his remarks were suppressed. Barnard was then at the height of his influence, and his memos were written on Premier J.-A. Chapleau's stationery.[86]

Barnard preferred farm inspections to exhibitions. His program involved a hierarchy of people looking, speaking, and listening, with facts about the condition of farming being transmitted upward to the department, and principles concerning good agriculture being transmitted down to farmers by lecturers. The lecturers would distribute honorific prizes because the money had been pre-empted to pay their salaries. The farmers' grant had been diverted to the bureaucracy. This system would permit centralization of knowledge, for 'l'enregistrement de ces diplômes au département de l'agriculture nous permettrait de constater, d'un coup d'oeil, l'état de l'agriculture dans toutes les paroisses du pays.'[87] The flow of information culminated in the person of Barnard, who would exercise surveillance over all agricultural operations. This system was a far cry from the Enlightenment's attempt to use public opinion to illuminate the workings of government. Exhibitions had been devised to vest public opinion with the power of vision, to ground it on real knowledge rather than mere opinion. In Barnard's scenario, farmers were to be looked at, and bureaucrats did the looking from the recesses of their offices.

It was not a repudiation of the exhibition principle but an extension of it. Exhibitions were a form of visual discipline, but an overly crude form. Like Jeremy Bentham's prison, the exhibition was highly 'panoptical': It created a special place where everything existed to be seen completely. It occurred only once each year, but its disciplinary effects were supposed to transform daily behaviour. Foucault argued that the significance of the Panopticon was as a 'generalizeable mode of functioning; a way of defining power relations in terms of the everyday life of men.'[88] In putting themselves on display, exhibitors advertised their submission to cultural and social norms. Exhibitions expanded the apparatus of surveillance so that farmers policed one another and, in true Benthamite fashion, 'each comrade becomes an overseer.'[89] But this was not good enough. Once-yearly inspection was too irregular, and there was no guarantee that the exhibition accurately represented the countryside. Slovenly farmers could win prizes with one good crop or one fine horse. Evans, and Barnard after him, insisted that agriculture should be scrutinized directly, that state-paid inspectors should survey the private property of citizens. (Model farm inspections were also introduced in Ontario, but only a handful of farms, run by the wealthiest agriculturists, entered the competition; the goal was to advertise the elites, after the hegemonic model, rather than to inspect the populace.)

The real problem with Bentham's theory was that it assigned too much power to public opinion. Comrades were not natural overseers, but were more likely to be aiders and abettors. Quebec reformers tried to combat this problem by undermining popular agency. Instead of the self-education of the exhibition, they restored the traditional form of lectures by experts and clerics. In both provinces, most agricultural reformers were liberal professionals to whom this model was natural, but the argument was more effective in Quebec, where professionals had more cultural clout. For Quebec journalists, agriculture was first and foremost a system of knowledge to be promulgated by expert teachers. By defining it as an exchange of information, rather than of money or trade goods, as at a fair, they justified the state's expansion into this area. Ontario agricultural journalism was dominated by outspoken liberals like William McDougall at the *Canadian Agriculturist* and William Weld at the *Farmer's Advocate* who insisted that agriculture, as an economic activity, should not be subject to government interference. George Brown also opposed the state's intervention in agriculture, but he left the editing of *Canada Farmer* to Rev. William Clarke, who advocated the professional model.

Ontario journalists were also more attached to the ideal of rational self-interest that exhibitions had been designed to encourage and that Barnard tried to short-circuit. In complaining of bad faith, Barnard showed an unfamiliarity with the liberal principles on which agricultural societies were predicated. Strictly speaking, bad faith was no objection. Montesquieu had remarked that it is in the nature of things for men to abuse power: 'C'est une expérience éternelle, que tout homme qui a du pouvoir est porté à en abuser; il va jusqu'à ce qu'il trouve de limites.' James Mill popularized this view in England with an article on 'Government' in the *Encyclopædia Britannica*,[90] and it guided the American constitution. The Clear Grit William McDougall advocated it in the *Canadian Agriculturist*: 'The fact is the abuse of power, or in other words oppression, is a tendency of man's nature,' to be prevented only by mass education.[91] If all exercise of power corrupted, only popular participation could check this abuse because, as Bentham stated, the public can have 'no sinister interest' against itself. Properly constituted with a system of checks and balances, any system of government, down to the agricultural society, would be self-regulating, or it would be flawed. This system was what was meant by the 'liberal principles' on which the Upper Canada societies were established in 1830. Barnard presumed that these societies might have sinister interests and should be guided by an observer with regulatory and punitive power. Best of all, to Barnard's mind, would be *cercles* guided by priests. Agricultural societies in Quebec failed to thrive because liberalism had too many enemies.

By the end of the century, classic liberalism was waning in Ontario as well. Ontario began to fund farmers' institutes during the 1890s, sponsoring lectures by professors and graduates of the Guelph College, and a few years later these experts were stationed around the province. Douglas Lawr remarks, 'These college men brought a new viewpoint to the question of agricultural education. They based their authority not on empiricism, but on their scientific training. Their methods did not favour the rivalry and competition of the show ring or the formality of the essay, but looked instead to the informal lecture and the actual demonstration of new techniques and equipment.'[92]

The process had occurred a few years earlier in the dairy industry, when direct methods replaced exhibitions in training farmers. There was a progression from conventions to travelling inspectors to schools. American dairy professor L.B. Arnold described it as a shift from abstract to practical instruction,[93] but it was more than that. Conventions, like exhibitions, brought practical farmers and dairymen together to exchange experience. With travelling experts, the farmer was now merely a pupil. Moreover, exhibitions were designed to prove that factories, not farm wives, should be turning the milk into cheese.[94] In 1876 the Ontario dairymen sent an exhibit of cheese to the Philadelphia Exhibition which won the highest awards. The Dairy Association attributed the victory to the factory system, arguing that even farmers who had opposed it 'are now believers in the universal law of progress, and see for themselves that in a factory superintended by men combining practical and scientific knowledge, a better article can be produced than by private efforts. It was owing to the intelligent management of the Factory system that we were able to compete so satisfactorily at the International Exhibition, which is a strong proof that we have made a progressive step in the right direction.'[95] Though they had been designed to help farmers acquire skills to improve their farming, exhibitions now persuaded them to relinquish skills and leave matters to the experts.

The exhibition was a forum for competition among social groups. As an organization designed along liberal and democratic lines, the agricultural society reflected the fault lines in nineteenth-century Canada. The larger the exhibition, the more likely it was to concentrate prize money among the prosperous farmers and breeders who had already improved their farms. The interests of the rich and the poor diverged in this respect, and these two informal groups competed for control of the exhibition. Neither won a decisive victory, just as neither capitalists nor labourers decisively defeated the other politically, even though the wealthy usually finished with a disproportionately large slice of the pie.

The exhibition, as a forum for confrontations, was itself under attack. In Ontario the agricultural elite tried to restrict unimproved farmers to observing rather than exhibiting. Their ideal was Debord's society of the spectacle, one of 'separation, estrangement and nonparticipation.'[96] Still, the township shows persisted, as occasions where tens of thousands of Ontario farm families came together rather than remain separated and estranged, and where they participated rather than merely observed. Democratic forms in these local agricultural societies, and the public voice they provided, enabled farmers to maintain control of the fairs. In Leeds in 1872, for example, membership shot up from 106 to 750 when the rear townships tried to secure the county fair.[97]

Similar confrontations occurred in Quebec. In Argenteuil, in 1883, the quarrel was between farmers and townsfolk, the old county town of Lachute versus the new railway town of St Andrew. The decisive meeting was held on a fine harvest day when farmers had to be in their fields and could not vote.[98] There were also ethnic quarrels, as seen in Chateauguay during the 1860s: Scottish Canadians garnered the most votes in 1862, French Canadians beat them out in 1863, the Scots took control again in 1865, and the French triumphed in 1867. Deputy Henry Starnes remarked that such dissension 'raised a spirit of competition among them which could not but produce the most beneficial results.'[99]

But in Quebec and, later, Ontario, another social group found that its best interest lay in replacing the competition and confrontation of the exhibition with smooth and continuous administrative action. These reforming quasi-intellectuals sought to replace political methods with bureaucratic methods of engineering progress. They did not simply side with one class against the other, but found their own best interest to lie in the expansion of the state. Gramsci theorized that this development would occur wherever economic progress was fuelled by international rather than internal developments, so that 'the group which is the bearer of the new ideas is not the economic group but the intellectual stratum.'[100] By arguing that the exhibition was a symbolic, educational re-enactment of market activity, agricultural reformers extended the state's reach into quasi-market activity. Self-styled intellectuals were permitted to enter into new activities redefined as 'informational,' and they secured jobs as bureaucrats, professors, and writers of manuals that the government sponsored by purchasing and distributing them. Hardly a year went by without William Evans petitioning the legislature for money to sustain his improving activities, all of which were, he insisted, good for the public and the state. By the end of the century, Catholic newspapers unabashedly affirmed that all initiative should rest with the government: 'Si on veut que l'esprit d'association s'affirme parmi nos cultivateurs il faut cesser de compter sur le travail stérile de l'initiative privée. Il faut que l'encouragement, le mot d'ordre et la propagande lui viennent de la sphère gouvernmentale.'[101]

Agricultural exhibitions struck an unstable balance between public and private interest. They were developed by social elites and supported by the state as instruments of indoctrination, at a time when, as seen, secular indoctrination was something of an innovation. But the indoctrinators had to 'stoop to conquer'; they had to resort to tactics more interesting and attractive than the delivery of dry technical information and moral homilies. In both form and content, then, exhibitions were manifestly popular institutions. As democratic associations of farmers, businessmen, and citizens, they were, for better or for worse, what people made of them. If they lent themselves to hegemonic propaganda by social elites or bureaucratic experts, there were also safeguards – elections and reliance on public subscription and public attendance – that gave the people a say. In Quebec, where farmers' control of the societies was undermined, the farming public withdrew its support, rendering the exhibitions and other reforming initiatives ineffective. In Ontario, where the public took control, exhibitions changed, becoming less instructive and more amusing as the century drew to a close.

4

The Provincial Exhibitions and Economic Development, 1846–1893

From mid-century on, there were a number of large-scale provincial exhibitions. They were a logical outgrowth of the county fairs and encouraged exhibitors to raise their sights, but they also changed the dynamics of the fair. The exhibition expanded to accommodate economic and cultural diversification and became more than a simple farmers' festival. Exhibits now included industry, fine arts, natural history, fancy-work, and many oddities to bemuse the judges and amuse the audience. The spectators were no longer humble farming folk anxious to raise plumper porcines: By the end of the century, hundreds of thousands queued behind the turnstiles leading into the big fairs. Some of these shows were run as quasi-private enterprises, and they posed a threat to the project of edification by ranking profit before improvement. This chapter sets out the framework within which the large provincial and private fairs developed. Once again, the account is one of sustained growth in Ontario and of stuttering, halting development in Quebec. In Ontario, the system of exhibitions successfully aroused a spirit of competition among regions, each anxious to attract the most exhibitors, the largest crowds, the wealthiest purchasers from across North America. In Quebec, regional rivalries had the opposite effect. The different districts were unevenly represented on the provincial exhibition's management committee, and this disparity created such resentment that many breeders and producers did not trouble to exhibit outside their district.

Quebec's exhibitions lagged behind those of Ontario, just as its economy lagged behind that of Ontario. However, it is impossible to distinguish cause from effect, to determine the extent to which a successful exhibition merely illustrated or actually caused economic development. Where prosperity and a flourishing exhibition coincided, there is no way to prove that the exhibitions had a significant role in causing the prosperity. Historians have always found sufficient explanation for economic development without resorting to exhibitions. The fact that farmers abandoned grain production for dairying can be attributed to market conditions rather than exhibitions, a symbolic reflection of the market. Although many late-nineteenth-century farmers did behave like rational economic agents of the sort the eighteenth-century improvers had sought to create, there is no way to measure the role exhibitions played in this development. But if the role of exhibitions in winning farmers to a capitalist mode of production cannot be measured in any reliable way, there are indications to suggest that they did play a role in developing agricultural industries. The final section of this chapter will examine the economic impact of exhibitions. This impact took two forms. First, exhibitions encouraged specialization within particular agricultural industries, and, second, they fostered the principle of economic management through state intervention.

The provincial exhibitions had local and international antecedents, beginning with the development of national and state agricultural shows in the United Kingdom and the United States between the 1820s and the 1840s. Scotland had national exhibitions beginning in 1822, and Ireland, in 1831. Yorkshire began its northern shows in 1837, and, two years later, the Royal Agricultural Society began to hold annual exhibitions around England.[1] Many Canadians bought their breeding stock at British and Scottish fairs. Americans began to hold exhibitions in Massachusetts in 1811, when Elkanah Watson displayed imported Merino sheep, and by 1817 that state was voting $200 for exhibitions. A New York State Agricultural Society was formed in 1832, holding its first State Fair in 1841.[2] Canadians bought and showed breeding animals in New York, but, without a comparable institution at home, could not hope to rival American standards. They also attended en masse: in 1848, 150 Canadians camped at Niagara Falls when the trains could not carry them all home.[3]

Calls for a provincial exhibition in Canada West began to appear in the early 1840s in the *British American Cultivator* edited by W.G. Edmundson, an ex-teacher, farmer, and vendor of implements.[4] In 1845 the Home District Society organized meetings in Toronto and Hamilton, an Agricultural Association was soon formed, and in the fall of 1846 the first provincial exhibition was held. Prizes worth $1100 were distributed among the 1150 entries, and receipts exceeded admissions by $408.[5] Edmundson declared the competition 'both spirited and creditable,' and reported high prices paid for animals: £57.10s for a three-day-old heifer. John Howitt won three of the seven Durham prizes; there were few other breeds. As well as the beasts and grains, visitors could admire implements, fruits, flowers, fancy-work, stuffed birds from Henry Croft, and 395 entries of factory and artisanal work. At the banquet that evening, conservatives and reformers gaily toasted one another, and keynote speeches were given by high Tory Chief Justice J.B. Robinson[6] and Clear Grit Adam Fergusson, an early importer of purebred Durhams. He praised the sturdy yeomanry, advanced their claims upon the public, and invited Canada East to join in the movement. Fergusson became vice-president, Edmundson became secretary, and Colonel E.W. Thompson, president of the Home District Society, became the president of the provincial association.

In 1847 the Agricultural Association was incorporated, though it was refused a grant until 1849. At the second exhibition, held in Hamilton, torrential rain transformed the grounds into an enormous mud puddle. Organization was chaotic and banquet tickets were not issued to journalists, who repaid the snub with silence. The Streetsville *Review* threatened that 'in time to come prize bulls will roar, and *diploma* pigs grunt in inglorious silence so far as our newspapers are concerned.'[7] In 1848 the fair travelled to Cobourg, and in 1849 to Kingston, where a small crowd of 'well dressed and well behaved people' heard Scottish agriculture professor James Johnston lecture and watched a less seemly group of 'Phisiogs,' 'dressed and painted in the most outlandish style,' parade with torches through the grounds after dark.[8] Niagara had its turn in 1850, Brockville in 1851, and the exhibition returned to Toronto in 1852. That year entries exceeded 3000, the prize list amounted to nearly $5000, and visitors numbered 30,000. One told a friend: 'On Thursday the City presented the appearance of one dense moving mass of human beings. The Exhibition was a creditable one to the Province, but there were too many people to see anything satisfactorily & they had to look at each other.'[9] The exhibits, sprawled over seventeen acres, ranged from a fine array of cattle (191 head) and horses (212), to fancy-work, art of a 'distressing mediocrity,' daguerreotypes, American-made reapers, carriages, furniture, and even a $500 grand piano.[10]

Canada East lagged behind Canada West in developing provincial fairs. In 1850 Montreal hosted an industrial exhibition where articles were selected for the Great Exhibition in London in 1851. However, not until 1852 were a Board of Agriculture and an Agricultural Association created for Canada East, after the model of Canada West, and in 1853 the first regular provincial exhibition was held on Sulpician land on Sherbrooke Street.[11] A prize list of $6000 drew about 200 cattle, 50–60 horses, 300 sheep, and 100 pigs, and about 500 other entries, less than half as many as the Provincial Exhibition in Hamilton attracted that year with a comparable prize list. Ayrshire cattle and swine were good, but the sheep were 'as bad as they could be.' A total of 35,000–45,000 visitors paid £1375 in admissions. Attractions included a torchlight procession, fireworks, a lacrosse match, a soirée, and a fancy ball. Exhibits included mowers and threshers made in the province, fire engines, cloths, soaps, dyes, and a sewing machine. Horticulture was magnificent, as were the oats and barley, but not the wheat. English Canadians, some of them from Upper Canada, dominated the prize list, winning four times as many prizes as the French Canadians and all but four prizes for pedigreed cattle and sheep, as well as most of the awards for manufactures.[12] French Canadians won most of the grade cattle and sheep awards and three-fifths of those for fancy-work.

Agriculture was the sturdy foundation of these provincial exhibitions, but industrial and other artifacts rounded them out. Two influences on their development were the mechanics' fairs of the 1840s and the international exhibitions held from mid-century on. In Canada the two developments were closely related, for in 1857 the legislature created a Board of Arts and Manufactures to oversee the mechanics' institutes and to organize industrial displays in Canada and abroad.

During the 1820s mechanics' institutes had begun to appear, first in London, then the United States, and they reached Montreal and Toronto at the end of that decade. They were sponsored by social elites, and they purveyed technical instruction to working men, along with moral and political education. The Toronto Institute defined as its task in 1857 'to direct and develope the moral and intellectual energies of the industrial classes of society.'[13] To attract more members, the institutes began to hold literary and musical soirées, excursions, and exhibitions. The Montreal mechanics' institute had its first exhibition in 1843, Toronto in 1848, and Bytown in 1853.[14] English improvers praised the Canadian exhibitions for their usefulness and their abstention from 'the firecloud and phantasmagoria.'[15] The institutes defined their mandate broadly: 'While the chief design of these Exhibitions is to encourage the manufacturers and artizans of our country, by furnishing them with opportunities of submitting specimens of their skill and ingenuity to the public, it is also sought by them to foster a taste for the Fine Arts; and by placing within easy access of all classes objects of curiosity, excellence or value in nature and art, to diffuse throughout the whole community a desire for the examination of whatever may tend to elevate, instruct and refine the mind and thus promote and illustrate the progress of Industrial Art in Canada.'[16] Previous connoisseurs of the arts prided themselves on the exclusivity of their judgment, but these men tried to popularize their tastes. This form of cultural hegemony promised not only to increase the value of manufactured goods but to 'civilize' the urban industrial classes, fulfilling the task that a love of flowers, nourished by horticultural exhibitions, was supposed to perform in the countryside.

The speeches were probably the most impressive part of these events. At the Toronto exhibition of 1848, the 195 exhibits were mostly mechanical or manufactured goods (42%) and fine art (37%), with a sprinkling of needlework, natural history, and Indian curios. In 1850 the exhibition brought in £56, but in 1851 receipts dropped to £12, and only 13 of the Institute's 340 members exhibited. There were no more exhibitions until the 1860s, but even in 1866 it was remarked that 'a small steam-engine, a few models, and the one each of Steinway's and Fox's pianos comprise the principal articles in machines and manufactures.'[17] Profits peaked at $800 and visitors at about 9000, but by the end of the decade these exhibitions lost money and were abandoned once again. In 1876 another attempt was abandoned when women members refused to help.[18] The Montreal Mechanics' Institute festivals began to earn a profit by the end of the 1840s. At a typical festival, the walls were covered with banners, 'appropriate mottoes,' and the names of great engineers; sometimes prize essays were read on the uses of education. Among the miscellaneous exhibits in 1853 were a working printing press, three sewing machines, scales, daguerreotypes, biscuits, and even grains and implements.[19] The February 1850 festival was crowned by a drunken scuffle, as anti-annexationists tried to destroy a portrait of Washington hanging on the wall.[20] The Montreal fairs soon began to lose money again, and the institute found other sources of amusement and profit. In 1886, however, it nostalgically recalled the days when 'the Mechanics' Institute was the only place of public meeting in Montreal, and the annual "Mechanics' Festival" was the great event of the year.' The Toronto Institute had closed its doors three years earlier. Others in Ontario and Quebec survived largely as libraries, and if they held exhibitions they lost money or showed only art, curiosities, and natural history, as at Elora and Paris in 1874.[21]

Mechanics' exhibitions needed art and needlework to succeed. Mechanics took little interest in them, leaving the field open for capitalists. With the decline of artisanal work, mechanics were only anonymous contributors to factory production. Urban artisans and working men preferred presence to representation, marching in the trade processions that accompanied important public events.[22] To some extent, disinterest in the exhibitions reflected disinterest in the mechanics' institutes, which were middle-class organizations designed to improve productivity and prevent disorder. The *Mechanics' Magazine* objected from the beginning that 'men had better be without education than be educated by their rulers.'[23]

In the mid-1850s the Montreal institute began to organize the industrial section of the provincial exhibition, but their *ad hoc* intervention drew attention to the fact that farmers had a funded board for such matters and industrialists did not. The institutes lobbied for their own bureaucracy and, in 1857, were rewarded when a Board of Arts and Manufactures was created in each section of the province, after the model of the boards of agriculture, to oversee the local institutes and organize industrial displays. The effect was disastrous. The boards demanded control of the institutes' grants, listing abuses as a reason; instead, the grants were abolished and many institutes closed their doors. The Lower Canada board, composed of delegates from the Montreal Mechanics' Institute, now existed mainly to oversee this same institute. Objecting to their paltry grant of $2000, one-tenth of what the agricultural boards received, the boards of arts refused for several years to co-organize the provincial exhibitions. They became involved in organizing night classes, though their efforts were never effective. As late as the 1880s the Quebec board, now the Council of Arts and Manufactures, fulfilled its mandate of technical education solely by offering drawing lessons, the fruits of which it proudly exhibited locally and internationally.[24]

The boards' activities were economically and educationally marginal. There was a board clique that included some able businessmen, but the bureaucrats ran things from the beginning. Ostensibly standing for an industrial interest, these men represented neither mechanics nor manufacturers, as was evident when labourers and manufacturers formed their own groups to press for tariff and workplace reforms. Almost the only reforms sought by the boards of arts concerned their own constitution. This narrow interest was largely true of the boards of agriculture as well, though they did influence quarantine legislation and encouraged new industries like hemp and beet sugar. In each case, the quasi-official status of these boards prevented them from airing political concerns.

The Great Exhibition of 1851 and its successors gave an impetus to the development of provincial exhibitions. They bestowed grandeur and importance on the activity of exhibiting and convinced local governments of the usefulness of having an exhibition bureaucracy. The great period of growth for this bureaucracy was during the 1850s, when Canadians sent exhibits to London, Paris, and New York. Local exhibitions were thought the best preparation because they not only inspired producers to make a better product but also judged the outcome. As well, the world's fairs established the genre of the exhibition as a way for societies to reflect upon their condition. The mid-nineteenth-century Canadian obsession with measuring progress through exhibitions might be compared with the late-twentieth-century obsession with measuring national identity through commissions and panels. Canadians needed constant reassurance that history was following its proper course, and they found it at exhibitions. North American exhibitions were more encyclopedic than those of England, where the Royal Agricultural Society exhibited only livestock and implements. In England there remained an enormous gap between agricultural and international fairs, but in Canada the same men organized both displays and encouraged the same exhibits. The provincial exhibitions were accorded an epistemological importance as a total representation of colonial society, from farmers and fishermen to manufacturers and mechanics. English exhibitions represented either agriculture or the entire world, but the Canadian exhibitions were understood to measure and affirm national identity. This was true of the other British colonies: at Nova Scotia's first industrial exhibition, held in 1854, Attorney General William Young declared that the purpose of the display was 'to make us better acquainted with our country, to make Nova Scotia more familiar to its own people. If to "know one's self" was, according to the ancient moralist, the perfection of human wisdom, it is of equal importance that the inhabitants of a country should understand its capabilities and be able to appreciate its progress.'[25]

Across British North America, international exhibitions provoked provincewide exhibitions and permanent exhibiting bodies. Nova Scotia's first Provincial in 1854 used the four classifications of the Great Exhibition: Mineral, Vegetable, Animal, and Manufactures kingdoms. It was a commercial leviathan indeed, for the exhibits were draped around the government buildings, with the council chamber housing fine arts, and the Legislative Assembly 'filled from top to bottom ... with the handiwork of the fair wives and daughters of Acadia.'[26] The industrial display was disappointing, and a more traditional classification, based on livestock and field products, was later adopted. New Brunswick's first industrial exhibition was held in 1852, while Prince Edward Island's was organized in 1850 to collect an exhibit for the Crystal Palace. On Vancouver Island, when a public meeting was held to consider a representation for the London Exhibition of 1862, journalist, politician, and booster Amor de Cosmos urged his listeners to organize a regular provincial exhibition as well. The British Columbia mainland had its first industrial exhibition in 1861 to select goods for display in London, but it began inauspiciously: 'The Local Exhibition was opened on the afternoon of the 13th instant, at 2 P.M.; but, in consequence of a sad accident which occurred during the firing of the salute, the opening address by Col. Moody, R.E., was not delivered until the evening.'[27]

By mid-century the provincial exhibition was a major event in Canadian life, and a great deal of economic, political, and cultural importance was attributed to it. Over the next few decades, it was gradually stripped of its importance. The remainder of this chapter traces what happened to these fairs: their development or devolution and the rise of private fairs, as well as their economic significance.

The travelling Ontario provincial exhibition fostered the growth of rivals. When it visited small and eastern towns such as Cobourg, Brockville, and Kingston, the Provincial invariably lost money. It was more successful in the west, where the livestock trade was centred, or in Toronto, where huge crowds turned out. After 1858, when organizers and exhibitors were 'fleeced, peeled, and skinned' with inflated accommodation, lumber, and train prices at Brantford, the association determined to limit the Provincial to towns with permanent buildings: Toronto, Kingston, London, and Hamilton, which rushed to build a Crystal Palace, and fell into bankruptcy.[28] In 1875 Ottawa joined the roster, and, in 1883, Guelph was added. Permanent exhibition grounds were expensive: They required a Crystal Palace for manufactures and fine arts; extensive shedding for livestock; separate buildings for horticulture, dairy, and implements; a restaurant; and a horse ring or grandstand – all spread out on several acres of grounds that had to be fenced and kept up. Toronto spent $24,000 on the grounds in 1858 and $100,000 when they were rebuilt thirty years later. It seemed wasteful to go to this expense when the site would be used for only two weeks every four years. London, where the fair was most profitable, resented subsidizing failures elsewhere, and in 1865 the *Free Press* suggested starting a fair of its own.[29] In 1868 the Middlesex Agricultural Society held the first Western Fair, offering prizes of $2500 (a quarter of what the Provincial offered). In 1870 the Western Fair Board offered $6000 in prizes and took in close to $10,000, and in 1877 prizes topped $10,000 and 75,000 visitors paid $18,776. Londoners magnanimously wished success to 'all minor exhibitions like the Provincial and county fairs.'[30]

Others followed. Guelph held a Central Exhibition from 1871 which, by mid-decade, was awarding more than $6000 in prizes and taking in $4000–5000 in admissions. Here, as at London, 'the great majority of visitors first pay their respects to the livestock.'[31] Hamilton Central Exhibition was held from 1872 on. During its first decade, the Central paid $4000–5000 in prizes and earned a profit of $2000–$3700 each year. Peterborough, Northumberland, Victoria, and Durham collaborated on yet another Central Exhibition from 1875, and its receipts ranged from $2200 to $3000; in 1885 it settled in Peterborough.[32] The Ottawa Agricultural Society held successful exhibitions through to the mid-1870s, paying prizes of $1800–$2500: admissions covered two-thirds of that cost, and subscriptions and grants the rest.[33] However, after building permanent grounds to host the Provincial in 1875, the society fell $10,000 into debt; the next Provincial, in 1879, raised this amount to $26,150, for which the directors were personally liable. This crisis coincided with the collapse of the Ottawa lumber trade and the bankruptcy of some directors, such as lumberer James Skead. These men finally convinced the city, which had invested $38,000 in the grounds, to take over the site in 1888 through the Central Canada Exhibition Company – which has run the Ottawa Exhibition ever since.[34] By the 1880s, competition between exhibitions was beginning to reduce receipts. Above all, there was the Toronto Industrial Exhibition, which set the standard for the private show.[35]

Toronto always drew the largest crowds at fair time, and, from mid-century, Toronto papers began to insist that the Provincial be permanently established there. In 1858 the city built a Palace of Industry on a 20-acre site south of the Lunatic Asylum on Queen Street West[36] and, after failing to retain the Provincial in 1859, decided to organize a regular fair. The first, held in August of that year, drew about 200 cattle and 80 horses, as well as some relics from the War of 1812 and handicrafts shown by a Mohawk chief. Several thousand city folk came, but farmers stayed in their fields, waiting to see if buyers turned out. By October the fair was clearly 'a failure in every sense of the word.' Livestock consisted of fourteen horses and cattle, while the Crystal Palace was empty but for a 'venerable matron' selling apples and cakes.[37]

In 1878 the grounds were moved west along King Street to their present site and occupied a parcel of 60 acres. A group of property owners defeated a bylaw to raise $150,000 in taxes to pay for this land, but, faced with the humiliating prospect of losing the Provincial, the city council inserted the cost into its budget for the year and postponed educational spending. By this time it was May, and crews slaved to finish the work in time. When the Crystal Palace was opened in August, it bore a sign saying, 'The Exhibition Committee of the City of Toronto exhibit this building as a pretty fair specimen of what the mechanics of Toronto can do in three months.' To recoup the costs, Toronto decided to organize an annual show. A board was set up, representing the city, manufacturers, merchants, fruit growers, poultrymen, stockbreeders, and the Ontario Society of Artists. The organizers sought public money, but Premier Oliver Mowat flatly refused to fund Toronto at the expense of the rest of Ontario.[38]

The first Toronto Industrial held in 1879 drew about 4000 entries and 100,000 paying visitors (London and Hamilton at the time had over 7000 entries, and Ottawa close to 10,000). Total receipts were about $52,000, more than four times what Ottawa took in ($11,656) and more than the previous two Provincials, at Toronto ($23,488) and London ($21,734), together. They did not cover expenses, which included a $20,000 prize list, a similar sum in operating expenses, and $25,000 in improvements to the grounds. In 1880 the Industrial receipts and expenditures balanced at $35,000, and the next year it earned a profit of $1576. Soon, the Toronto show outdistanced all the other fairs: in 1884, while the Western Fair drew 53,000 visitors during its week-long show, the Toronto Industrial accommodated that number in a single day. In 1888 approximately 300,000 visitors paid $60,000 to pass through the gates at Toronto. In 1891 the gate receipts at Toronto were $64,090, and at London, $18,470. These were enormous crowds, swelled by trainloads of visitors from the countryside. The other fairs could not compete with this behemoth, but all benefited from it. In 1884, as the Toronto Industrial closed its doors for the season, twenty-seven train carloads of exhibits and livestock left for the Western Fair, and the next week twenty-nine loads were trundled from London to Ottawa for the provincial exhibition.[39]

The Provincial, meanwhile, declined. Toronto and Hamilton refused to host it, resenting the intrusion of the board. London accepted it every four years, but otherwise it was restricted to Ottawa, Guelph, and Kingston, where it annually lost thousands of dollars. Because the board represented regions across Ontario, it met rarely and at great expense, and it lacked the support of private citizens who rallied around the local and private fairs.

One cause of the Provincial's decline was a change in exhibition culture. It received money because it was serious and improving. But after Confederation, the key to a successful exhibition became attracting crowds, so that large prizes could draw the important breeders. The Toronto Industrial could offer $25,000 in prizes, twice as much as the provincial exhibition. To bring the crowds to Toronto, manager H.J. Hill offered fireworks, races, acrobats, balloon ascensions, even wild west shows re-enacting battles of the American West. Sideshows and patent medicine hucksters proliferated. Many objected to the riff-raff and to the sale of liquor, which the city and the organizers winked at. Women performing in scanty clothing shocked conservative visitors, who did not expect to find them at ostensibly 'improving' public events frequented by women and children. Fair boards defended these amusements as harmless, but the Provincial's organizers could not spend public money on them: 'Suppose $1,000 extra came in through means of attractions, and that *one* young man is started on the down grade to ruin as a result ... with his morals contaminated, hopes blasted, and moral decay written upon his temples. Has the country gained or lost? ... The state is undoubtedly not guiltless if she knowingly aids in any such work, or even winks at it.'[40] The Agricultural Association became an apologist for the purely agricultural show, free of vulgar amusements. At the International Fair and Exposition Association meeting in Milwaukee in 1889, Agricultural Association secretary Henry Wade 'championed the sort of exhibition that was thoroughly educational and which drew people by that fact alone. He was fearful, he said, of the tendency of late to make exhibitions part of what was known as "show-business." He believed that an exhibition ought to draw without the employment of acrobats.'[41] Hill responded that this ideal was fine in theory, but exhibitions must also stay in the black. This exchange, coinciding with the death of the provincial exhibition, marked the failure of the project to elevate public taste by means of exhibitions. Public taste proved more transforming than transformed.

The Agricultural Association tried to nobble its rivals by demanding a legislative monopoly. In 1879 its president hinted at legal action to squelch the private shows, and ten years later the association demanded that county exhibitions be abolished because they detracted from the Provincial.[42] The provincial exhibition, designed to encourage competition in all things, was run by men who sought laws to protect them from the free market. In the legislature, one member denounced the Provincial for exercising a 'retarding influence, as it tends to prevent a free competition in agricultural exhibitions.'[43] In 1889 the grant was cancelled and the last provincial exhibition was held. The Agricultural Association continued to sponsor competitions for best farm and for essays, but the farm competition never drew more than a handful of entries and was criticized as awarding prizes for fancy buildings, while the winning essays, usually penned by journalist Thomas Shaw, were exercises in moralizing that blamed the decline of agriculture on farmers' lack of knowledge, receptivity, diligence, and manliness.[44] The association funded an annual fat stock show, but probably did more harm than good by trying, unsuccessfully, to develop a show at Toronto to rival that of Guelph. While the Guelph Agricultural College and agricultural societies flourished, this centralist branch of the program to encourage agriculture withered. In Ontario, then, bureaucratic organizers like Henry Wade failed to wrest exhibitions from public control. The immense profits made from the 25 cent admissions paid by hundreds of thousands of visitors led fair organizers to cater to popular opinion.

In Quebec, bureaucrats exercised greater power over the organization of agriculture, but the result was an inefficient system that lacked public support and lost money. Beginning with William Evans, these men worked to undermine exhibitions: Evans staved off the first provincial exhibition for several years by diverting the public grant to his journal. Another obstacle to the development of provincial exhibitions was the fact that only Montreal could draw crowds sufficient to cover expenses. But unless the fair circulated around the province, the French Canadians would hardly benefit from it. Thus, a quasi-geographical, quasi-ethnic rivalry, which sprang up even within the agricultural bureaucracy (between the Montreal-based Board of Agriculture and the Quebec-based director of agriculture, Édouard-A. Barnard), impeded the development of large-scale exhibitions in Quebec.

The pattern of rivalry was firmly established in the early years of the Provincial's existence. After a successful first year in Montreal, the exhibition travelled in 1854 to Quebec City, where a modest collection was gathered, described as equal to an Upper Canada township show. The sheep were 'such as a beggar would steal if he wished to create compassion, their appearance being that of animals in the last stage of a galloping consumption.'[45] Poor communications, cholera, general elections, and rain were blamed.[46] The next Provincial was in the heart of the English district, the Eastern Townships, and attracted few exhibitors from any distance. In 1856 the show moved to Three Rivers on the north shore, but even here Britons won almost two-thirds of the prizes, though the French did well in butter, swine, horses, and Canadian cattle (the Montreal *Gazette* complained that prizes of £5 went to scraggy beasts not worth £2).[47] Half the stands and pens stood empty, leading to complaints that the meagre display discredited the province. The Montreal show incurred a debt of £180, that of Quebec a further £468, and the Sherbrooke fair of 1855 another £388. At Three Rivers, admissions brought in £100, but the fence to keep non-payers out had cost a larger sum.[48]

The Provincial returned to Montreal in 1857 and remained there for two more years. Montreal had the best grounds, drew the largest crowds, and attracted the best breeders, and the Grand Trunk Railway provided free passage. The other towns fought back: In 1859 the mayor of Quebec, Hector Langevin, canvassed the delegates to support Quebec, but

Montrealers again outnumbered them twenty-five to twenty-three. Defying the votes of the agricultural society delegates, the board cancelled the show on grounds of financial embarrassment and held a meeting of the association in Three Rivers. It was attended by only twenty-two delegates, and they voted for Quebec.[49]

That year, 1860, was an unfortunate year to boycott Montreal. Victoria Bridge across the St Lawrence was newly completed, and Queen Victoria sent her young son Albert, Prince of Wales, to open it. The Lower Canada Board of Arts and Manufactures obtained a grant of $20,000 for a 'Great Exhibition' in Montreal, and it invited farmers and industrialists from across British North America. All three of the other boards refused the invitation. The Lower Canada Board of Agriculture was determined to have its exhibition in Quebec City, while the Canada West boards were determined to have their own show and quickly applied for their own grant. Closer to home, the English-dominated Board of Arts alienated the French-dominated city council and forfeited a $20,000 vote by insisting that the Crystal Palace be built on McGill University property, in the fashionable English district, rather than in the eastern, French-speaking part of the city.[50] (The councillors proved to be right, for the site was too congested for agricultural shows, and the palace became a white elephant.) Then the English elite decided to boycott the city's reception for Prince Albert and to stage their own. After they subscribed $10,000 for this event, they learned that the prince could not accept a private invitation, and they were forced to come to terms with Mayor Rodier. But the wealthy citizens had exhausted their goodwill and did not subscribe to the palace.[51] The board finally borrowed $12,000 from McGill, saddling itself with a debt it could not repay.

Many people denounced the idea of an exhibition on principle. One newspaper insisted that the prince should be shown the land and its people in a natural state, 'and not got up like a children's tea-party, with magic lantern effects – and small gingerbread efforts at astonishment.' Moreover, Canada's strong point was its 'boundless' forests, fields, and rivers: how could any collection of isolated rocks, grains, and planks convey boundlessness?[52] The editor of the Montreal *Gazette*, Brown Chamberlin, who was also the secretary of the Board of Arts and the exhibition's main organizer, replied that the prince could learn more from an exhibition than from touring the land: 'Not our towns alone, for they are but pigmies; not alone our people in holiday attire, and the beautiful scenery which he will encounter in his travels, but the fruits of the handiwork and toil of our laboring men, and the marks of a gradual development of a taste for the refinements of cultivated life.'[53] In the end, this claim – that a representation could speak more powerfully than the thing it represented – was rejected. Exhibits at world's fairs gave Canada some genuine physical presence, but the same display, moved to Canada, was separated and distinct from the world outside the Crystal Palace and was more obviously a representation standing in for something greater. In any case, the prince finally did see this real world for himself on tour.

Albert opened the exhibition with the gracious remark 'I am not ignorant of the high position obtained by Canada in the Great Exhibition of 1851,' and he walked around the exhibits with McGill principal J.W. Dawson as his guide. The *Times* correspondent was less gracious, remarking, 'There was not a great deal in it to delay this part of the ceremony.'[54] After the prince left, more exhibits arrived, among them a pyramid of cutlery, a velocipede, and a 'pretty bust of Her Majesty in spermaceti.' Of 460 prizes, French Canadians took less than one-fifth, Canada West took one-tenth, and Nova Scotia won a few for minerals.

The Hamilton show, meanwhile, was a puffed-up provincial exhibition with an inflated prize list of $15,000, which *The Times* pronounced equal to the best county shows at home. The provincial exhibition in Quebec, which the prince did not visit, was a smaller event that drew 110 horses, 300 cattle, 120 sheep and swine, and a smattering of agricultural products. The governor general braved the rain and mud, but others were less hardy. Locals blamed the Montreal-based Board of Agriculture for the disorder.[55]

The Crystal Palace affair in Montreal was far from over. The board spent most of the 1860s pleading with politicians for more money, and it barely averted a forced sale in 1862 when the sheriff seized the palace for debt.[56] In the meantime, it was used as an armoury and suffered damage from horses' hooves, while faulty girders caused the corners to bulge and a leaky roof let in the rain. Well into the 1870s the federal government, McGill University, and the Board of Arts all claimed legal title to the dilapidating edifice.[57] In 1873 the university agreed that, for $12,000, the government could have the building if it removed the palace, for the land had been sold. The Board of Arts was in no position to take advantage of the offer, for it had fallen into complete disarray and divided into rival French and English factions, each holding its own elections and claiming to be the true and legitimate council.[58] Finally, an order-in-council named a new Council of Arts with French and English members.

There remained the problem of where to put the troublesome palace. The Council of Agriculture had bought land from the nuns of the Hôtel-Dieu on Mount-Royal at Mile End, but it had designs on a more spacious terrain on Sherbrooke Street East. It refused to improve the Mile End property until forced by law to do so in 1875, when it signed a $12,000 contract for buildings, using money voted by the city.[59] The Crystal Palace idled three more years until 1878, when McGill insisted on immediate removal of the structure. Premier Joly de Lotbinière, an enthusiast of exhibitions, took decisive action. That fall the eighteen-year-old Crystal Palace was dismantled and moved, 'iron carcass,' bricks, and all, to the grounds at Mile End, where, rebuilt, it did yeoman service until it burnt down in 1896.

Sherbrooke had the Provincial in 1862, but it returned to Montreal for 1863, 1865, 1868, and 1870. J.E. Turcotte, mayor of Three Rivers, protested 'l'esprit d'envahissement de la grande Babylone dont les tendances et l'habilité à tout monopoliser pèsent lourdement sur les intérêts agricoles.'[60] But the meagre Sherbrooke display received scathing reviews: 'The contents of the gallery would hardly have been accepted as a gift by a Pawnbroker – indeed, it seemed as if all shows of that description had been ransacked. Old prints, dilapidated Indian curiosities, worsted work and drawings, which any bread and butter Miss, who had just entered her teens, would have blushed to acknowledge – all things at our Agricultural Show, largely subsidized by the provincial Government.'[61] The Provincial did not return to Sherbrooke for a quarter of a century. Quebec was offered the exhibition in 1867 and 1868, but the city council failed to vote the necessary funds.[62] In 1871 Quebec City finally hosted the Provincial by a vote of $4000, and again in 1877 by voting $6000.

Even though the Montreal shows were not stale and flat, they were unprofitable. Those held in the 1860s racked up debts of $3157, $2322, $1886, and $6009.[63] Nonetheless, they bolstered Quebec egos by drawing prominent Ontario breeders. In 1865 George Miller and John Snell of Ontario competed neck and neck with James Logan of Montreal, the Dawes of Lachine, and Townships breeders like H.D. Moore and Stevens Baker of Missisquoi. But French Canadians also won many prizes: Adolphe Ste-Marie of Laprairie won 8 prizes, mostly for sheep, and C.A.M. Globensky won sixteen, mostly for swine. Joseph Perrault noted proudly that French Canadians accounted for 648 entries out of 1763, or 37 per cent. They showed three-quarters of all agricultural products on the grounds (grains, roots, dairy) and one-quarter of the livestock, where they won one-third of the prizes for horses, 13 per cent of cattle prizes, 21 per cent of sheep prizes, and half the pig prizes.[64] Ontario breeders won more than half the cattle prizes and one-third of the sheep prizes, but showed no swine and few horses. In 1868 French Canadians accounted for 32 per cent of the agricultural prizes (534 of 1687), and at Quebec City in 1871, 35 per cent

(148 of 420). French Canadians were strong in Leicester sheep, native horses, tobacco, and cereals, but Britons won the most valuable prizes for pedigreed stock.

With the Crystal Palace erected on the new fairgrounds, Montreal hosted the Provincial three years running, in 1880, 1881, and 1882. Year after year, the costs of these exhibitions rose while their receipts fell (from $27,874 to $10,000). Another, in 1884, was 'un fiasco sur toute la ligne. Peu d'exposants, infériorité des articles, et enfin déficit dans la caisse du comité.'[65] Exhibitions in Quebec City were smaller: in 1871 and 1877 the admissions and the prize list balanced each other at $6000. In 1885 smallpox struck Montreal, and city authorities converted the Crystal Palace and other buildings into emergency hospitals.[66] Sherbrooke seized this opportunity to reclaim the Provincial in 1886. The bid was organized by the Eastern Townships Agricultural Association, formed in 1885 to build up a cattle market that would attract American and English buyers. The dominion minister of agriculture, J.H. Pope, himself a Townships breeder, secured a federal grant for it.[67] The first show, held in 1885, was a great success, with $5000 of prizes awarded.

French-Canadian nationalism brought the Provincial to Quebec City in 1887. Honoré Mercier, the newly elected premier, personally insisted it go there and subscribed $50 towards the budget.[68] Managed by Joly de Lotbinière, the show was a success, with more than 50,000 visitors paying over $10,000 in admissions. Organizers toured the district to bring out exhibitors and visitors, and the clergy seconded their efforts: the bishop of Chicoutimi told priests to 'engager vos paroissiens à envoyer à cette exposition les produits de leurs fermes et de leur industrie, et de s'y transporter eux-mêmes.'[69] Six hundred *colons* came from Lake St-Jean and Charlevoix bearing exhibits. The affair was seen as an important expression of French-Canadian culture.

During the 1890s, private companies in Montreal and Quebec organized shows that replaced the Provincial. Extensive documentation records the stormy beginning of the Quebec Compagnie d'exposition in 1894.[70] It was formed by a group of politicians and businessmen, led by Senator Philippe Landry, agronomist and Tory stalwart, who invested several thousand dollars in it. The wealth of manufacturers and merchants on the board led one reporter to remark that 'les directeurs pourraient prèsque faire l'exposition à eux seuls.'[71] Joly subscribed $200, but refused to do more because he thought exhibitions should be run by the state. The company had secured a promise of $15,000 from the provincial government and $10,000 from the city, but a reforming mayor, S.-N. Parent, was voted into office, at a time when reform meant municipal control of private utilities.[72] Parent demanded that the provincial government run the exhibition and, when the Conservative government refused, he withdrew the funding, though he finally returned most of it. In Montreal, S.C. Stevenson tried to discourage this rival to his own Montreal Exposition Company, which, he remarked, had not yet paid a penny to shareholders. 'Our experience last year,' he said, 'was disastrous and not such as to encourage any new Association in undertaking similar work.' In 1893 the Montreal Exposition had taken in $46,042.97 and spent $45,961.63, leaving a profit of $81.34.[73]

Lieutenant Governor Sir Narcisse Belleau, chair of the first Quebec Exhibition forty years earlier, opened this new one but did not live to close it. Twenty thousand visitors gathered at the Plains of Abraham on opening day alone, and the $8000 prize list attracted 800 entries, more than the 625 that graced the previous Montreal show. John Featherstone sent swine from Ontario, Thomas Irving dispatched Ayrshires from Montreal, the Oka Trappists and deputy minister of agriculture Siméon LeSage exhibited Canadian horses, and J.D. Guay sent Canadian cattle from Chicoutimi. It was an exhibition run by French Canadians for French Canadians. Nearby was a midway, with amusements ranging from balloon ascensions with parachute jumps, athletics, a naval review, and horse and bicycle races to strongmen, photographers, panoramas, a circus, and an encampment of Lorette Indians. A bear that escaped and wreaked havoc among the pigs had to be shot. A waxwork show of instruments of torture through the ages was well appreciated, while General Wolfe's sword occasioned remarks that the field of war had become a field of peace. In 1898 the company ambitiously bought grounds in St-Charles and put up buildings far beyond what it could afford, landing it $40,000 in debt. That year, a surviving list of entries shows 90 per cent of the 435 exhibitors came from Quebec, with one from New Brunswick, four from the United States, and the rest from Ontario. In 1899 receipts were $11,757 and costs, $10,892. In 1900 the exhibition attracted 132 horses, 407 cattle, 399 sheep, and 208 swine, as well as hundreds of industrial entries.

In Montreal two rival companies, one English and one French, began to negotiate for the grounds in 1889. The French-Canadian press demanded special representation for French-Canadian capitalists.[74] The company that took over the grounds had many French-Canadian officers, but not enough for Joseph Perrault, who made a scene at the shareholders' meeting in 1892, arguing that this respresentation alienated the farmers.[75] In fact, there were extensive displays of French-Canadian cattle and horses at the

Montreal shows, thanks to help from LeSage. Moreover, the French press provided extensive coverage of the event (*La Minerve* sent six reporters in 1880), and the crowds of 60,000–70,000 per day in 1891 and 1982 belie Perrault's remark. Visitors included farming families distinguishable by the sight of women suckling their infants.[76] However, the company never thrived, and the death of S.C. Stevenson (also the long-time secretary of the Council of Arts) in 1898 proved a fatal blow. In 1901 the company was liquidated.[77]

There was also a series of dominion exhibitions in the 1870s and 1880s: provincial shows bolstered by federal grants, half to be spent on prizes and the other half paying for transportation from distant provinces. The first grant was bestowed on the Ottawa Exhibition in 1879, when the local Agricultural Society asked for aid, trying to prevent the losses incurred in 1875. In 1880 the grant subsidized the Provincial at Montreal, but Nova Scotia and New Brunswick refused the offer of free rail transport. In 1881 Halifax received the dominion grant, in 1882 it returned to Ontario, in 1883 it went to Saint John, and in 1884 it was shared between Quebec and Ontario. The project was a failure, rarely attracting serious exhibitors outside the host province. Only Manitoba benefited substantially, sending exhibits of enormous grains and vegetables, Indian artifacts, and even the occasional buffalo. Winnipeg and Vancouver vainly sought the grant in the 1880s, although the government did sponsor an exhibition in Regina in 1895, at a cost of $40,000.[78]

Exhibitions were neither an English-Canadian plot nor a panacea for a troubled community. They did not subsume differences of ethnicity, class, or industry in the name of a common good, but instead intensified local feuds. In Ontario the Provincial was, for several crucial decades, a symbol of unity and prosperity for the whole province, but in Quebec it represented only the district that held it. Moreover, public enthusiasm, measured by admission receipts, was never so great in Quebec as in Ontario. Judged as instruments of hegemony, exhibitions in Quebec were manifest failures. In Ontario, nobody criticized the idea of the exhibition, though interested parties did dispute who should control individual fairs. In Quebec, critics denounced the very idea of exhibitions and laboured to abolish them; most farmers refused to have anything to do with them; and those that were held attracted far fewer paying visitors. A politically charged climate was one reason for this lack of success, as priests and journalists jealously scrutinized all cultural events. Another school of thought blames French-Canadian 'mentality,' which, liberal scholars such as Fernand Ouellet have argued, was reactionary and hostile to innovation and enterprise. But this theory only adds a level of abstraction to the more obvious material explanation, and it invokes an element of choice which, Ouellet's critics have suggested, is inappropriate, given difficult economic conditions in Quebec. These economic conditions alone are sufficient to explain the sorry state of Quebec exhibitions. Even in Ontario, exhibitions succeeded best in the rich farming land to the west, where the livestock trade flourished. Provincial exhibitions held in eastern Ontario lost money; agricultural societies there had some of the lowest memberships in the province, and they held fewer exhibitions. Exhibitions most helped the economy where the economy could most help itself. If land was poor, so were displays, and they discouraged exhibitors and audiences alike.

Exhibitions could not transform an economy, but they could reinforce its pattern of development. Most notably, exhibitions encouraged specialization and concentration within the agricultural economy. On the one hand, they distributed prizes unequally to the more prosperous farmers, breeders, and manufacturers. On the other hand, they provided these producers with an annual gathering point that facilitated the establishment of formal organizations to defend common interests. Exhibitions fostered the growth of 'captains of agriculture' and, it was argued, when these leaders benefited, everyone benefited.

It has already been shown that local exhibitions concentrated wealth by concentrating the prizes. The larger the exhibition, the more marked the concentration. Thousands of participants might take away prizes from a provincial exhibition, but a handful of breeders and producers were certain to dominate the prize lists year after year. Taking livestock as an example, a survey of prize lists for the Ontario provincial exhibition reveals that less than 10 per cent of exhibitors won from one-third to one-half of the prize money (see table 5). Livestock breeding was an expensive and risky business, and probably unprofitable without the subsidies and advertising that exhibitions provided. Large sums were involved: over the five years 1861–5, F.W. Stone, who imported cattle and sheep, won $1869 at the Provincial, and in 1865 he sold $1700 in stock at the Provincial and $1000 at a local Toronto show. Breeders like Stone showed and sold their stock throughout the East and Mid-West, and in 1864 a Conservative paper remarked:

> Canadian breeders have obtained a reputation that reaches far beyond the limits of the Province. They have made themselves known in the far west, and when they have chosen to compete have carried everything before them at the fairs of the central and eastern states. Politicians cannot compare with

> them in wide-known fame. For one who has heard of the existence of a John Macdonald, or a Cartier, or a Brown – who has a dim idea that these gentlemen are respectively leading public men, and a demagogue in Canada – a dozen know all about our Stones, our Snells, and our Millers, whose quiet labors for the improvement of farm stock have gained them wider credit and done infinitely more good to the country than the combined work of all the politicians for twenty years.[79]

In 1867 David Christie sent his animals to the New York State Fair at Buffalo rather than to the provincial exhibition, while F.W. Stone sold six Cotswold sheep at Buffalo for $900, and Canadians accounted for 300 of the 800 sheep at the Illinois State Fair. Canadians won all the Devon cattle and long-woolled sheep prizes at the Ohio Fair in 1872. In 1873 John Craig took 150 swine to the St Louis Fair, where he won a $300 sweepstake prize and sold his animals for $100–$200 each.[80] The famous Shorthorn steer Clarence Kirklevington of Bow Park won every time he competed at Chicago in the 1880s. During the 1870s, Canadians began to sell breeding animals back to Britain: at one sale in 1877, forty-five Canadian Shorthorns brought in £36,350. By 1886, Canadian-bred bulls beat out imports at Toronto.[81]

Most of the leading breeders were from Ontario, but many Quebec breeders earned a good living in Quebec, and some acquired international prestige. In 1888 A. Mousseau reported on twenty years' experience: he bought Cotswold sheep for $782 at a Provincial and subsequently sold $1202 worth of sheep and earned $357 in prizes, leaving a profit of $777, half of which came from prizes.[82] An Eastern Townships breeder, M.H. Cochrane, entered dozens of animals at the Provincials in Hamilton and Montreal in 1868 and won $744, causing 'real, practical farmers' to 'complain bitterly of being unable to secure prizes.' Senator Cochrane became Canada's most prominent breeder, with enormous sales in England and the United States, including $40,000 worth of cattle and sheep sold at the Chicago Stock Fair in 1882, and $20,000 worth in Kansas in 1884.[83]

All this success was good for Canadian breeders, but what did it do for Canadian farmers? It made available first-rate breeding stock and increased the value of grade cattle at a time when farmers turned to livestock husbandry. Shorthorns put on flesh easily and cheaply, and had a low ratio of offal to flesh. J.L. Gourlay estimated that one Shorthorn bull introduced to the Ottawa valley by the Agricultural Society in 1851, then passed on to farmers around the county, increased the value of stock by over $10,000 in

The champion shorthorn steer Clarence Kirklevington of Bow Park, Ontario. Clarence won the sweepstakes at the Chicago stock shows in the 1880s. When he was finally slaughtered, hardy Ontario cattlemen were seen to shed a tear.

the thirteen to fourteen years it served.[84] Ontario farmers were well positioned to take advantage of the growing urban demand for meat in Canada and Britain from the 1860s, although British buyers urged them to raise the standard still higher: 'Unanimously they all speak out loud, "Why don't your people use better bulls?"'[85] Reformers and breeders throughout Ontario waged war on the 'scrub' bulls that continued to predominate, especially in outlying areas, and the average in 1880 was still only one purebred bull to 3000 grade animals.

In 1880 an inquiry into the state of agriculture in Ontario found that, in 459 townships reporting, 53 had no improved stock, and 113 had slight, 126 had considerable, and 117 had general improvement.[86] Often reporters did not distinguish between common native cattle and grades with purebred sires, or they registered only a few purebred stock but did not say if farmers had good grades. Nonetheless, the report suggests that purebred animals were less likely to be seen in rear townships, most of which had no agricultural society because they were too recently or too sparsely settled. Seven townships without pure stock did have agricultural societies, usually with only a few members. A few reported improved stock, but had no agricultural society. On the whole, townships with strong agricultural societies had better livestock than surrounding townships. In Prince Edward County, Hallowell and Hillier had 62 per cent membership in the societies and numerous purebred and good grade cattle, while the other townships reported limited improvement. In Lennox, South Fredericksburg had no society and no purebred stock; North Fredericksburg had much improved stock and a society with 34 per cent membership. Drummond in Lanark attributed the good livestock there to the agricultural society. Nearly half the farmers in Colborne, Huron County, belonged to the agricultural society, and the township had $25,000 of imported animals. In East Gwillimbury, York County, close to 20 per cent of farmers belonged to the society and 20 per cent of livestock were purebred. In Wellington, the heart of livestock country, reported membership was usually above 30 per cent and the report noted that 'Wellington possesses better herds and more good stock than any county in the Dominion. Durham blood is largely diffused and a poor beast is becoming a rarity.'[87] The presence of a breeder or a market led many farmers to improve their stock, but exhibitions certainly facilitated this process.

In Quebec, Barnard convinced the government that dairying was more profitable than the meat trade, and he sought to keep the Canadian cow (a fine milk animal, but too small to breed for meat) free of foreign blood. In 1871 Quebec had half as many cattle, excluding milk cows, as Ontario, and by 1901 the number dropped to 43 per cent. During the same period, the proportion of milk cows in Quebec rose from 63 per cent of those in Ontario to 72 per cent.[88] Exhibitions probably encouraged farmers who raised beef cattle, which looked good on showgrounds, but had less impact on dairy cattle, and perhaps discouraged dairying as an industry. In 1861 Ontario judges remarked that there was a fine lot of Ayrshire grades, 'but there ought to be a separate class for them as milkers, as they are the best adapted for that purpose, and do not show favourably with the short horn grades.'[89] Instead, over time, 'Shorthorn grade' became synonymous with 'grade.' No breed other than the Shorthorn was widely used in Ontario until the end of the century, when the dairy breeds began to spread. The provincial exhibition did much to advertise Holsteins to breeders.[90] But this change required a complete reform in judging criteria, with the introduction of milk tests. As late as 1886 the *Livestock Journal* mocked the Canadian cow for its concave back and convex belly, bones so prominent one could hang a hat on them, and a scrawny udder which 'looked as though it would afford at least tea-milk for one family at every milking.' When they appeared on prize lists, 'the sun went backward ten full degrees in the sky of Lower Canadian agricultural progress.'[91]. Based on appearance, the editor scorned one of the finest milking breeds in the world.

Because size was excessively prized, enormous bloated animals took the top prizes, animals too fat to breed. 'The symmetry of an animal is scarcely ever looked for; if they are big and fat it is all they care for; and there is no doubt if Barnum's woolly horse was shown among the Cotswolds he would get a prize.' The top breeders in England sacrificed the best animals in their herds by overfeeding them, with the prizes won serving as an advertisement for the rest of the herd, which was not so highly fed.[92] When David Christie showed some imported 'Athelstane' Shorthorns, their 'immense proportions evolved some amusing criticisms from the spectators.' Christie stopped exhibiting to keep his animals in breeding trim. Noting this contrast, the *Canada Farmer* declared that 'breeding and showing are conflicting interests.'[93]

Fat stock shows encouraged the raising of monster steers, 'useless hulks of blubber, entirely beyond the digestive capacity of the average consumer.'[94] The Groffs, Germans from Pennsylvania, regularly won hundreds of dollars at fat stock shows with steers weighing from 2400 lb to 3200 lb, twice the ideal weight. At the Chicago Fat Stock Show, where the best breeders on the continent competed, thirty of forty carcasses in 1886 were too fat to eat. In 1887 not one carcass prize went to an animal that had won a prize while still alive.[95] By the 1890s leaner stock with finer lines were winning. Swine competitions also advertised pigs as big as cows, 'mere fat-making machines, with about as much consciousness as jelly-fishes.'[96] But British tastes changed; workers chose to take their fat as butter, not lard, and pork packers like William Davies urged farmers to raise leaner hogs weighing 150–225 lb.[97] Sheep like John Snell's 450-lb 'small moving mountain of mutton'[98] were also slimmed down. In short, exhibitions had a structural bias in favour of large animals, to the detriment of the animals themselves as well as to smaller but economical breeds. However, by the end of the century, organizers had begun to correct this bias, with explicit instructions to judges not to reward obesity. It was also alleged that grain prizes went to heavy varieties that looked good, but were too vulnerable to the midge or other depredations to be raised profitably.

This distinction between what was good for the showground and what was good for agriculture marked another important change in the history of exhibitions: the decline of emulation. The pampering methods used by breeders were very far removed from useful barnyard techniques. Moreover, the shows encouraged prohibitive prices beyond the reach of most farmers. In 1875 Cochrane sold a cow for $18,000, and the next year, at an auction held by Cochrane, Simon Beattie, and John Hope of Bow Park, two cows were sold for $21,000 and $23,000 each to a Kansas stockman.[99] Only the richest farmers could afford even one highly bred bull to improve their grades; others might pay occasional stud fees, but most simply used grade bulls. Exhibitions tended to encourage a division of labour between breeding and farming proper, and, as a result, small-scale farmers could find little to emulate in the prize winners. Exhibitions encouraged specialization in many agricultural industries, and they fostered professional associations and markets for these industries.

Exhibitions facilitated association among agriculturists by bringing them together regularly. Most breeders' and growers' associations in Ontario and Quebec were first formed at a provincial exhibition, and they continued to hold their annual meetings at the large shows. Dairy associations first appeared in connection with the large shows, but then they resorted to conventions instead. Dairy products were not suited to exhibition: they could only be judged by taste, so spectators could not form their own judgment, and they absorbed the 'fetid emanations' of visitors. Thus, one observer remarked of the Ontario dairy associations, 'it was not till they threw up their mammoth exhibitions, and began teaching their operatives direct, that they made any perceptible advance. Since then they have moved on with strides that have put them ahead of all competition.'[100] But if exhibitions were not instructive, they still served as advertisements. Dairymen continued to exhibit, and the number of farm wives showing butter grew during this period. The dairy associations organized model cheese and butter factories at the large exhibitions and sent collective displays to the international exhibitions.

Poultry, fruit, and honey producers also formed associations to organize conventions and exhibitions at home and overseas. Fruit could be displayed to advantage, although the growers complained that 'the people at the county fairs count too much on size.'[101] The growers introduced a points system so they could obtain reliable judgments and direct efforts towards the most marketable fruit. The Fruit Growers' Association of Ontario was formed at the Provincial of 1857 with the goal of 'collecting, arranging, and disseminating useful information.'[102] Honey was brought to public attention by means of exhibitions. At the Toronto Industrial in

1879, apiarist D.A. Jones showed 30,000 lb of honey. It sold so well that he doubled production the next year. The Beekeepers' Association (formed September 1880) called the exhibit a 'complete revelation' by which 'beekeeping made itself known as an industry in Ontario.'[103] In 1883, $30,000 worth of honey was exhibited.

Exhibitions aided the thriving agricultural implements industry that emerged in Ontario, permitting manufacturers to study innovations and to advertise to farmers.[104] In 1850 most of the implements exhibited in Ontario were sent in by an American firm, at the Agricultural Association's behest. By the end of that decade, Canadians accounted for every entry, and reporter William O'Brien boasted that 'an American-made machine is now as great a rarity as a Canadian one was a few years ago.'[105] In 1851 Hussy and McCormick reapers could be seen at the Upper Canada Provincial Exhibition, as at the Great Exhibition of London, where they caused great astonishment. The British judges gave McCormick first prize; Canadian judges, without putting them to practical test, preferred the Hussey in 1851 and again in 1852. Informal tests were held elsewhere, as at Port Hope in 1853, when, in the absence of judges or prizes, farmers judged the machines for themselves.[106]

Rarely were implements properly tested at the Provincials. Partial tests were introduced in 1856 and 1857, while full trials at Brantford in 1858 drew eight mowers, six reapers, and five combined machines. The next trial was not held until 1871, with the result that 'the owner of each machine declaims loudly upon its superiority and there is no means of eliciting truth out of this confusion of claims. The implements must be judged either from appearances or by the most imperfect accounts of what they have severally done.' In 1859 the *Leader* remarked that the lack of trial was observable 'in the extra gilt and polish every where to be seen. And it could not be otherwise, where machines are judged by appearances, the makers will strive to outdo one another in that particular.'[107] When trials were finally restored, the implement makers rebelled against the prize system, complaining that judges were biased or ignorant and that verdicts changed from one show to the next. A trial placed every machine on an even field, which was not in the interests of the large companies. In 1872 these companies asked the Agricultural Association to stop the trials. The association asked other makers for their opinion: of the forty who responded, more than half opposed the trials,[108] and the association began to offer prizes 'for Exhibition only.'

During the recession of the 1880s, smaller companies closed their doors and the larger companies consolidated, expanding their market share at home and creating markets abroad. The leading firms began to use international exhibitions to advertise their wares. John Watson of Ayr sent a collection to the Paris Exposition of 1878 and personally attended to explain his machines to the judges. He earned a gold medal, sales worth $287.14, and an order from Normandy for 1000 Meadowlark reapers.[109] In 1867 Hart Massey sent a mower-reaper to the Paris Exposition, but without expecting any orders, for the company could not meet even home demand. In 1885 Massey sent a binder to the Antwerp Exhibition, again without any particular expectations of success, but when that binder won the top prize, beating all the American, British, and European machines, he began to think seriously about developing foreign markets.[110] At the Colonial and Indian Exhibition of 1886 and the Chicago World's Fair of 1893, Canadian implement makers exhibited in force.

The true test of the economic impact of exhibitions is found in the international exhibitions, which will be examined in Part Two. Here I merely wish to show that exhibitions did play a clear, if not easily quantified, role in fostering Canadian agriculture and industry. The result was fairly modest, as were the sums expended. The effect was probably more measurable on the upper strata of agriculturists and industrialists than on the general farming public. Exhibitions did not redistribute wealth among farmers so much as they concentrated it, a result that made them appropriate and representative instruments of late nineteenth-century capitalism.

Exhibitions also encouraged state intervention into the economy. The late nineteenth century, no less than the early nineteenth century, was an age of state formation. Exhibitions were always the thin edge of the wedge: they provided justification for a much larger program of intervention. The development of governmentality in Canada was an eclectic and gradual process with many different instigating factors, of which exhibitions were only one. Without attributing excessive influence to them, however, it is possible to identify at least two means by which they exerted some influence over the development of a state-managed economy.

First, exhibitions encouraged a vision of the marketplace as constituting a free process of economic competition and exchange, mediated by an external and very visible hand standing in for the invisible one. Although individual judges and organizers might be corrupt or incompetent (as they often were), they collectively bolstered the principle that objective management and judging could be performed by individuals who were outside and above the competition. Applied outside the showgrounds to political economy, this principle encouraged a perception of the government as composed of disinterested men who could manage the economy and the nation impartially. It gradually supplanted the traditional liberal view that governments were composed of men who might govern in the interests of their class, rather than of the nation at large, and should therefore intervene as little as possible.

Second, by instituting this program to encourage competitiveness, the Canadian government established a precedent that was, paradoxically, later used to defend economic protectionism. Its intervention in the struggling agricultural sector before *laissez-faire* became economic orthodoxy was, in the 1870s, imported into the manufacturing sector at the expense of farming interests. As Ben Forster has shown, this new political influence of the manufacturing sector reflected its economic power.[111] However, manufacturers justified protectionism on the grounds of their economic weakness rather than their strength. From mid-century onwards, they appealed to a principle of 'infant industries' which was first elaborated in rural Upper Canada in the 1830s. In that decade, when the Upper Canadian legislature began to fund agricultural societies as Lower Canada had done for a decade, John Rae, a Scot living in Goderich, wrote an economic treatise that defended state encouragement of the arts and sciences so as to transfer discoveries from elsewhere and to promote new inventions. The whole community, he argued, should share the burden of initial costs in new industries. He instanced importing seeds and animals from abroad: Where individuals could not assume this cost, they banded together, and if it was too costly still, the state aided them.[112] This example must be a reference to Canadian agricultural societies, for those in England and Scotland did not import livestock, but encouraged better breeding among existing species. In Canada, however, every cow, every horse, every sheep had to be imported, at least at first. Rae followed the fortunes of Canadian agricultural societies and engaged in an interesting polemic with William Evans: Value existed in nature, according to Evans, while Rae contended that it lay in the surplus of production over consumption.[113] As Robin Neill argues, Rae's economic theories, which constituted a major revision of Adam Smith, reflected the nature of the economic activity around him. They were formed through arguments with his contemporaries, but they proved to have a lasting influence. Forty-five years later, when he advocated a policy of tariff protection, Sir John A. Macdonald drew upon mid-century debates and policies, but he also directly invoked John Stuart Mill's 'infant industry' theory, which Mill had taken from Rae.[114] Farmers, who had benefited from state subsidies for decades, could not object to the principle of intervention. Agricultural societies relied on liberalism, economic as well as political, to function well, but they undermined that liberalism and sowed the seeds of their eventual decline.

5

Exhibition Culture

The story thus far has chronicled the struggle for control of the developing exhibition bureaucracy in Canada, with farmers, politicians, journalists, and professional improvers all vying for power. But the organizers of exhibitions were not their authors in any historical sense. The paying visitors determined the success or failure of these events. Historical agency, in other words, lay with the audience rather than the organizers of exhibitions. Recent work in reception theory rejects a dichotomy between autonomous cultural texts and passive readers or consumers of the text by collapsing the distinction between cultural production and consumption and by declaring use itself to be a creative act. Michel de Certeau remarks, 'The presence and circulation of a representation (taught by preachers, educators, and popularizers as the key to socioeconomic advancement) tells us nothing about what it is for its users.'[1] He argues that Foucault's disciplinary techniques met resistance which neutralized their effects on practices in everyday life.[2] Applying his theories, one might read resistance in the acts of fair-goers who, told to admire a prize cow, instead laughed at its obesity and subverted the intended meaning. Or they might flock to the races instead of the lecture hall. The sum total of these actions altered the shape of the exhibition, installing the popular reading as the determining one.

Exhibitions changed as their source of funding changed, and went in turn from elite subscription to government grant to admission fees. Exhibitions tempered instruction with amusement because the elites realized that if they wished to popularize improving ideas, they would have to speak a popular language. Only the form was supposed to be popular, not the message, but the one collapsed into the other. Even before admission was charged, exhibitions had to be popular, for a show with no audience was a sad joke. The use of public moneys established a principle of accountability to public opinion which the growing dependence on admissions entrenched. The directors of the Western Fair explained in 1884: 'It is self-evident that the people can be neither instructed or entertained unless you can induce them to attend, and it is from the crowds that are attracted by these amusements that the Society gets the money to improve the exhibition by offering larger prizes to the more important classes [of exhibits] ... we recognize the fact that your officers are not placed in their positions as rulers or censors of public opinion, but merely representatives and servants of the public and their highest merit is implicit obedience to the voice of the people when clearly expressed.' A sum of $2000 spent on grandstand amusements brought in an extra $3000 at the gates: The people had spoken clearly.[3] Cultural agency passed to the popular classes, in flagrant contradiction to the original idea behind exhibitions which the financially troubled Toronto Agricultural Society continued to espouse in 1883: 'that of educating the public mind up to a higher standard of the true and the beautiful in Nature and in Art.'[4] The provincial exhibition run by improvers died out in 1889, and the Toronto Industrial Exhibition, 'the people's fair,' went from strength to strength, with admissions rising above 200,000 in 1885 and above 400,000 by 1900, almost one-fifth of Ontario's population.[5] If exhibitions partook of 'the war between play and purpose,'[6] play won, for amusement came to outweigh instruction for most fair-goers.

Reception theory only partially debunks Guy DeBord's view of a society composed of passive spectators, for late nineteenth-century fair-goers were more passive than their earlier counterparts. Exhibitions were designed to stoke farmers' pride and cupidity and turn spectators into participants, so that '"lookers-on" be induced to be "lookers-on" no longer, but be up and doing that which Providence has given them the abilities to perform.'[7] By the end of the century, visitors merely stared at paid performers. The cattle were foreign objects to city folk. Manufactured goods were shown by retailers, rather than manufacturers, as store wares for sale and not for technical education, as the Society of Arts intended at the first Great Exhibition.[8] It has been suggested that consumption is itself a kind of spectacle.[9] By the end of the century, against the intentions of their founders, exhibitions became a metaphor for impotence. Nietzsche spoke of history as a 'world exhibition' at which man 'has become a spectator merely enjoying himself and strolling around.' Nietzsche and the Frankfurt school criticized the mass culture that developed as giving too much power to the vulgar.[10] Some scholars worry that a 'carnival culture' threatens morality and art, while others give a sprightly defence of popular culture.[11]

Exhibitions and the popular culture they expressed expanded together. The work of the best farmers and artisans could still be seen in the fond hope that they would plant desire and knowledge in the hearts and minds of producers: 'Where the leaders are so well abreast, the rank and file must follow.'[12] But the range of exhibits and activities grew enormously, as organizers gave free rein to their imaginations to attract crowds. The exhibition used to represent culture, as Matthew Arnold understood it: the best of all that was said and done, although a purist like Arnold might disdain the philistine excellence of iron-mongers and grain-growers. But the idea of culture itself broadened to become 'almost identical with our whole common life,'[13] and exhibitions came to reflect this range. Canadian exhibitions were more encyclopedic than those in England, as one English farmer remarked on visiting the Toronto Exhibition in 1880: 'The exhibits in the central building put me in mind of the exhibition in London in 1851, on a small scale.'[14] Canadians designed their exhibitions so that an 'intelligent stranger' could observe 'at a glance as complete a representation as possible of its entire resources and products, natural and artificial.'[15] Some resented this encroachment on the farmer's festival, complaining that the 'show of finery and fixings interfered with the real character of the Exhibition and its objects,'[16] but the encyclopedism had been present all along; there had simply been less extra-agricultural material to exhibit. In this new land without other traditions, an emerging community would identify itself with this public celebration of its strengths. As industry and the arts developed from the agricultural economy, they naturally appeared at the fair.

While the fair began to look more like a microcosm of the world, the world began to look more like the fair. Cultural osmosis worked in both directions. Department stores and modern advertising appeared during the nineteenth century, and the exhibition was midwife to both,[17] as well as to the twentieth-century mall: It is no accident that the Toronto Eaton's Centre resembles Paxton's Crystal Palace. In Philadelphia, John Wanamaker renovated his department store to resemble the Centennial Exposition, and in Hamilton, Thomas C. Watkins, an aficionado of world's fairs, built a 'palatial store' of cast iron and crystal, the interior opened up like an exhibition hall with eighteen-foot-high ceilings, no interior partitions, and mahogany show cases.[18] In Montreal and Toronto, permanent exhibitions were established, the one organized by Perrault to display French-Canadian handicrafts, the other a showroom for manufacturers organized by W.H. Howland and Frederick Nicholls. The Canadian Manufacturers' Association praised the endeavour, which would 'afford many manufacturers superior facilities for the display and sale of their productions' and 'tend to educate the general public in regard to the material and constant progress of domestic manufactures.'[19] Stores called themselves 'permanent exhibitions' or put up signs that read, 'The Exhibition is Free, call and see.' The Massey Company turned delivery day into a spectacle, with a parade and a marching band. In 1860 general merchant Henry Morgan of Montreal was horrified to find his nephew placing merchandise in the store window: 'Will you have us display all our goods so that our competitors will see what we have?' he demanded. Morgan's became the first store in the city to display merchandise openly with prices attached.[20] Before long, window-dressing with pyramids and bunting became an elaborate art.[21] Meanwhile, the Toronto Industrial tried to resemble an established store: 'The great feature about this exhibition of ours is its *permanent* appearance. You don't see any pine board business about it. It is a community in itself, with an everlasting hue hanging over it. Its buildings look as if they were erected to last, the drives resemble city streets, the refreshment booths have the appearance of stores, everything you see has a solid permanent aspect. Things are not put together to last a week, or a season; things are there for keeps.'[22]

Because stores resembled exhibitions, exhibitions had to do more: 'The day is past and gone when live stock, machinery, agricultural implements, fruits, vegetables, dairy, ladies' work, photographs, pictures, etc., are likely to attract a crowd again. These things are seen in every shop window in the city or at every depot and emporium in the country. They will always do their share towards drawing a crowd, but only their share.'[23] This 1881 writer recommended a military tournament, musical competition, monster concerts, Caledonian games, fireworks, baby prizes, bike races, lacrosse match, tug of war, encampments of oddfellows and masons, and a trades procession, which 'if properly conducted and mixed up, so as to embrace something of the carnival or masquerade element, could be made at once instructive and amusing.' At the Dominion Exhibition in 1880, just such a procession of 'Tradesmen and Terribles paraded the principal thoroughfares. The Terribles consisted of demon-like creatures, who discharged fire-works from their perch on the top of a covered wagon. Following them came the tradesmen, barbers, carpenters, tinworkers, blacksmiths, picture-framers, shoemakers, &c., who gave practical illustration of their various businesses.'[24] There was a logic to this display: Exhibitors and organizers tried to be more spectacular than everyday life, so as to create a festive feeling and bring in the crowds. Because these methods worked so well, businessmen adapted them to daily business, which, in turn, pushed the exhibition to greater innovations. The fair's 'liminality,' or distinctness from everyday life, was never absolute, but was constantly being renegotiated.[25]

Culture inside and outside the exhibition grew more commercial and more spectacular. In 1886 the secretary of the London Society of Arts mused, 'The inauguration of an age of commercialism may or may not have been an unmixed blessing; anyhow, the exhibition inaugurated such an age.'[26] Exhibitions were machines designed to transform artifacts into commodities, to turn the display of objects into a market for objects, by bringing together potential buyers and sellers and encouraging farmers to sell their surpluses. No matter what one brought to the exhibition, that object became a commodity, integrated into the commodity culture. The officious organizers of the Ontario Provincial banned coffins as unseemly, but in the Toronto Exhibition, undertakers had a special building. Not only 'tasteful shrouds' and upright coffins appeared but even skeletons, sold by a medical student at $80 apiece.[27] The art wing at the Toronto Industrial succumbed to the prevailing influence: 'One could wish that the distinction between the commerce and the art were more marked. We still feel somewhat of the essence of the market as we step into the Art department.'[28] The ladies' department was the exception that proved the rule, for it was intended to defy the market system. Ladies spent a lifetime piecing together a quilt from ten thousand bits of silk, to show that they were protected from the market by wealthy husbands whose goods could be found further along in the mechanical department or out in the stables.

Along with the new commodity culture came advertising. Advertising and spectacles had existed before, and P.T. Barnum, for one, had grasped their combined potential and made a flourishing business of circuses and permanent exhibitions. But before mid-century, Barnum's respectability was suspect: He was denounced from pulpits and had to obtain certificates of approval from well-disposed clergymen.[29] Exhibitions had the approval of the crown, the peerage, and the middle classes, who dreamed of harnessing the spectacle, the commodity, and the advertisement, and thereby effectively legitimized them. Just as no artifacts at the fair could escape becoming commodities, no communicative acts could escape being promotional. It hardly mattered what the signs said, so long as they drew attention. At the stove department of the Southern Counties fair at St Thomas, 'every available space between the exhibits' was filled with cartoons, one company showing 'the heathen Chinese, the warlike Indian and representatives of all the other nations' dancing and singing with extended arms, while another showed 'Sir John, Mackenzie, Mowat, Blake, Selby, and other political leaders who perform vigorous cellar flaps, hoe-downs, and *entre-pas* around the burner.'[30] What had this to do with stoves? the bewildered reporter asked. What indeed? But no matter, so long as visitors came.

Exhibitors boasted of prowess at visual rhetoric: 'That phase of advertising which addresses the eye, and, through trained expositors, the intelligence, has made extraordinary advances since the Exhibition of 1850.'[31] At Montreal in 1880, 'the most simple and prosaic collections of confectionary, and knickknacks are built into structures of the most graceful design and proportion, and even stoves and furnaces are so disposed as to be actually picturesque.' That year, Oriental pagodas graced many a stall, and there was even a pyramid of corsets.[32] The most imposing displays were by Johnston's Fluid Beef Company, which built castles made of beef tins or showed a giant clubbing a boa constrictor to death, 'to convey an idea of the wonderful strength-giving properties of Johnston's fluid beef.'[33] Other exhibitors used spectators as advertisements. Uxbridge Organs put up a sign at the Colonial Exhibition in London in 1886: 'A Fact. An Uxbridge Organ was being played at the Canadian Exhibition where a great sturdy stout-hearted Briton stood listening, and as the player brought into effect the grand magnificence of the Organ, a tear stole down his swarthy cheek and he exclaimed, "They can't beat that in Heaven."' Heintzman Organs quoted a farmwife: 'O, sor, but them pianos of yours can be heard all over the building away above everything else. We have big crops this year, and if my ole man don't get one of them for my Elizabeth Ann I will, and don't you forget it.'[34] Christie Brown Company made artistic displays with hundreds of biscuits (prompting the exclamation, 'It is astonishing that such an effect can be produced with a prosaic biscuit') and, on children's day, gave out free samples: 'With eyes dancing, mouths open and hands raised they clamored for the toothsome biscuits like a swarm of bees.' A paid press review enhanced the effect.[35]

The daily press latched onto exhibition advertising quickly. Newspapers puffed exhibitions and sold special exhibition numbers by the tens of thousands. Journalists knew that reviews would serve as advertisements, so, in the 1880s, they began to charge by the line, often simply inserting notices supplied by the company. Then the exhibitor would boast that the *Globe* or the *World* had favourably reviewed his wares. Occasionally farmers objected to the flagrant dishonesty.[36] The *Canadian Manufacturer* protested the fact that only businessmen had to pay for copy, and vaunted its own objectivity in reviewing exhibits;[37] in fact, its interests parallelled those of exhibitors – advertising for business. The *Canadian Manufacturer* resembled the exhibition in this respect: In no single instance could it be accused of corruption because there was no part of it that was not a piece of puffery; this was its purpose. There was no *hors-promotion*.[38] The exhibition became a symbol for a promotional discourse that critics feared was seeping into daily life. In 1901 a popular book denouncing materialism complained of 'the abuse of showing everything, or rather putting everything on exhibition; the growing incapacity to appreciate that which chooses to remain hidden ... One sometimes wonders if society will not end by transforming itself into a great fair, with each one beating his drum in front of his tent.'[39]

In sum, the exhibition was at the vanguard of modern mass culture, in all its commercial and carnivalesque aspects. Thus far, the focus has been on the production and management of exhibitions. The remainder of this chapter will describe the exhibition itself as it was experienced by visitors. The first part describes the objects on display, with some remarks on visitors' reactions, while the second part examines fairground activities: the judging, the improving lectures, speeches and banquets, and the amusements – those provided by fair managers and the rough-and-ready subculture of the fair. I have deliberately adopted a descriptive approach to illustrate the difficulties in subjecting the myriad diffuse choices of the crowds, which shaped the exhibition and mass culture, under a traditionally analytical approach that attributes cause and effect to reasoned decisions and actions by individuals. To reduce this rich popular culture to bloodless academic categories would bely the argument of this chapter – that on the fairgrounds, popular culture was able to defy the intellectual imperialism of the Enlightenment.

One had first to get to the fair. For some, this meant a long trip on a crowded train, perhaps a cattle car fitted out with board seating. Others rode buggies, as at Vaughan Township in 1865, where 'from an early hour in the morning, buggies, wagons, and almost every vehicle on wheels began to pour into the village, and to deliver their freight of human beings at the doors of the several taverns.'[40] Farmers raced their horses, as the excitement took hold of them. At Binbrook Township, 'many a staid old team of plow horses, driven by equally staid, but temporarily frisky old farmers, was excited into hilarious impromptu races with passing rigs or driving in an aggravatingly zig-zag fashion along the road to prevent the fiery steeds of impatient and haughty young bloods, coming up behind, from passing.'[41] This bravado, plus free-flowing liquor, wreaked havoc upon conveyances: 'On the day after Ridgetown fair, the road leading from the town to the grounds was literally covered with smashed rigs.'[42] In the city, streetcars would be overloaded, conductors unable to collect fares, and horses sometimes collapsed under the strain. By the 1890s, electric streetcar service in Toronto ran every minute along King Street to the grounds. If it rained, the roads often became impassable, while the exhibition grounds would turn to mud, and visitors would jam into the halls. In 1882 a storm in Montreal knocked down buildings, blew implements across the racetrack, and shook the Crystal Palace until it swayed and the windows shattered, sending the crowds scurrying back into the rain.[43]

Small exhibitions were set up in makeshift fashion in a field, while manufactures would be in the town hall. Large cities had crystal palaces and smaller ones used the drill shed, with the result that, during the Fenian Raids in 1867, exhibitions were curtailed by troop movements. Most settled shows put up fences and turnstiles, charging 10 or 25 cents admission. There were frauds galore, as at London in 1869, when a man claimed to have bought 350 tickets from a gate-keeper at $10 per hundred.[44] There were also scores of gatecrashers, particularly boys who sat in rows on the fence, waiting for the constable to turn his back.

Once in the door, visitors often separated, the men heading for the livestock or machinery and the women for the galleries of fine art or fancywork, if they did not go immediately to the horse-ring. It was a rare exhibition that had no horse race, whether in a track with a grandstand or simply through the main street of the village. Where horse racing was banned from exhibitions, the restriction was got around by awarding prizes for 'horse in harness, speed to be considered.'[45] Thoroughbred horses were never numerous: The Provincial in Kingston in 1867 drew only one fifteen-year-old stallion. The 1881 Agricultural Commission found that there was no market for race horses in Canada because of a law against gambling, so breeders sold to the United States.[46] Trotters were more popular, especially the small French-Canadian horses, which were 'rum 'uns to look at, but good 'uns to go.'[47] After the horse races came the donkey races, as small boys careened around the fair grounds. There were also prizes for all-purpose agricultural horses, including draught horses. At the smaller shows, prizes went to different people every year, a doctor with a handsome roadster or a widow with a fine pair of carriage horses. If highly bred horses were exhibited, their owners kept them locked in stables under a blanket to protect them from the crowds. The winning horses, cattle, sheep, and swine could be seen on parade around the horse ring or through city streets. In 1852 a 'grand procession' of livestock from the York County show marched to the Provincial grounds in Toronto, along King, York, and William streets. First came the Toronto Brass Band in a horse-drawn carriage, then 'the noblest and favorite horse, Clyde, led off, as certainly the largest specimen on the ground.' The earth shook beneath their hooves. Then more horses, a jackass, Durhams, Devons, Ayrshires, ponies, a monstrously fat and plodding steer looking 'more like a rhinoceros than an ox,' and, finally, thirty carriages and a crowd of horsemen.[48]

The noble horse was the bane of agricultural reformers, who complained that French Canadians pampered their horses excessively and awarded them exorbitant prizes.[49] In fact, the proportion of horse to cattle prizes was not above that paid out in Ontario, and horses were a lucrative industry throughout Canada. Reformers also objected to the racing because it encouraged the breeding of horses unfitted for labour. Moreover, exhibitions were supposed to teach slack farmers that hard work and discipline led to success, but the gambling that accompanied horse-racing promised instant profit, won by chance, not effort.

There was little scope for chance in the cattle competition. Breeding cattle required such great investment and large herds to keep up the bloodlines that the same exhibitors won year after year, decade after decade. The cattle sheds were the true heart of every agricultural show. On Farmers' Day at the Toronto Industrial, 'in serried ranks they trod manfully to the turn-stiles, paid their quarters, then rushed straight ahead to the cattle departments.'[50] Townspeople also visited the cattle, poking them with sticks, or 'with daintily gloved fingers or with the point of natty parasols, simply because poking of some sort or other is supposed to be essential in the examination of every prize animal,' and they judged the animals for their curly tails or angry glares.[51] Durhams, 'the greatest breed the world has produced,' were the king of the show in Ontario, while Ayrshires predominated in Quebec. Reporters confronted with these magnificent cattle 'rose to heady lyrical heights': David Christie's 'Oxford Lad' was 'an animal of large dimensions, great length, wide-chested, expansive ribs, full, clear eyes, with a powerful masculine expression.'[52] Bulls were praised for their 'good loins and thighs,' while a prize cow was 'an attractive dame, with a sweet eye and finely-rounded carcass.' At the early Ontario exhibitions, Shorthorn rivals were Herefords and Devons; in later years, new breeds disputed their hegemony, such as the Angus-Aberdeen or Galloways, whose 'black, woolly coats and hornless heads arrested the attention of passers-by, who generally asked, "What do you call these?"'[53] Their gathered owners squabbled noisily about which breed was best, but the dispute could not be settled at an exhibition: small Devons, for example, had finer meat and brought a higher price per pound than the bulkier Shorthorns. Sometimes the bulls ran free, as at Grey County in 1862, but usually they were tethered to posts or shown in rings. Judges would run their hands over the animal, feeling the texture of the skin and flesh, sometimes straining to find the curve of the bone under inches of fat. Even the way fat sat mattered: The finest animal was kept in high condition from its birth, as opposed to those fed on grass and stuffed in the last few months, resulting in protruding knobs of fat over rangy, tough beef, rather than the marbled meat that Britons prized.[54]

Nearby would be sheep pens filled with a variety of breeds, with black faces, white faces, sometimes painted faces. Sheep judges fought a constant battle against fraud of all sorts, the most common stratagem being premature shearing to make the animals look larger and their wool more luxuriant. At the Provincial in 1866, nearly every competitor was disqualified. Some breeders showed their animals unkempt, while others sheared to make them look blockier or dyed the wool bright colours. The long-woolled Leicester and Cotswold sheep always made the best appearance, for Canada excelled in these breeds and often swept all the prizes in American fairs. Farmers began to cross these two breeds to get Leicester wool with Cotswold size, but as a result the sheep around the country deteriorated. This led to more fraud, as exhibitors hid the Cotswold blood with 'ill-concealed traces of the scissors, razor and branding-iron used to obtain the smooth, bare face' of the Leicester.[55]

Sheep had come a long way from the 'old scraggy lank-sided, razor-backed race' seen at the beginning of the century, as had the swine: 'Tall, gaunt, slab-sided animals they were, with high arched backs and long legs and snout, with hair along the spine of nearly six inches in length.'[56] Sometimes scrub stock appeared at exhibitions, but, in swine, the trouble became rather the reverse: overbred, obscenely fat animals, unable to support their own weight, 'living pork barrels, cylinders of meat with vestigial snouts at one end and tails at the other.'[57] In 1881 pork-packer William Davies told the agricultural commission that the improved breeds needed an infusion of blood from the old 'alligator' pigs of yore.[58] Swine were latecomers to the exhibitions because they were difficult to transport, and they appeared in number only when the pork-packing industry developed around Toronto. In 1868 empty pig pens housed the overflow of sheep and slumbering attendants. A good pig, like a noble Clyde or a majestic Durham, was a thing of beauty: In the township of Ailsa Craig, near London, Mr McArthur's Berkshires were 'a perfect picture to behold. As a short old gentleman present remarked, "You couldn't whittle finer models out of bass wood."[59] Names appeared over the pens: McArthur's included the '10th Earl of Balmoral,' and one old boar was called Sir John A. Macdonald.

Other staples of the large exhibitions were dog and cat shows. Their howls and cries added to the din of machinery in motion; competing organs playing everything from hymns to Annie Laurie; the hammering of late arrivals; and the cries of hucksters, patent medicine men, and bootblacks who travelled around the exhibitions in packs and sang 'Black your boots and make 'em shine. T'will only cost you half a dime.'[60] Neither the least noisy nor the least competitive department was poultry. Farmers' wives brought geese and ducks, and sometimes a rabbit or two, to small fairs, while wealthy fanciers showed dozens of birds in various hues. In 1877 the Provincial degenerated into a scene of chaos when the chickens escaped and ran about the grounds and into the street, with caretakers and gamins in full pursuit and spectators shouting helpful advice. Fanciers could be a perfidious lot, snapping tail feathers off rival birds or even poisoning them, as occurred at one Provincial. All eggs laid were shaken up and made infertile so no prize bloodlines could be stolen. Another poultry fixture from the late 1870s was the Glass Hen, or incubator. Visitors paid 10 cents to get in, and another dime to take a 'motherless' chick home with them. At the Quebec Provincial Exhibition in 1871, cockfighting was held.

Between the animate and the inanimate displays was the once-animate: natural history. Some of Canada's leading naturalists, including David Boyle and John Macoun, first attracted attention to their work by exhibiting their collections of useful and harmful flora and fauna. Natural history cultivated a scientific and utilitarian attitude towards the world, all the while inspiring admiration for God's handiwork.[61] In 1860 a farmer complained that prizes in this department were too high, but a decade later they were so low that no specimens appeared at the Ontario Provincial. Journalist Kit Coleman described one sorry display: 'The fish are repulsive, and the stuffed birds have a tired air, as though they have done this sort of thing once too often and would like a rest.'[62]

Other inanimate objects ranged from field produce to manufactures. The Agricultural Hall was not the most spectacular, but it was pregnant with symbolic meaning, second in importance only to the animals. Here the visitor went to admire 'the captivating cabbage, the marvellous mangold wurtzel, the whopping wheat, the primitive pea, the prize-procuring potato, the amazing apple and the 777 other articles necessary to occasions of this kind.'[63] Early on, wheat was king, and dozens competed for the Canada Company prize at the Provincial, worth £25, and the proud title of the best grain in the land. Unscrupulous exhibitors put the best grain at the top of the sack: one man lost the prize for this sin. Henry Moyle's prize-winning wheat, at 66½ lb to the bushel, was said to owe its weight to its 'flinty character.'[64] Abbé Pilote set the students at his school in Ste-Anne-de-la-Pocatière to work handpicking grain samples. These stratagems show how intensely the farmers competed in wheat, but this intensity eased with the decline of wheat growing in central Canada. By 1888, 'for the Canada Company's prize for the best 25 bushels of fall wheat there was no competition, there being only one exhibit, and it not of very fine quality.'[65] The wheat was flanked by other cereals: rye, oats, barley, and the forage and green crops that reformers advocated. At larger shows, seedsmen put up large displays, as did millers: in Montreal, Ogilvie's erected a pyramid of flour and gave out freshly baked bread.[66]

Vegetables were subdivided according to their producers (farmers and gardeners) and their uses (fodder and table food). Roots to stall-feed cattle in winter were one of the crops that agricultural improvers worked hardest to introduce. Some societies, as in Wellington and Wentworth, held competitions to see who could raise the most (at Blenheim Township in North Oxford, the winning acre had 22 tons of turnips), while Ontario South had an annual turnip-hoeing competition. The winning roots were gigantic and, although tasteless, signalled a fertile soil and kindly climate. To prove the agricultural capabilities of western and northern Canada, land and railway companies, municipalities, and agricultural societies sent the largest specimens they could find to the big fairs. One such collection, proudly displayed in Toronto in 1881, prompted the cynical remark that if it 'came out of Muskoka, it was taken there first.'[67]

Fruit did not have to be grotesquely large to attract visitors, for piled-up plates of luscious apples, pears, peaches, or berries were an attractive sight. Canadian apples showed well because the alternating warm days and cool nights made them a brilliant red colour. In the early enthusiasm for fruit growing, the largest prize usually went to the collection with the most different varieties, correctly named, but in later years fruit growers preferred to reward successful specialization. The fruit department was the staple of fairs in Niagara, but at Ottawa in 1869, 'a two bushel basket would have held the whole of it.'[68] Fruit growers did not have to complain of being neglected by visitors; on the contrary. One year sneak thieves stole all the grapes and pears shown by one hapless exhibitor at the Dresden fair. In Burford Township and at the West Brant show in 1860, 'outrages were committed' by 'a set of marauders' who devoured every fruit in sight. The next year at the Provincial, when visitors were too crowded to move or be policed, 'a general onslaught was made upon the cheese and everything edible, and even roots & vegetables were consumed by Vulture-like chaps.'[69]

A more platonic appreciation of fruit could be enjoyed in the women's department, where modelling waxwork fruit was an honoured art. Preserved fruit was admitted, either in jars or as jellies and jams. Home-made grocery exhibits swelled the female presence on fairgrounds after mid-century. In 1892 domestic work and dairy products made up 34 per cent of entries at Markham fair. Some of these entries were factory-made cheeses, but women showing dairy butter accounted for most. Canadian butter had a bad reputation as fit only for axle grease: One English dealer was fined for possessing 350 lb of it.[70] Reformers urged a switch to factories to obtain butter of more uniformly good quality, but the very best 'gilt-edged' butter, for which top price was paid, was made by farm women who had won a reputation for themselves, such as Mrs John McClurg of Lobo, who won first prize at the township, county, and Western fairs from the 1870s on. She exported her butter and sold it locally for as much as 30 cents a pound.[71]

Cheese-makers entered into the spirit of the fairs, beginning with Hiram Ranney, the 'prince of cheeses,' who established a reputation during the 1850s with the mammoth cheeses he and his wife produced. In 1852 he showed a 700-lb cheese destined for England, and another, in 1857, was as 'big as a cartwheel.' In 1861 both Ranney and James Harris made 1200–lb cheeses. In 1865 A. Smith of Norwich stole the honours in Ontario and New York with a 4000-lb cheese, but it was badly pressed and gave off a foul odour.[72] In 1866 Ranney and Harris together made a cheese weighing 3½ tons. Too large for ordinary transport, this 'greatest wonder of the dairy age' was paraded through the streets in a specially made car, drawn by six grey horses. It was shown in the United States and then Britain, at 6 d admission. Finally, 180 lb was sent back to Oxford County and eaten with relish.[73] It lived on, immortalized by James McIntyre of Ingersoll, who penned an 'Ode on the Mammoth Cheese':

We have seen thee, queen of cheese,
Lying quietly at your ease,
Gently fanned by evening breeze,
Thy fair form no flies dare seize.

All gaily dressed soon you'll go
To the great Provincial show,

To be admired by many a beau
In the city of Toronto.

Cows numerous as a swarm of bees,
Or as the leaves upon the trees,
It did require to make thee please,
And stand unrivalled, queen of cheese.

W.A. Deacon remarks that 'his poet's eye rightly saw it as a symbol of the glory of his community.'[74] Another poem explained the lesson conveyed by the big cheeses issuing from that region:

Big cheese is synonymous name,
With fertile district of the Thame,
Here dairy system's understood,
And they are made both large and good.

Giant cheeses became a part of Ontario folk culture, a source of great pride and amusement for country and city visitors to the exhibitions.

The dairy department was one where men and women competed directly; horticulture was another. This attractive department was very popular with visitors. At Brockville in 1851, a central tent was 'gaily decorated with evergreens and flowers and surrounded by a beautiful floral cupola.' Peaches were posed around the cupola, and grapes hung from it.[75] The same visitors rarely attended purely horticultural shows: At a larger fair, the floral hall made a pretty diversion, but as an event in itself, a horticultural show was too respectable to draw crowds.

The ladies' department proper did not advertise commodities but served as an advertisement for the exhibition itself: Women were a hook to lure in the men. Only when this department was well filled did townspeople take an interest in the country fair. 'Whenever ladies' are encouraged to bring out their useful and ornamental productions there is always a larger attendance, and apparently greater animation,' remarked a paper in Barrie. Fairs with sagging admission rates would reverse their fortune by raising prizes in the ladies' section: in South Leeds, prize money for fancy-work rose from $125 in 1889 to $390 in 1891, more than was spent on livestock. Exhibiters found their reward not in the prizes, which were small, nor in advertising their commodities, which were meant only for display, but in advertising themselves: At the Asphodel, Belmont, and Dummer fair in 1864, 'the ladies did their duty nobly; and not the least attraction at the show was the many fair faces, which certainly won admiration, and will doubtless get prizes (of husbands) too.'[76] The ladies' department is discussed in detail in chapter 9.

Women dominated the amateur art department. Some amateur painters showed marked talent, whereas others sent hideous copies of coloured lithographs. One gallant reporter remarked: '"The Holy Family," an oil painting, is the work of a lady amateur, and is therefore not open to severe criticism.' Of a landscape labelled 'self-taught,' another wrote: 'In a kindly spirit we would recommend the author to retain such specimens at home, and to apply to some really good instructor.' Journalist Hector Fabre commented of the Montreal Exhibition in 1880: 'Il faut certainement avoir de la bonne volonté pour croire qu'il y a un département des beaux-arts dans le Palais de Cristal.'[77] Reporters and visitors had great fun passing judgment on the efforts of amateurs and professionals alike.

> It was clear from the observations made that the visitors were not highly critical, but they were, nevertheless, generally just in their observations, the invariable test being one that Aristotle would endorse – was it like nature? The children's portraits were great favourites, and a man would remind Bill or Tom how well a Billy or a Tommy that was at home would look in oil. One poor lady discovered a resemblance to a lost child, and there was much loving reminiscence; the heart was too full for reticence – landscapes, figures, flowers – all had to yield the palm to the children in the opinion of these good, simple souls. How English all this is![78]

Many noted Canadian painters exhibited at the fairs. Every year, Robert Whale and his sons 'would cart crates of oil paintings to the Provincial Exhibition, always submitting the maximum number of pictures allowed in each of the categories. They relied on the prize money, for there was no steady market for art.'[79] By the age of sixteen, Paul Peel had already distinguished himself at the Western Fair in London. Another exhibitor, Charles Chapman, showed 'some of his old standard pictures on hand, but these have received some masterly finishing strokes, which make them look almost new.'[80]

Specialized art exhibitions had been held since the 1830s and were organized by artist societies in Montreal and Toronto from the 1860s onward. In later years professionals began to abandon the fairs, where pictures were badly hung – too high or in dark corners – and judging was idiosyncratic. At the Southern Counties show in St Thomas, wooden frames were nailed directly to the wall, and painters received their pictures back with the glass smashed and nails protruding.[81] Artists resented their work being

The Quebec Provincial Exhibition, shown crowded with people and artifacts of all shapes and varieties.

mixed in with specimens of penmanship or amateur works 'deficient alike in drawing, colouring and originality.'[82] A sketch of a Quebec provincial exhibition shows a cluttered building with paintings crowded closely together, stretching up to the ceiling, while quilts hang from rafters, and dead birds and beetles occupy the tables.

Fine art and fancy-work were at one pole of the manufactures department; at the other were machines and agricultural implements. At times, when the implement makers resorted to polish, gilding, silk, and embroidery, they were not so distant. At local fairs, there were usually a few hoes and ploughs or perhaps a buggy. Manufacturers were, as a class, less attached to exhibitions than the breeders and farmers. During the 1860s the *Journal of the Board of Arts and Manufactures of Upper Canada* complained of poor industrial displays at the provincial exhibitions.[83] Like implements makers, manufacturers began to object to the prize system: A farmer who did not win prizes sold his wheat all the same, but a failure in the industrial department injured sales.[84] By the end of the century, many fairs no longer offered prizes for manufacturers. Though this exhibition-only system was in their interest and at their request, the manufacturers who did participate attributed it to their benevolence. The *Canadian Manufacturer* explained: 'What they have done and are doing is for love of Canada; and although some may affect to think that the days of sentiment are past, and that in this stern utilitarian age only sordid motives prevail, this certainly is not the fact as regards Canadian manufacturers.'[85]

A few machinery companies entered large exhibits. Waterous Engines of Brantford, which sold $220,000 of steam implements in 1875, exhibited widely. In 1868 the company fenced off nearly half an acre on the provincial fairgrounds and set up seats for 2000. When the whistle blew, crowds flocked around to see the machines in motion, including an 18-horsepower portable steam engine (worth $1700), shingle and lathe machines, and a sawmill capable of sawing 2000 feet per hour, which won first prize of $20.[86] Waterous also sponsored fairs, offering a chopping mill for a prize at the Southern Fair in 1879. McKechnie and Bertram of Dundas in 1872 put up a building measuring 80 by 24 feet to house $10,000 worth and 30 tons of metal-working and wood-working machinery.[87] In 1869 the most striking sight at the Provincial in Ontario was an imposing array of threshers; in 1872 it was a 40–foot windmill. Other machines shown at fairgrounds were Henry Bulmer's brick makers, drainage-tile machines that could produce 2700 tiles per hour, ditch diggers, a suspender-making machine, milking and bread-making machines, centrifugal separators, and ever larger steam engines, such as the 80-horsepower Wheelock engine shown by Goldie and McCulloch in Toronto in 1892. Portable steam engines were driven around the grounds, carrying visitors and occasionally running one over. Much blood was shed in the machinery departments: Visitors were forever crushing their fingers or having arms torn off.

The agricultural machines were pitched at farmers, but many of the largest machines could be sold only to firms. Manufacturers attended the fair to sell their goods, so suppliers had a rare opportunity to reach this audience. But manufacturers probably enjoyed showing off their powerful and imposing inventions to the crowds, in the same spirit that prodigious quilts were proudly hung from rafters. These exhibits publicly displayed substance, as did the large factories dotted around the Ontario landscape. Engravings of factories and medals were the letterheads of choice in the late nineteenth century. The age of triumphant capitalism was shot through with showmanship.

Many household manufactures were, according to *La Minerve*, also of platonic interest for most visitors.[88] Furniture makers outdid themselves with the luxuriance of their exhibits. Jacques and Hay were zealous exhibitors in Canada and abroad, with magnificently carved suites. In 1852 the company built a cottage, carpeted and hung with crimson and damask; inside was walnut furniture, including a magnificent bedstead priced at £60, bedecked with carved wreaths and garlands, birds, fruit, wheat, wild flowers, a Cupid, and a Madonna. A later exhibitor was John Hoodless of Hamilton, who, in 1882, showed a drawing-room set of carved walnut with scarlet satin puffed covering, a Turkish suite in plush mohair, inlaid tables of ebony and walnut, chairs, and fire tables. This sort of exhibit prompted self-satisfied musings: 'The collection of manufactured goods is eminently suggestive of the real condition of the Canadians; luxury amongst a few, solid comfort and increasing prosperity amongst the many.'[89] How different was the exhibition from the world of undernourished working-class families.[90] These visitors could barely afford the admission, and the ubiquitous commodities would be, for them, artifacts for scrutiny only, museum pieces.

Perhaps women doing piece-work at home could afford the sewing-machines found at every exhibition after mid-century. Sewing-machines made a stir at the fairs, their salesmen being 'terrible fellows for talking and "blowing."' Wanzer, Singer, and other companies worked hard to outshow their rivals at provincial, county, and township shows, sending dozens of machines as well as women to run them. Sometimes a blind girl was employed to show how simple the machines were to use. Wanzer offered a free machine for the quickest shirt made on the spot: a Woodbridge woman won, taking forty-two minutes.[91] Piano companies also had competitions and brought in professional performers. Visitors who could not buy a piano could at least enjoy the music at the fair.

Scattered throughout the exhibition could be found quirky exhibits devised by ingenious inventors. At the Provincial of 1883, these items included a refrigerated coffin equipped with bells, and a perpetual motion machine. In 1871 a 'sick minister' sent in a model bedstead that could also sleep a child in a drawer, a wash-stand that could double as a lady's work table, and a walking stick that could serve as a camp stool. The next year, J.P. Merritt of St Catharines exhibited a 'universal Chronographer,' which showed how all the events of human history were connected but was 'not very intelligible to an ordinary spectator.'[92]

It is impossible to list all the objects exhibited at the fair. Such a list would, in the end, amount to a description of Victorian material culture. And yet, so long as there was a backwoods, there persisted small agricultural fairs that attracted a handful of livestock and a few samples of farm products, wrung from a harsh land. Gaspé, unable to sustain an agricultural society in the 1820s, managed it in 1847: 'There was a Show of Cattle and Vegetables which, for the first time, drew together the Farmers of the Neighbouring Parishes'; the next year, 'some fine specimens of pigs & sheep were seen, as also an imported Alderney Bull and Cow – the Potatoes, Turnips, Onions, Parsnips, Carrots and Cabbage were of large size and fine form, while the Butter & Cheese – especially the former, proved that the female portion of the population were second to none in their department.'[93] Four decades later, the first fairs were appearing in northern townships, as at Idylwild, Muskoka, where twenty-three prizes for grains, roots, fruit, butter, flowers, and domestic work were distributed among six exhibitors.[94]

The formal structure of showground activities did not change a great deal over the century. They can be roughly categorized as judicial, officio-political, commercial, and entertaining. The judging was the most important; without it, an exhibition was just a bazaar. The judging introduced a note of formal education, for it would draw visitors' attention to the very best of every species of exhibit. This was, to some extent, an attack on private judgment, but judges had little latitude, for visitors made up their own minds which animal or machine or quilt was the finest, and a judge who defied their collective wisdom would only discredit himself and the institution of judging. As Horace Greely remarked of American judging, 'The popular verdict often reverses that of the committee, and fully atones for the injustice by the heartiness of its condemnation.'[95] One Ontario fruit grower also found that 'the public in the aggregate, are pretty good judges of a fruit show and they are not backward in giving their opinions freely when gross errors have been made.'[96] At the county show in Woodbridge in 1877, one horse received only second prize, 'to the dissatisfaction of many spectators, who cried to the driver to "throw it back." At first he shook his head, but the cries being repeated he thought worse of it, and separating the blue ticket from the reds in his hand, threw it on the sod and drove away.'[97] At a ploughing match in New Brunswick in 1851, affronted competitors elected their own judge, who overturned the official verdict.[98] Exhibitors were always ready to challenge a judge's verdict, often using their right to appeal to the Fair Board. Most boards upheld judges' verdicts, though they might reform the judging the following year. By 1873, litigation, 'threatened or actual,' was a serious problem, and one fruit grower successfully sued the Provincial Association for bad judging.[99]

Much of the judging done in nineteenth-century Canada was very poor. Often no expert could be had, and judges were simply prominent men who knew nothing of the business: 'Many are so ambitious of the honour and so covetous of the fee connected with a judgeship that they will accept positions for which they well know that they are unfit.'[100] One farmer at the Western in 1877 groused: 'It's an awfu' peety when a mon gets up a lot of good sheep, and has to axheebit them afore a thunderin' jackass – a felley in a white westcoat, hair oil, an' gold chain, instead o' a plain farmer wha has been amang sheep a' his days!'[101] At the Provincial in 1881, the judge of Southdown sheep pointed to a pen full of them and asked: 'What do you ca' these black-faced beasties?'[102] When experts were found, they usually had exhibits on the grounds or had sold their stock to an exhibitor, so the glory would redound on their herd. In the early years, this conflict did not seem to matter, for judges often awarded themselves prizes, but in later years it was considered scandalous.

One solution was to find experts in related fields, but this strategem skewed the results: Durham men judging Ayrshires would look for beefing qualities, so that, in 1867, the biggest Ayrshires, infused with Durham blood, most pleased the judges.[103] Much more satisfactory were experts brought in from another township, another county, or the United States. Unfortunately, this solution was costly, for visiting experts demanded expenses paid and an honorarium. Two late innovations aimed at securing more reliable verdicts were points systems and the single judge. It was hard to find three good judges, and they hid behind one another's verdicts, whereas one judge would be more careful in his decision. Although no reform in judging would ever please everyone, standards of judging improved, especially as scientific apparatus came into use. Milk tests were notoriously unreliable for many years: There was no record of what the cows ate or of the average yield, and some cows were milked to death. As well, the early tests measured only milk quantity and not its richness, an important consideration for butter production. Over the years, exhaustive tests were introduced and a better idea of yields was obtained, but not before many a breeder had withdrawn his herds from competition amid harsh words.

Judging was a serious business, as the rituals showed. Sometimes exhibits were anonymous so as to ensure impartiality, but because anonymity hampered publicity and business transactions, the larger fairs abandoned the practice. Exhibitors were not permitted to approach the judges; nor could crowds jostle them, so the judging went on before the doors were open to the public. Sometimes exhibits arrived late on the grounds or the judges would have to criss-cross the grounds to find the objects randomly situated wherever the exhibitor found a few spare feet, so the judging might take up most of a one-day show.

As the experts became more proficient at their task and better equipped to perform it, their judgments often became counter-intuitive. It was no longer the biggest ox, but the ox with the highest proportion of meat to offal – which required close study – that won the prize. It was no longer the biggest, most colourful fruit, but the sweetest that judges preferred. Fat pigs were passed over in favour of lean ones; tastier, but less spectacular. The best chicken might have damaged its feathers in the cage but still win a prize for its splendid form, which only a breeder could appreciate. In all cases, prizes went to goods most suited to the market, but public opinion could no longer follow the judges as it once had, for lack of expert knowledge. This was especially true at the large fairs, where most visitors knew almost nothing of what they saw in the various departments. By the Enlightenment model for the exhibition, spectators were always judges on their own account, but now they became mere spectators.

Expert authority also paraded itself before spectators in the form of lectures by dignitaries, who might urge improved farming or protectionism on the auditors. Ministers of agriculture lectured copiously, often undeterred by a complete absence of agricultural knowledge. D'Arcy McGee gave patriotic speeches, as at London in 1861, where he reflected on the American Civil War, or at Montreal in 1863, where he hailed the 'generous rivalry in the arts of peace' and the British Constitution.[104] Sir John A. Macdonald, occasionally called on to open shows, told jokes about the calves of lady dancers being suited to an agricultural show.

Departing Governor General Dufferin's farewell speech, in favour of loyalty to British institutions, was given at the Provincial of 1878 and printed in newspapers across the land. He remarked: 'It is not the departments of a mere Provincial Show which lie mapped out beneath my feet but the territories of our great Dominion, whose wealth and capabilities these courts exhibit. Nor is it in the presence of a detached crowd of casual sightseers that I seem to stand, but face to face with that entire population with whose destinies I have been so long associated.'[105] By identifying the exhibition as a microcosm of the entire dominion, Dufferin seemed to say that Canadians could actually meet together as one people. Organizers at the Toronto Industrial (known after 1904 as the Canadian National Exhibition) argued that it was not a local affair, but 'one of the Dominion's largest as it is one of its most important annual events,' where 'the pulse-beats of our national life can be felt as at no other time or place.'[106] An identity was being built up that, though indigenous to Toronto, was endowed with national significance by the sort of metonymic projection that the exhibition tempted speechifiers to make. George Brown's Upper Canadian community was conceptually superimposed on the entire country.

Historians have a 'boy scout's fondness for Fourth of July orations,'[107] but for spectators, the speeches were probably the least attractive part of the fair. As Warren Gates has remarked of Grange picnic exhibitions, 'Although platform speakers were highly visible and widely reported, individual speakers were probably seldom heard by any significant fraction of those attending the Picnic.'[108] Lectures given at Montreal in 1850 and Kingston in 1849 were ill-attended, and in 1856 Baron de Longeuil's address was 'voluminous, and, as usual, far from being interesting to the public generally. The chief topic touched upon is the horse.'[109] Educational lectures disappeared for a few years, later reappearing when the Ontario and Quebec governments paid professional agronomists to speak to Farmers' Institutes and *cercles agricoles*. At the larger shows, breeders and manufacturers held conventions where papers were read, but the lack of a wider audience was generally regretted. Traditionally, speeches at the exhibitions were supposed to be apolitical, but many were full of rousing campaign rhetoric. When George Brown, who owned a stock farm, was asked to speak on agriculture at the Brantford Southern Fair, Tories were outraged and cancelled the event. Even D'Arcy McGee overstepped the mark with a partisan harangue at the Missisquoi Fair in 1867, a year when political feeling ran high: 'Something like a riot occurred and for some years no more addresses were given at the fair.'[110]

Usually there was an inaugural speech or two, several more at the prize-giving ceremony, and extended toasts at banquets to various interests, not forgetting the press and the ladies. In Quebec, banquets were prohibited as a waste of public money. At the Toronto Industrial, organizers held daily lunches and published the proceedings: In 1888 Oliver Mowat was the guest of honour; in 1889 it was Sir John A. – whose response to the announcement that lunch was conducted on temperance principles has not been recorded. Sir John and his wife came to inaugurate the fair, an operation performed by pressing a button, which then sent an electric signal to the machinery hall and set the machines in motion – whistles shrieking, bells ringing, and crowds cheering. Usually the visiting dignitary said the requisite words of praise while his wife, sitting silently beside him, pressed the button. This time, before Lady Macdonald could put finger to button, John Leys, MP for Toronto and a Holstein breeder, set his hat on it and triggered the cacophony.[111]

Inaugural ceremonies were rarely held on opening day, for visitors knew better than to come when few exhibitors would be ready. Still, a new governor general was always an attraction. Governors' autumns were largely taken up trailing from fair to fair. When the Prince of Wales came in 1860, enormous crowds pressed close to catch a glimpse of him, and 'the swaying of the turbulent mass, under the opposing influences of the terrible cavalry and their own inclinations,' almost proved disastrous. Hoop skirts and crinolines suffered badly.[112] When Prince Arthur visited a few years later, 'the crowd was too large and too feverish with excitement to settle down to business. The royal pageant, too – which in comparison dwarfed even the interest in the Exhibition – absorbed people's time and attention.'[113]

Few dignitaries created as much excitement, but there were other distractions that the serious-minded despised. The Victorians were of two minds with regard to entertainment. Some thought harmless fun a social necessity: 'The denial of fitting recreation to the poorest orders, and even the not providing them with means to gratify the natural appetite for amusement, is one of the grossest oversights of an enlightened administration. If ... we had places open to entertainment, invitations to innocent walks and enjoyment, games, fairs (not proscribed, but well regulated), and even public spectacles and shows, provided for those who have not the means of purchasing the smallest pleasures, it would do more to extinguish discontent, and make a satisfied people, than the bitterest penal statutes, and the most tiresome labours of theoretical philanthropists.' For this Tory writing in 1833, amusement would make reform unnecessary and nip popular politicking in the bud: 'Deprived of more gentle relaxations, men are driven to the gin-shop; they talk over their superiors – and whoever talks of others in order to praise them?'[114] His comments answered complaints against the circus published in the Reform and Methodist press, Mackenzie's *Colonial Advocate*, and the *Christian Watchman*, and they were capped with the insult that Reformers and Methodists were themselves mere theatre, 'mountebank pranks and droll exhibitions of mummery and quack piety.'[115] Fifty years later the Salvation Army was the target of such epithets,[116] and the war on low amusements continued. Agricultural fairs were not low amusement in 1830, but when they took on the trappings of circuses, moralists complained.

Objections to the amusements at the fair varied because the amusements varied. Some were simply spectacular, like the fad for torpedo explosions. Some critics wanted fairs stripped of these entertainments, scaled down, and returned to the farmer. Others fought only those distractions deemed demoralizing: anything involving women or gambling. Horse-racing was bad, lady racing worse, and no language was too strong to denounce gambling stalls: 'They are accursed fountains sending forth bitter waters that smite with blasting and turn into an arid desert this land over which they flow.'[117]

The gambling and the amusements appealed to desires for profit and for visual display that the exhibitions had sought to cultivate, and succeeded in only too well. The early improvers tried to channel these desires in such a way that greed, ambition, and pleasure-seeking would expend themselves usefully in the pursuit of fine cows and grains. In fact, exhibitions fuelled these forces, as fair boards cast about for new ways to satiate them. To traditionalists, the expansion of exhibitions after Confederation seemed a cancerous growth on the healthy, honest countryside.

The early amusements had, after all, been wholesome. There was excitement enough in the cattle competition or the ploughing match:

> The day was fine, the land was dry,
> With steady hand and eager eye,
> They spent the day in keen contest,
> To prove which plough, which man was best.
>
> ...
>
> But every ploughman on the ground,
> Laid up his ridge so smooth and round;
> Those who were judges all did say,
> T'was done in a superior way.
>
> Then next a banquet was prepared,
> At which a Monarch might have shared;
> Their day's work finished in they went,
> And feasted to their heart's content.[118]

To mark the larger exhibitions, stolid Victorian entertainments were held: regattas, balls, parades, and tugs of war. These were public festivals in which citizens took part, not professional entertainments. Lacrosse games and dances performed by Indians were only semi-respectable, especially when they became rowdy, as at Montreal's first Provincial in 1853. Band competitions were another popular entertainment at the fair and an occasion for community pride, or community malevolence, for bands took defeat badly. In 1853, when a judge disqualified the manifestly superior Kingston band and gave the palm to Dundas, newspapers were full of the scandal. American bands came north, Ontario bands competed in Quebec and Quebec bands in Ontario, Indian bands did the rounds, and so did lady bands, complete with velvet uniforms and plumed helmets. *Grip* tittered at the prospect: 'Oh, how the manly bosoms of the Londoners will palpitate when the girls toot their little horns. How they will squirm when they see those ducks of girls pressing their ruby lips to the mouth-pieces of the brass instruments – how they will wish they were trombones and cornets!'[119] Children were often included in the entertainment. At Toronto, between races, the public watched school cadets do military drill, girls perform calisthenics, and a 'broom brigade' of little girls, each armed with a broom carried as a rifle and a dustpan strapped on like a knapsack.[120]

Public subscription paid for these sideline activities, which grew ever more elaborate: In 1881 a Citizens' Committee raised $11,581 from Montreal businessmen and paid $4525 for entertainments, including a torpedo explosion that threw a Man of War 200 feet into the air (straw men were added to heighten the effect), lacrosse games, a tug of war, balloon ascensions, fireworks, electrical illuminations, and horse and bicycle racing. Performances were given by rower Ned Hanlan and channel-swimmer Captain Webb; the latter smoked cigars and set off fireworks while swimming. David Morrice presided over the Citizens' Committee and personally donated a fountain to the showgrounds.[121] This entertainment was still wholesome, yet the frontier of respectability was being pushed back. Balloonists and acrobats were amusing, but lady balloonists and lady acrobats were demoralizing: 'For can one imagine a wife or sister sunk much lower than to seek applause from the multitude dressed in glittering spangles and flesh-colored tights with loose flowing hair?'[122] By this time, Toronto Exhibition organizers had thrown caution to the wind and even engaged female dancers. In 1889 a troupe of Hungarian dancers who kicked their legs into the air to revealing heights excited the opprobrium of Methodist ministers.[123] Many people were more amused by the Methodists' outrage than offended by the dancing. Other attractions brought in by H.J. Hill and J.J. Withrow that year included a re-enactment of the burning of Moscow, fireworks, trapeze artists, and the grandstand show with its races. Entertainment tended to the martial. In 1888 the seige of Sebastopol was performed every evening before an enormous painting of the fort, and in 1890 cowboys and Indians rehashed old enmities. Another excitement was the introduction of electricity in 1883: Amidst evening gloom, 'suddenly a light like sunrise shot through the building: on the instant everything recovered its form and beauty; the building was again thronged with people; and another day, the day of Electric Light, began.'[124]

As the entertainment grew more spectacular, the informal sideshows grew more numerous, including 'many of Barnum's old cast-offs': the fat woman 'of a ponderosity beyond description,' the human skeleton, bearded woman and strong man, as well as transparent frauds: mermaids, missing links, fortune tellers, and ubiquitous Zulu warriors who had never seen Africa – 'their war dance and yells were very tame, civilized affairs.'[125] A Toronto reporter toured round the fakir shows on Yonge Street during fair week and found such scams as a 'World's Museum,' a piece of canvas with fourteen peepholes, through which one spied pictures cut from the *Illustrated London News*.[126]

Disaster often struck, creating more excitement. Things went wrong with balloons and parachutes. One year a young Ottawa butcher, Tom Winsley, helped a balloon lift off but did not let go the rope, and he rose up in the air with it until he could grip no longer and fell to his death. 'Down he came like a rocket, executing A SERIES OF SOMERSAULTS in the air,' gasped the press.[127] At London that year, strongman Signor Lawanda held a barrel of water with three heavy men on it in his teeth, but a misstep landed 1000 lb of barrel, water, and men on him, sending him to the hospital.[128] In 1890, at the Toronto grandstand, a horse broke loose and charged around the track, upsetting half a dozen horses and vehicles. It was making for the fence, beyond which were thousands of schoolchildren, when 'California Tom,' one of the cowboys performing with Captain Horn's Wild West Show, leapt on the beast, swung over its neck, and tripped it to the ground. A cap passed around for California Tom netted him $50.[129]

Demonstrations, sermons, and speeches provided impromptu entertainment. In 1863, when a revivalist preacher warned of the coming Napoleonic world régime, his words were drowned out by a band of minstrels.[130] In 1887, at the Hamilton Central Fair, a labour demonstration was interrupted by a band that 'began to toot sturdily the tune Let Us Be Happy Together.' A Tory reporter mocked the speaker's efforts: 'Amid the cries of the fakirs and hokey-pokey peanutmen, the guying of the small boy, and the bustle and noise of a constantly shifting and wholly uninterested crowd, the silver tongued orator of Michigan struggled down the homestretch of his peroration.'[131]

There were gaming stalls, set up at the fair gates. Catch pennies at the Guelph Fair included pitching coppers into a bowl, rolling pins at a board, and shooting pine arrows for a cigar.[132] At Montreal, 'one could win canes and penknives innumerable if he only could drop the rings in the right places; he could win doubtful looking cigars from a doubtful looking shuffling of dirty cards; or he could throw away five cents while throwing soft balls at the hammer hardened skull of an Ethiopian that only needed washing to be white.'[133] Many young farmers confident in their skills were wide-eyed in astonishment when they lost, but if they suspected skulduggery, they took justice into their own hands. When a walnut-shell man took $5 from a farmer at the Stoney Creek fair in Saltfleet, 'the yokel then pulled off his coat and made the light-fingered individual disgorge.'[134] To many, these stalls were a kind of thieving: they stole money from country boys who knew no better, and if the rube occasionally won, this was also a kind of theft. When one boy wagered a penny and won six, a passer-by sternly advised him to trade them back, 'and then stand with the world an honest boy again.'[135]

Theft of a more blatant sort flourished. The exhibition was a paradise for pickpockets and con artists because, so long as the fair served as a market, farmers strolled the grounds with huge sums in their pockets, ripe for the plucking. Time and again, newspapers reported hundreds of dollars stolen from visitors: Ottawa fair organizers nailed stolen purses to the gates as a warning. Thieves concocted stories to fool visitors, selling 'gold' watches at inflated prices year after year. They came from miles away to converge on the exhibition, as did detectives, who descended in swarms to help local police meet the problem. Many of the thieves were well known, and no sooner were they spotted than they were hauled up for vagrancy, fined, and put on a train out of town. Dozens of constables strolled fairgrounds, watching out for light-fingered thieves, but they could not police the entire, overcrowded town, and the largest thefts took place outside the grounds. Thieves raided the homes of townspeople who had gone to the fair, stole money from under pillows in hotel rooms, slit purses at train stations or on crowded streetcars, and sometimes they went further. In 1880 a Markham farmer who had come to see the Toronto Industrial accepted a lift in a buggy, only to be chloroformed and lose the $500 hidden in his clothing.[136] In 1871 Thomas Secord, returned from prospecting in California where he had made his fortune, attended the South

Oxford Fair to buy horses. He visited old friends, toured the bars, and showed off jewels and money; his 'mangled' body was found in the woods outside the town.[137] Drink was implicated in this crime, and also in the rape of a woman working on the Montreal showgrounds who was assaulted as she went home by men gathered at the liquor shanties around the gate. Drink was also at the bordellos, which did a roaring trade during fair week. Constance Backhouse has described a raid on one such establishment in the last days of London's Western Fair.[138]

Refreshments, like amusements, were either innocent or sinful. There was always a restaurant, sometimes several at the large fairs. At Missisquoi in 1874, C.J. Phelps paid $25 for the concession and promised to serve a meal of cold meat, hot potatoes, hot tea and coffee for 25 cents.[139] The London *Advertiser* was scornful of the fare, 'half-cold and flabby roast beef and altogether cold potatoes.'[140] Banquet food was often no better: In 1862 the Provincial Association offered sour beer and foul fish. Often charities and churches organized the restaurant: The Woman's Christian Temperance Union made a good deal of money this way. There were many other specialty vendors on showgrounds, touting 'the cheerful, colic-producing melon, the substantial bologna sausage, the soul-cheering soda-water.' A visitor to the Montreal fair 'was invited to partake of diminutive oysters from overgrown tablespoons out of a larger-sized whisky glass, into which vinegar had been plentifully poured and pepper plentifully popped and as a test for the human stomach he was offered a choice between whitey-blue milk and cider that never knew the odor of an apple to wash down the infantile bivalves and the well developed vinegar.'[141] Vendors paid a fee to erect a stall or strolled the grounds. At the Provincial in 1854 there were just nine refreshment booths, but by the end of the century there were dozens, selling everything from pigs feet and jellied turkey to pumpkin pie and 'villainous whisky.' Outside the fairgrounds, hastily erected shanties lined the road. They bore names like 'The Hole in the Wall' and were crowned by a Union Jack flapping in the September breeze, and they sold everything from mild, creamy lager to 'tangle-leg' strychnine whisky.[142] Some were licensed, while others paid a fine in lieu of a licence, after which the police let them alone. Lager was often permitted and whisky banned, but lager booths kept a jar of whisky under the counter. The *Globe* was disgusted when the Toronto Industrial Exhibition Association took out a liquor licence, reasoning by analogy with educational institutions: 'It would be monstrous to have schoolmasters licensed to sell spirits, porter, and ales, and a whisky booth in a college quadrangle would be justly looked on as an outrage.'[143]

Drinking and fighting were a staple of fairs from one end of the century to the other. For one confirmed drinker, the Toronto Industrial marked the beginning of a binge that lasted until his arrest more than a month later at the Peterborough fair: lacking a dollar to pay the fine, he went to prison for a week.[144] But at the Bruce County exhibition of 1878 the reporter complained: 'We have rarely, if ever, seen so many young men of decent and respectable appearance the worse for drink as on this occasion.'[145] At Burford, where whisky booths were permitted inside the grounds, 'the usual number of fights were on the programme and carried out, and anyone who was presumptuous enough to enter the crowd was quite liable to come out with a broken nose or a discolored optic, whether he spoke or not. For several hours it was a perfect pandemonium, lights were extinguished in the general melee, and men with their blood fired by the miserable liquor they had poured in all day were as wild as cowboys in the West.'[146] When Colonel By declared a holiday and fair in 1829 at Bytown, on the square of land south of Wellington and west of Bank streets, he was unprepared for what followed:

'Twas not to buy or sell they came
From far and near, the blind, the lame,
The grave, the merry, and the gay
Upon that old eventful day:
They all assembled wild and free,
To have a ranting, roaring spree.[147]

While farmers raced their horses, canallers drank rum with bay leaves sold at the refreshment stands and bet heavily on the races. One race was so close that a fight broke out among the Tipperary and the Connaught men over which animal had won, and for three hours the battle raged.'[148] Some newspapers ignored the fights; others entered into their minutiae. A Milton paper reported a 'fairly paralysed' Salvation Army man at Halton county fair getting the worst of a fight and being dragged off by his wife.[149] Men fought most often, but some women entered the fray. When Mrs J.F. Buckingham returned from the Western Fair with her husband, a man came up, made immodest remarks, and seized her. Mr Buckingham grappled with the lout, while his wife, in her own words, 'hammered him for all she was able.' Spectators watching them roll about on the streets declared it just like an old-time Donneybrook.[150] Not all the fights took place on the showgrounds: Organizers had their own share. When the Missisquoi county society tried to decide where to purchase showgrounds in 1860, one rowdy, son of the society's president, disrupted the vote when he shouted 'Hold up both hands.' Stymied, the officers had members file through the building one at a time to vote. They put a guard at each door, but the westerners rushed in through the windows and overwhelmed the eastern men. Westerners stoutly insisted that the easterners had thrown the first punch.[151]

The drunkenness was, in fact, declining, both relative to the growing numbers of visitors and in absolute terms. The London police chief described the Western Fair of 1888 as the 'lightest' ever held: '"The biggest number of cases we had any day was 20 or 25. They were mostly simple drunks. Now, in old Fair times it was nothing to run in 30, 40, and even 50 in a night." "Why," said Sergeant Jenkins, "one Fair I put in 36 myself alone."'[152] And, they added, had they formerly been as 'particular' as they had become, the jail would not have held all those arrested. James Young observed the same decline in drinking at the Galt fall fair.[153] London police attributed the change to a decline in drinking and to the fact that saloon keepers watered the whisky. A greater police presence was undoubtedly another reason.

This control signals an important development in the history of exhibitions. The early fairs had been small, stern events, with the hegemonic ambitions of the ruling classes clearly visible. But this presence was only a show of strength to hide the weakness of early administrative and police apparatus. After the riot in Bytown in 1829, Colonel By prohibited fairs, for he could not maintain order at them. By the end of the century the state was able to enforce order at the fair, and citizens by and large submitted to the restraint. That subversive element which Bakhtin calls 'the carnivalesque' was attenuated in the course of time.[154] Hysterics like Thomas Shaw might envision social collapse from the gambling booths and belly dancers, but it required a feat of imagination to follow him; no such feat had been necessary in 1829. Gradually, the cultural influence of the popular classes prevailed at the fair, but their culture had been transformed no less than that of the middle classes, tempered by decades of cultural contact and interpenetration at the exhibitions and elsewhere.

The most important exhibit at the fair was the audience. Many accounts of mid-century shows hailed the sight of Ontario's sturdy yeomanry: In 1847 Lord Elgin spoke before 'six or seven hundred substantial Upper Canada yeomen – a body of men not easily to be matched. It is indeed a glorious country!'[155] The London *Advertiser* bragged that 'no agricultural country on this fair earth can send together upon such an occasion a more uniformly well-clad, energetic, observant, intelligent and progressive gathering. They represented the basis of the country's stability and progress.'[156]

In 1874 Charles Foy, the Canadian agent at Belfast, published a description of a Canadian fair: 'The scene presented a fair picture of prosperity, content, and advancement. The stalwart, burly, jovial farmers would have compared favourably with any similar gathering in the old country. Their wives, plainly dressed, absorbed in examining improved butter churns and wringing machines, looked cheerful and well to do. Their daughters, plump and rosy-cheeked, daintily trimmed from feather to shoe buckle.'[157] Or, as one Toronto paper said of 'Young Canada from the farm,' 'he may have dandruff on his coat, but there is independence and integrity under it.'[158] Farmers' Day and Citizens' Day drew the largest crowds at the fairs. The Grange collaborated on Farmers' Day, in exchange for 20 per cent of the profits. Farm labourers expected to have the day off: John Clarridge, a labourer for David Smith in Caledon Township, received his $5 salary on September 27, 1895, and took the day off to attend Brampton Fair.[159] On Citizens' Day the mayor called a half-holiday, and the working class came to the fair. There were also Children's Day, Women's Day, Germania Day, American Visitors' Day, and many others.

It was a matter of local pride to draw a large crowd. Every year, fair organizers surveyed the small opening-day crowds with pursed lips and scrutinized the sky for threat of rain. By the third day, success or failure would have been decided, the directors beaming with pleasure amid a merry crowd of holidayers or skulking around half-empty showgrounds. In Montreal, which had the biggest showgrounds of all, a crowd of 2000 'seemed like a very small toad in a very big puddle,' but a few days later 60,000 squeezed themselves into the grounds.[160] The huge crowds that exhibitions attracted had seemed more remarkable early in the century, before the rise of cities, but the infectious excitement that a large crowd produces never disappeared.

People exhibited themselves. There were innumerable baby contests. At the Peterborough Central, a $10 cashmere dress went to the mother of the best baby, a $10 suit to the father, and a $5 gold coin to the infant.[161] A reporter satirized such proceedings by suggesting classifications like 'Ittle oody-boody-woody-poody-um.'[162] The baby competition was held in good fun, but by the twentieth century a study of the Iowa State Fair shows that it was used to advance scientific parenting.[163]

Most descriptions of exhibitions were really business advertising. Perhaps the most heartfelt response to them survives in poetry. Even livestock journals published poetry by readers, usually sentimental musings on dead children or autumn foliage. But fairs, too, came in for their share of poems; a staid *Canadian Agricultural Reader* would leave aside superphosphates to versify with shaky scansion and dubious rhyme:

Ye husbandmen, both far and near,
Up, up, stir round, prepare,
With sons, and wives, and daughters too,
To attend the Farmers' Fair.

Bring wheat and corn of various kinds
Bring all that's new and rare,
And barley, oats, rye, buckwheat, millet,
All to the Farmers' Fair.[164]

Most of these poems would run through the list of exhibits, delighting in the simple enumeration of the objects that exhibitors had worked to bring forth from the land, as in this appreciation of 'our own Township fair':

Canadens cucumbers, crabbed and curious
Canadens cabbage, globular, great –
Patters of butter so golden and glorious,
Churned in the morning by Charlie and Kate; –
Squashes in plenty, and, each one a barrow-full, –
Carrots and Mangolds as long as a rail,
Praties that roasted, would make the most sorrowful
Laugh at the Quacks, as he weighed down the scale.[165]

The lovingly bestowed adjectives were themselves crabbed and curious. The produce piled up in heaps – a blaze of red tomatoes or capsicums, green lettuce, orange carrots, yellow corn, and purple eggplant – stirred visitors with the instinct to honour and preserve this noble sight in words. Nor was the larger message lost sight of. By enumerating the fruits and grains and animals found at the fair, their material weight was asserted, and with it the material weight of rural life. The objects on display were ponderous, fleshy, and substantial. They gave agriculture substance, a place in the cosmos that could not be overthrown by mere professions like the law or medicine, for whoever saw a good exhibition from lawyers? The *Agricultural Reader* heckled these city classes:

Ye clergy, teachers, students, come,
Come taste the bright blue air;
Pale, sallow, sickly, 'feeble folk,'
Turn out to the Farmers' Fair.

Ye Lawyers, Doctors, Merchants too,
Come gather round – for where
Shall *non*-producers learn *their place!*
Save at the Farmers' Fair.

These poets were not just lecturing to others, but reassuring themselves that

There's health, peace, and wealth on the farm,
And with ours no life can compare;
But young folks and old need holiday whiles,
So you see we are off – to the fair.[166]

The plough match, which covered fields with arrow-straight lines, brought out the same thrill of pride. A Nassagaweya man urged his compatriots:

> Do not be satisfied merely with a name to live, but raise yourselves in your profession or occupation, whatever it may be.
>
> 'Tis infamy to die and not be missed,
> Or let all soon forget that thou dids't ever exist.
>
> Married men, come out next year and plough! Bachelors, be not afraid! young men and boys! come! and take the prize from both, and then I tell you, we will have such ploughing matches, as will be an honor to us.[167]

In crumbling yellow newspapers across Canada, page after page of names survives, listing the winners at the local fair or the Grand Provincial Exhibition. Newspapers sometimes baulked at printing them. In 1886 the editor of the Brantford *Expositor* exclaimed that none but prize winners glanced at them, and there were too many half-rate township shows anyway, so he refused to print the prize lists.[168] But, under popular pressure, the lists soon reappeared. Prize winners would not be denied their glory, readers the pleasure of seeing familiar names. The exhibits and the exhibitors have crumbled or died, but for those who seek out the lists of names, perhaps looking for an ancestor, the fame survives, celebrating the inglorious arts of peace.

PART TWO

Canada at the International Exhibitions

6

International Exhibitions and Canadian Nationality, 1851–1867

The international exhibitions provided the developing colonies of British North America with an opportunity to construct a self-identity and to broadcast it to the world. Exhibitors dreamed of astonishing visitors by their spectacular displays of natural wealth, manufacturing skill, and cultural and scientific genius. With elaborate arrangements of artifacts, pushing pamphlets, and grandiloquent attendants, they concocted descriptions to beguile immigrants and capitalists or simply to impress the popular audience. But if exhibitors and organizers had a great deal to say to the world, they also had a lot to learn from it. The sight of a neighbouring colony's economic strengths and weaknesses might prompt reflection on trade treaties and political relations. To the historian, these exhibitions are a unique form of mythologizing. They are stories that reveal contemporary ideas about material culture and national destiny. They were continually rewritten, to meet the needs of the next great international exhibition, to obliterate past defeats, and to win new victories. And, because they were collaborative productions, written in the works of many people rather than the words of one writer, exhibitions can illuminate collective aspiration and self-definition.

To no small extent, Canadian participation in the international exhibitions helped to generate a belief that collective aspiration and self-definition could be expressed. This belief resulted from an unusual degree of government intervention in the creation of these displays. The first Great Exhibition was initiated and largely sustained by private enterprise in Britain. At a time when laissez-faire doctrines prevailed, and public opinion held that the political sector was a trammel upon market forces, the Crystal Palace gathering was a celebration of all that market forces could accomplish in the way of material and fraternal progress. For Canada, however, the history of the early world's fairs soon becomes a history of government policies and practices. English and French organizers of the international expositions encouraged this policy by refusing to negotiate with any but official – that is, political – representatives of the colonies. But it was also manifestly obvious that private interests in Canada could not afford to send goods across the Atlantic to London without some sort of subsidy. At least one Canadian patriot proclaimed that Britain would benefit from a strong display by its colonies and that it should itself subsidize and transport a Canadian exhibit.[1] The British were not moved by this argument. However, the Canadian government did come to believe that a good display was in the national interest and should be undertaken by the legislature. As Canada was but a colony of Britain, lacking diplomatic powers itself, exhibitions provided the first opportunity for the legislature to represent a Canadian national interest abroad. The story that Canada told to visitors at the international exhibitions was a story narrated by its government. These displays initiated an enduring faith that the country has a national identity that the government can authoritatively decipher and set down.

This chapter will focus on the period leading up to the confederation of Canada, New Brunswick, and Nova Scotia in 1867. There were many reasons for and against this union of the provinces, which have been examined thoroughly elsewhere. I will ask whether participation in the international exhibitions had any influence on this course of events. Without invoking some metaphysical national identity, it is possible to show ways in which successes or failures in the exhibition hall shaped some of the terms of the debate. I do not try to provide a complete account of this long political negotiation, but merely wish to establish that exhibitions were one of many factors involved. I will show that the fairly narrow range of goods shown at these exhibitions both raised and lowered expectations as to what Canada could achieve through economic and political activity. These expectations, in turn, helped to shape the form of the new government and the allocation of powers.

London, 1851

The international exhibitions marked a new chapter in international trade and in education, but not, as was hoped, in diplomacy. The scheme was developed by the Society of Arts, a voluntary society that encouraged arts and manufactures. When the society raised the idea of national exhibitions, similar to those held in France, Prince Albert convinced it to invite the world, in keeping with England's recent adoption of free trade. As the scheme grew ever more grandiose, a Royal Commission took over and raised subscriptions for the building – Paxton's Crystal Palace – and a prize list of £20,000. The Great Exhibition of the Works of Industry of All Nations was supposed to prove that Britain led the world in industrial development and to enable British workmen to study manufactures from other countries. The experience proved something of a humiliation to Britain, for the cheap industrial goods produced by that nation paled beside the more finely crafted works from the continent. One result of the exhibition was the founding of schools of design throughout the country.[2]

Many were sceptical about this daring new scheme. *The Times* predicted disaster, and many Britons feared the influx of foreigners with their revolutionary ideas. The Americans were slow to realize the exhibition's significance, and they did not send important manufactures until their paltry collection had first been ridiculed by visitors. One who was well pleased by this humiliation was Canadian commissioner Alfred Perry, who poked fun at the Americans: 'So miserably meagre is their appearance that the people laugh when they look at it.'[3] Perry was from Montreal, where the exhibition had quickly obtained enthusiastic support. In February and March, while the provincial legislature pondered its response to the royal invitation, Montreal's leading men assembled to plan their own course of action.

Despite the loss of imperial preferences, which provoked the election of an annexationist mayor in March, Montreal's promoters and politicians rallied to the British project. Surveying the crowded public meeting, Papineau exclaimed: 'Il n'y a pas deux opinions sur le sujet qui nous occupe.' During his exile in 1839 he had visited and admired the French industrial exhibition as a means of scientific education.[4] Christopher Dunkin, a Conservative lawyer from the Eastern Townships, argued that a good display at the exhibition could procure capital and emigrants, develop the country's resources, and provide an opportunity for 'bringing together the people in more kindly unison.'[5] Papineau and A.N. Morin, the Speaker of the House of Assembly, also hailed the opportunity for fellowship, but William Badgely, another Conservative Townships lawyer, scoffed at such sentimentality: 'The exhibition was intended not so much for uniting the art of different nations as for improving it,' he said, by stirring up competition, just as the agricultural fairs had done. But where agricultural exhibitions were intended to create a competitive spirit among peasants, industrial exhibitions would redirect existing political passions into more productive avenues. Economic competition could, thanks to the invisible hand of the market, be channelled towards a common goal of prosperity. Political rivalry promised no such happy outcome. The meeting anticipated the optimism and good feeling, cemented by cooperation on public works, among French and English Canadians which characterized public life during the 1850s.

Somewhat bravely, the assembled public men of Montreal declared it 'highly desirable' to send a display to London, and they determined to hold a Grand District Industrial Fair that fall to collect exhibits. They sent invitations across Canada asking for materials that could be sold overseas.[6] Upper Canadian authorities and the legislature itself fell in with the Montreal plan, for lack of alternative arrangements, and £2000 was allocated to fund the gala affair. Premier Francis Hincks explained that, 'in a country where party politics run high, every action of the government is viewed with distrust. Government, therefore, had waited till the people themselves moved.'[7] Upper Canada's slow response to the invitation probably reflected a fear that its rudimentary manufactures would not be up to standard. Even in Montreal, many feared that 'Canada is too backward to cut a respectable figure in an Exhibition open to the whole world.'[8]

Similar fears paralysed efforts in the Maritimes. In Prince Edward Island the Royal Agricultural Society held a preliminary exhibition, but decided that the exhibits were 'so very trifling' that they should be kept at home. Only when members objected that the colony's interests would be seriously injured if it were the only British colony unrepresented did the society's officers forward the collection.[9] Newfoundland, to its shame, was the only British colony unrepresented at Paxton's palace in 1851. Only a few New Brunswick articles made their way over after the government decided not to exhibit: New Brunswick improvers enviously eyed the ninety packages sent from Canada and blamed their own abstention on apathy among the highest classes.[10] Nova Scotian authorities did collect some items, mostly stuffed animals, coal, Indian curiosities, and local cloths; the whole was, according to the lukewarm praise of Reform journalist William Annand, 'really unexceptionable.' He, too, blamed apathy among the affluent for any shortcomings and concluded that Nova Scotia could not expect to shine with lustre, but might attract passing notice. The effect was weakened when the British press mistakenly attributed Nova Scotia's fine exhibit of stuffed birds to Canada.[11]

The Montreal exhibition of 1850 was a tremendous success. Shortcomings in the display were more than compensated for by other splendours, including a regatta, gymnastics, parades, fireworks, dinners, and a fancy ball. Visitors included politicians and businessmen from New York and Boston who had come to negotiate railway deals, and it was declared that, after such a public celebration of friendship, the first of its kind, Canada and the United States could never be enemies again. Governor General Lord Elgin sent an account to the colonial secretary, Earl Grey, remarking: 'You will appretiate the importance as bearing on the interests of Peace and of the Connexion of such speeches in such a company as that of the Editor of the Boston papers.'[12] The exhibition was held in the new Bonsecours Market: the north wing housed the arts, manufactures, and natural history, and the south wing the agricultural produce and implements. Upper Canada took the top wheat prizes, while Montreal swept the vegetables, but there was little competition in oats, barley, spring rye or peas, or such crops dear to reformers' hearts as flax, hemp, turnip, and beets. Cheese and butter from both sections of the province won awards. A small display of agricultural implements included wooden and iron ploughs, threshers (one cost an *habitant* his arm), fanning mills, reapers, forks, grain cradles, and stump extractors, and nearby were steam engines, a portable grist mill, a model locomotive engine by a St-Hyacinthe lad, and a fire engine by George Perry. Leather work abounded, including porpoise leather by C.-H. Tétu, touted for durability and softness: an American took out a patent on the spot. An extensive mineral collection was displayed, courtesy of William Logan, the provincial geologist. Timber was less well represented. The Montreal *Witness* recounted:

> Whilst walking around taking notes, we were asked by two French Canadian *habitans* how they would know what articles had taken prizes? We pointed out the tickets, but they were still at a loss to know what were first and what second prizes. We showed them the words 'first' and 'second' on different cards, and as soon as they had mastered the distinction, they ran to a row of boards standing up against the wall. We followed them with some curiosity, and found that they had sent in some twelve boards, of as many different kinds of wood, and had taken nearly as many first prizes. Now, but for these two brothers, Parisault, of St. Martin, near Montreal, the great Canadian exhibition of 1850, preparatory for the World's Show of 1851, would have been almost wholly destitute of samples of Canadian woods.[13]

There were sleighs and pianos, furniture, brass, imitation marble mantles, firearms, and even 'a curious diving helmet' to be seen. Cornelius Krieghoff and Antoine Plamondon sent paintings, which were displayed alongside lithography and jewellery. Upper and Lower Canada shared the honours, although *La Minerve* complained that the English press ignored French victories.[14]

Traditionally, the task of interpreting an exhibition rested with an address by some dignitary on a 'lonely-wigwam-to-bustling-city' theme. In 1850 Justice C.D. Day gave a keynote address that marks the transition from agricultural to industrial exhibitions.[15] The Montreal exhibition was dignified, as the Upper Canadian shows had not been, by the spectre of the Great Exhibition. Instead of merely chronicling Canadian civilization, Day discussed the very nature of civilization. He began by arguing that an industrial artifact was, like a piece of art, a form of discourse: 'You come but now from the interesting display of these productions of man's ingenuity in another hall; you have surveyed them there: did you see in them nothing but wood and iron, and brass and leather? Look at them again, and they will tell you of hours of deep, laborious, persevering meditation – of weariness of body – of exhaustion of spirits – discouragement, almost despair – of revived hope and energy, and victories – glorious victories of mind, won inch by inch over the strong, although inert, resistance of matter and its laws.' Fine art, Day continued, was but the highest example of a process all artisans imposed on brute matter – and he pointed to Hiram Powers' sculpture *Greek Slave*, which thrilled crowds at the Crystal Palace the following year. The Great Exhibition represented the same process, now collectively applied to history. Historical eras, Day explained, declared themselves 'by general combinations and assemblies of the nations of the earth: sometimes at regular periods, during a succession of ages, and for a uniform object; and at others, on a single occasion, and for a temporary purpose, common to all.' The Greeks gathered for sport, medieval men to make war; but modern men gathered in a spirit of peace, as the Crystal Palace emphatically declared to the world.

Day dreamed of history being made as art was made. To the mid-Victorians, art seemed to provide the best example of intentional creation.[16] The art-manufacture was a microcosm of the exhibition, as the exhibition was to social processes writ large. It argued for a history that was not random but directed, as association united with the intentional activity of the artist and the artisan. Day's speech owed much to one given by Prince Albert earlier that year which heralded the coming gathering as 'a true test and a living picture of the point of development at which the whole of mankind has arrived in this great task, and a new starting point from which all nations will be able to direct their further exertions.'[17] This view of history was an answer to Tocqueville who, when touring the north of England, remarked that everything in Manchester's appearance attested to the individual powers of man, and nothing to the directing powers of society.[18] It was an answer to David Hume, who argued that history was not rational because intentions rarely determine outcomes, and to Rousseau, who declared the progress of civilization to be the progress of theft, violence, and immorality.

The Great Exhibition expressed Victorian religiosity, by intertwining natural theology – which catalogued God's design of the created world – with the pursuit of secular progress. To many Victorians, art was a genre within religion, its purpose being to elevate the mind towards the lofty and the spiritual. Exhibitions revealed the plan of God in the world, extending natural theology to industry. Whereas William Paley, in his 1802 work on *Natural Theology*, reasoned from design in a watch to design in nature, Day, quoting from the Bible, reversed this order: 'As there are tongues in trees and books in running brooks, which utter to the heart things of high import; so also in these evidences of human intelligence and labor, is there a language not to be mistaken. They are the enduring records of progress from barbarism to civilization; heralds of future growth and excellence; the tracks – to adopt the apt word of another – the tracks of thought.' One visitor to the Crystal Palace developed a full-fledged 'industrial theology': 'Art works with Nature's materials; and while adding to the number of lectures written in them, it cannot alter their general character and tendency. Industrial theology is therefore only an extension of the ordinary manner of teaching natural theology to the daily transactions of industrious men in one of its parts.'[19]

Commodities were proliferating at a monstrous rate in mid-Victorian England, and exhibitions were a place to contemplate them, to attempt to fix their place and function in a purportedly Christian society. The Crystal Palace was emblazoned with the words 'The Earth is the Lord's and the Fullness thereof,' to assert the omnipotence and omnipresence of God. Samuel Nelles, a Methodist educator, voiced similar sentiments in Hamilton in 1851: The Great Exhibition indicated that 'a more frequent intercourse, a more enlightened discernment of true interest and above all the increasing prevalence of a peaceful religion are about to usher in a better Era upon the world.' One Canadian visionary imagined that this great 'gathering together of all the nations, Jews and Gentiles, to praise Him, the just and good Governor of them all' was 'the forerunner of New Jerusalem.'[20]

Those reflecting on exhibitions tried to intertwine the two master-tropes of the century: Progress and God. But, Nelles reflected, it was difficult to see moral progress in material objects: 'If in such an Exhibition we could only include a picture of things *moral*; if we could but hold up before those millions of idolaters a faithful delineation of the pure & rational religion of Christendom.' There were other measures of godliness: Day's argument that the exhibition exemplified a new spirit of peace provided another criterion of moral progress. Unhappily, this argument was soon exploded for, at the time of the next great exhibition, held in France in 1855, England and France were at war with Russia. To the end of the century, the paraphernalia of Christianity enveloped exhibitions, from opening prayers to closing benedictions pronounced by bishops. But the moral progress that these orators invariably discerned was hard to distinguish from material proliferation. The Montreal Catholic paper *True Witness* (angered by the recent English Papal Aggression Bill) accused the Crystal Palace of encouraging mammon-worship, a rebuke that was probably accurate. The attempt to find a criterion of moral progress was abandoned in favour of analytic definitions of civilization, derived from the means of measuring it. Although blind to morality, exhibitions could measure the progress of civilization in a narrower sense by tallying the number and the quality of industrial exhibits from country to country and from race to race. But this project also proved incoherent.

Across British North America, packages of exhibits were boxed up and sent to London for opening day, 1 May 1851. Canadian commissioners bought a timber trophy that the *Illustrated London News* unkindly described as an 'uncouth sort of pile.'[21] The Canadians also secured agents in England, including Sir Randolph Routh, who had served the Commissariat in Upper Canada and, in 1853, was the moving force behind an Industrial Exhibition in Dublin. Isaac Buchanan, a Hamilton merchant living in Scotland, refused to serve as commissioner and denounced the affair for its free-trade leanings and misplaced idealism, declaring it 'at best a sentimental pageantry.' On reflection, Buchanan changed his mind and demanded a free ticket to the opening, only to learn that the position carried neither duties nor perks.[22] Also in London to represent Canada were William Logan and Alfred Perry. Logan acted as a juror in the mineral class and arranged the Geological Survey's exhibit. Perry went to London to advertise his brother's fire-engine. He pressed for a practical test of the machines and, to everyone's surprise, won first prize, beating the larger but less powerful European models. Perry also served as caretaker to the Canadian court and wrote long letters detailing Canada's triumphs to the Montreal press. When Queen Victoria visited the Canadian court first of all the courts, Perry exclaimed joyfully: 'For the first time since Lord Elgin signed that outrageous Rebellion Losses Bill, I felt as if I could have shouted

God Save the Queen,' and he thrilled to report that 'Her Majesty thought much of the beans, and the Prince pronounced the wheat and pease excellent.' The one-time annexationist Montreal *Gazette* pondered: 'It is possible that even yet, after all that has come and gone, the Great Exhibition may be the means of doing something to tighten the bonds between Canada and Britain, which have latterly been growing so loose.'[23]

Perry's letters trace the conservative re-enchantment with England, which had been troubled since 1849 when Elgin passed a bill indemnifying loyal and disloyal subjects for their losses in 1837–8. Tories ran riot in Montreal and stormed the Parliament Buildings. Perry himself threw a brick that shattered the gas chandelier and caused the building to burn to the ground.[24] A few days later he sparked an assault on Elgin: 'Unfortunately for the peace of the city, I kicked a brick-bat at my feet.' He picked it up and hurled it at Elgin, who ducked to avoid it, and the chase was on. With the mob in pursuit, Elgin finally pulled up at Government House and 'rushed up the steps with the agility of an athlete.' One can only imagine Elgin's feelings on learning that a £20 prize he offered to Canadian exhibitors in London went to the Perry engine. One Upper Canadian sneered that the Perry brothers would not value honours received from Elgin, but were 'sufficiently sordid to seek pecuniary reward from the same source.'[25]

Alfred Perry and Lord Elgin did agree that the Great Exhibition achieved tremendous good for Canada. Elgin thought it 'perhaps more serviceable to Canada' than to any other country represented.[26] According to Perry, Canada's true value was at last recognized: 'I was glad to hear several gentlemen, who looked like men of influence and standing, say to-day, that they had always been opposed to Colonies, and would have been very glad to get rid of them; but, that having come to the Canadian Division in the full persuasion of being confirmed in their impressions, they had seen enough to convince them, that they had been wrong.'[27] This story was probably an exaggeration. The *True Witness* reported that the Canadian articles 'consist in a great measure of raw materials, and of such manufactures and natural products as to make little show; therefore, but few visitors stop to examine.'[28] Reviewers were not slow to find fault: One complained that the fire-engine was garish and, as for the furniture, 'We like the beavers carved round the edge of the table; we cannot approve of the same animals crawling like rats on the crossbars of the legs.'[29] The woollens were very coarse by British standards and, when black walnut chairs were presented to the queen, Logan recounted: 'I understand Her Majesty rather smiled, and no wonder; for in my opinion they are not very good. Only think of their being adorned with yellow worsted fringe.'[30] The

Canadians also combatted errors in the *Official Catalogue*, such as 'During one-half of the year, the surface of the country is covered with snow and ice, and thus remains totally unproductive.' The commission published a pamphlet setting matters straight and reduced this figure to 'two to four months.'[31]

In fact, despite some sniping, the British press warmly praised the exhibit from Canada. Premier L.-H. LaFontaine wrote to Edward Ellice in London: 'Nous avons été fiers de l'attention qu'ont atirée les objets canadiens à votre grande et belle Exhibition.'[32] Minerals drew the most attention, especially the remarkable geological collection. The jury remarked: 'Of all the British Colonies, CANADA is that whose exhibition is the most interesting and the most complete; and one may even say that it is superior, so far as the mineral kingdom is concerned, to all countries that have forwarded their products to the Exhibition. This arises from the fact that the collection has been made in a systematic manner, and it results that the study of it furnishes the means of appreciating at once the geological structure and the mineral resources of Canada.' This passage was widely quoted in Canada for many years. *The Times*, also laudatory, reflected deeply on the minerals displayed by British North America as a whole. Nova Scotia's coal and iron ore were nobly represented, while New Brunswick's samples of iron ore, manganese, plumbago, pitch, and coal were 'most paltry as an index of the mineral wealth which really exists in that fine Province.' But the real lesson, according to *The Times*, was that the Maritimes possessed the coal that Canada lacked and, if that country wished to become an industrial nation, it should combine with the lower provinces. Here was grounds for thought. But when the British North American provinces thought about the world's fairs in the years before Confederation, they did so in a spirit of rivalry rather than of closer collaboration. In particular, the Canadian display of 1851 was later remembered to have an 'almost magical effect' in bringing honour and profit to the colony, which the other provinces sought to emulate.[33]

New York, 1853

The Americans were slow to rouse, but eventually they surpassed the English at exhibiting. At New York, in 1853, organizers erected a Crystal Palace, in imitation of Paxton's edifice, to house 'the choicest products of the Luxury of the Old World and the most Cunning Devices of the Ingenuity of the New.' 'Yankee ingenuity' was already notorious, and a modern

The Canadian Court at the Great Exhibition of 1851 emphasized the un-European aspects of life in Canada, with a colourful sleigh and gigantic canoe to attract the curious. The visitor could then turn to study the economic exhibits, including the minerals set out in viewing cases.

scholar, Robert Post, views the exhibition as a turning point for American science and technology, which thereafter enjoyed greater public support.[34] American industry was well represented by companies such as Singer, Corliss, McCormick, Colt, and Goodyear, but other countries, including British North America, were not so well represented. The American organizers hired an agent, James Whitman, to tour around and drum up exhibitors. He travelled from the Canadas to the Maritimes, arguing that participation would help secure a reciprocal trade agreement, which Canadians had sought since the repeal of the Corn Laws in 1846: 'What more favourable opportunity could possibly be presented to Canadian enterprise, of showing to the citizens of the United States, collected from all parts of the Union, the number, variety, and value of these articles with which they could favourably trade with them upon favourable terms, and of convincing *the people* of that country who have never yet been appealed to on this point, of the necessity of demanding from their Government immediate legislation to that effect.'[35]

In Canada the boards of agriculture twiddled thumbs, waiting for the government to send instructions and money, but this step had to wait until the new session. No action was taken until the spring of 1853, when the election of free-trade-minded Democrats in the United States made reciprocity more likely. Almost simultaneously, Quebec City organized a preliminary exhibition to select goods and, on 1 May, the Canadian government named W. Antrobus Holwell, an ordnance storekeeper from Quebec, as agent, to the annoyance of Whitman, who quarrelled with Holwell over control of the Canadian court and confiscated a stove in lieu of payment.[36] Everywhere Holwell went he met 'general apathy and want of cooperation'; producers were too busy meeting orders, or they dismissed the venture as 'speculation and humbug' or as 'without interest to Canada, especially after her great success in the London Exhibition.'

Holwell finally scraped together 153 exhibits plus minerals. They won sixty-six prizes, though none above a bronze, and mostly for agricultural wares and fancy-work. Other winners included blankets from Dundas, a thresher-separator shown by Van Brocklin of Brantford, craftwork from the Lorette Indians, Tétu's soft porpoise leather (touted as ideal for gouty gentlemen), sleighs, a plough, a seed sower, pottery, tiles, and a table top decorated with maple leaves. Canada was an agricultural country, but even the agricultural display was poor. If it sent any message to the United States it was that here was a market ripe for the taking, and that resources needing 'Yankee ingenuity' might be developed. Ontario Grits blamed the 'Quebec influence' for the 'confused and slovenly' display of goods that made

Canada seem rooted in a state of 'semi-barbarism.' Archibald McKellar, vice-president of the Kent Agricultural Society and later Ontario commissioner of agriculture, complained that things were 'badly arranged': 'For example – you see beautiful Blankets on a frame, which seem not to have been covered at night, or dusted, since the palace was opened, as they are now covered with a deep coat of dust – beside there is a table covered with leather, socks, cheese & &c. promiscuously thrown together as if they had been shook in a letter Box – under this same Bench is to be seen a number of Casks and Barrels, said to contain grain from Canada from the position these casks occupy the visitor would naturally suppose it was int[end]ed to conceal their contents.'[37] He feared that an opportunity had been lost to convince visiting Canadians of their country's brilliant future as a commercial and manufacturing nation.

To the east, the two colonies that had least distinguished themselves in London were determined to restore their good names. Newfoundland booked space in December, and the local press welcomed this opportunity of 'atonement ... for injustice done the country on a former similar occasion,' and of showing 'by tangible substantial proof that we have articles of export which may well induce the Americans to unbend their prohibitionary tariff.' The Agricultural Society, Native's Society, and Mechanics' Institute amassed a collection of grains, fish, oils, furs, leather goods, slate, and stuffed birds. There was even a magnificent model of a seal fishery, carved by William Knight of St Johns, showing a three-foot brig of mahogany, fully rigged, moored to a plastered surface representing ice, with miniature sailors slaughtering seals. Mrs M.S. Peace sent a book of poetry, *The Convict Ship*, which hailed the 'stern' and 'sea-girt' land as reminiscent of her native Scotland.[38] Prince Edward Island sent a small collection of agricultural products to the fair: the barley won a prize, and a Russian economist inquired about the oats.[39] Both these provinces were satisfied with the experience. Indeed, reciprocity was secured the following year, although the exhibition's role in securing the agreement or in the subsequent increase of trade is difficult to estimate, in view of a general upturn in trade.

The fair provided an opportunity to gain information as well as to impart it. Britain sent a team of practical men to examine American manufactures, with Charles Lyell to survey American geology. The Canadian minister of agriculture, Malcolm Cameron, commissioned a report from the agricultural journalist William McDougall, who had supported his ministerial aspirations. Now Cameron asked McDougall to examine and report on new seeds and implements useful to Canadian agriculture, 'with a view to action by this Department, and the Boards of Agriculture, in promoting their introduction into this province.'[40] McDougall responded with a list of reforms, ranging from a grant to the Toronto Model Farm, lest its failure discredit agricultural science, to bounties on flax-scutching machinery. On the whole, McDougall preferred private to public enterprise and he urged the government not to compete with business by importing animals or implements. 'To encourage, to stimulate, to diffuse information, and by means of societies and Boards of Agriculture among the farmers themselves, to promote *general* improvement, is the duty of Government. But special action must be limited to a very few cases, and such as lie beyond the reach of common means, or injury, and not benefit, may be the result.' When he spotted a self-raking reaper that might be useful in Canada, he persuaded the inventor to send a machine to the provincial exhibition at Hamilton. Other machines caught his eye, but 'all those of real utility are either already known to Canadian farmers, or soon will be through the medium of our annual Provincial Exhibitions.' McDougall saw only a limited role for government, which merely helped entrepreneurs to advance themselves. Holwell, meanwhile, advocated a much greater role for the government, for he believed that Canada could only appear to advantage at international exhibitions if the government took complete control of the display. At the next world's fair, the Canadian government took this advice.

Paris, 1855

By 1855 the novelty had worn off and dreams of peace had foundered. The Paris Exposition opened late, attracted fewer visitors than the Crystal Palace had done, and, like the New York fair, lost money, but it did earn the upstart Napoleon III a new respectability. Because France remained supreme in 'taste' and art manufactures, Paris claimed, inside the Palais d'Industrie, to lead the world.[41] That world in 1855 included Canada and Prince Edward Island, but not the other British North American provinces, which refrained from exhibiting in Paris. Prince Edward Island sent a few specimens collected by the Royal Agricultural Society, but the Canadian government decided to make a spectacular display.

The mid-1850s were prosperous years for Canada: Commerce had revived and railways and canals were being built and extended with great speed. The premier in 1855 was the moderate conservative Allan Napier MacNab, a huckster extraordinaire best remembered for the phrase 'all my politics are railroads,' which, according to J.M.S. Careless, 'was quoted with sage approval as the very essence of modern statesmanship.'[42] In 1854 MacNab formed a coalition with the Hincks–Morin liberals, rallying all but a few die-hard Tories and Clear Grits. The basis of this alliance, according to Peter Baskerville, was 'a shared desire for economic rationalization' that 'transcended sectarian differences.'[43] It took the form of investment in public works and railways (up to £3000 per mile of Grand Trunk track), so that, by 1855, the public debt was close to £4.5 million, with public revenues about one-third of that amount. MacNab's government now needed to fill the boxcars with immigrants and agricultural produce. The exposition of 1855 provided a splendid opportunity to woo immigrants, capital, and markets.

As well as serving economic interests, the Paris Exposition promoted political harmony. Exhibitions no longer symbolized universal peace, but this one solemnized a new cordiality in British-French relations. In the Crimea, the longtime enemies together battled Russia, and in Paris, Victoria strolled among the exhibits on Napoleon III's arm, the first reigning English monarch to visit France in over four hundred years. When she visited the tomb of the first Napoleon, one New York newspaper exclaimed: 'Great God! it was a scene to have brought Promethian fire to his ashes, to witness the sovereign of that England who had never relaxed her grasp of his throat till she had manacled him for life, drop a tear over his grave! ... The axe is laid to the root of international jealousies.'[44] Canada basked in this warmth, its court decorated with 'lovingly intertwined' French and English banners. As curator Alfred Perry proudly remarked, 'No other country could so appropriately symbolize the alliance between the two great nations.'[45] And if the Canadian display lured French immigrants to Canada, as French Canadians hoped, this influx might silence the calls from Upper Canada for representation by population which had recently begun to be heard.

The Canadian government voted £10,000 towards the display, and another £10,000 for French widows and orphans – to bribe the French into good humour with things Canadian. E.-P. Taché disingenuously told his fellow councillors not to be moved by such material considerations in voting the grant.[46] It also named a 200-man committee (nearly every leading public man served on it), which immediately confronted two problems. First, Canada East was not to be permitted to put Canada West to shame, as it undoubtedly would by virtue of its advanced manufactures and its population of French Canadians, who would take a greater interest in the French fair than English Canadians. This problem was resolved by the decision to give 'a national and general character to the Canadian section of the Universal Exhibition of 1855, which was wanting to the exhibition of 1851.' The government would purchase, ferry, and display the exhibits, so as to 'set the country in the place and stead of individuals.' The intention was to give a general idea of the country, 'without confining the information to one, or to another section, but being of an equal or general character.'[47] The government, by wholly subsuming private enterprise under its banner, revealed clearly that it was more concerned with the political aspects of the fair than the economic.

The second problem lay in capturing the public imagination. The kernel of the collection had to be rocks, trees, and grains, because this was what Canada could best sell overseas. How could the organizers make this sort of exhibit interesting and appealing? A subcommittee was appointed to address this question, and it included businessmen John Young and John Gamble, Geology Survey members William Logan and Count de Rottermund, and agronomist-politicians J.-C. Taché, William Rhodes, and John Langton. This subcommittee resolutely pondered and plotted a response. Studying the jury reports of 1851, they learned that the two criteria repeatedly mentioned by juries were completeness and the instructiveness of the arrangement. These attributes had earned praise for Logan's minerals and a Council Medal for George Lawson, for the 'admirably displayed, very complete, instructive and scientifically arranged collection of the alimentary products of Scotland.' (Lawson moved to Canada in 1858, taught natural history, chemistry, and mineralogy at Queen's and Dalhousie, and was secretary of the Nova Scotia Board of Agriculture.)[48] The committee quoted these remarks, as well as Robert Owen's verdict: 'Among the numerous samples of raw produce contributed by different countries, there are several collections of especial value which derive additional merit from their completeness and from the fact that they illustrate the trade and manufactures of an entire country. The importance of such collections not only in a commercial but in a statistical and scientific point of view, is very great.'[49]

With such clearly defined goals, the committee worked to achieve a complete and completely explained collection of natural resources. The Geological Survey set the example with an elaborate map, the fruit of fifteen years in the field. Logan had good reason to exert himself, for friends in the London Geological Society privately told him that he had just missed receiving their Wollaston Medal in 1854 because he had not published any scientific findings, and that a good map might be enough to secure him this highest geological honour. Logan was also under pressure to prove to Canadians that the survey was useful. A parliamentary committee, headed by John Langton, had concluded that Logan should strive to publicize his findings.[50] The mineral exhibit beat English and French displays to win a Grand Medal of Honour (Canada was the only colony to win one). The jury lauded both the map itself and the size and extent of the samples that had been gathered, 'offrant une suite complète, classée et étiquetée avec le plus grand soin, des minéraux utiles et des matériaux de construction que recèle le sol du Canada.' Canada also received a Medal of Honour for trees and for furs: 'Ainsi préparée et interprétée, l'exposition canadienne se présentait, en outre, avec la plus heureuse et la plus intelligible disposition.'[51] British Commissioner Henry Cole called the display 'the most interesting part of the Exhibition,' and *The Times* exclaimed: 'Through the huge mass of objects which the Paris Exhibition contains, there is not a single display so practical, complete, and strictly industrial.'[52] Canada lost the top agricultural award to Britain (British grains were in glass jars, the Canadian grains in sacks). But in timber and minerals, Canadians had found the key to success: packaging.

The Canadian timber trophy in 1855 was no uncouth pile, but a monumental structure sixty feet high. There were 200 specimens of sixty-four varieties of timber, finished and unfinished, as well as trimmings of grains, furs, manufactures, and models of public works. It personified Canada, according to its exuberant architect, Alfred Perry: 'The pyramidal Trophy of our country is seen rearing its head to the very aerial roof, and stands like a sturdy giant girded with spoil, his feet upon the fruitful plain, knee deep in golden grain, implements of ease and peace to feed the multitude; his head among the snows, clad in skins and lined with hewn woods, he holds forth to the admiring gaze of Europe the products of his forests, mountains, rivers, lakes, and mines.'[53] Visitors could climb an internal staircase and survey the building as a whole: 'From this gallery the view of the building is really magnificent; this extensive edifice, nearly 4000 feet in length, presents itself to the gaze of the visitor in all its varied aspects, with its numberless decorations and variety of colors, the fairy like confusion of all the objects displayed on the ground floor and in the galleries, and the iron and crystal vault of the immense industrial caravanserai. The complete view of the annexe, the aerial and indefinable prospect renders this gallery one of the most curious points of the Exhibition of 1855.'[54] This was a coup. *The Times* derided the exposition as 'a mighty jumble,' and later described the 1867 exposition as 'a hive' that the visitor could follow 'in detail bit by bit,' but without any overarching '*coup d'oeil*' or 'grand vista.'[55] The *Gestalt* experience of comprehending whole and part at once was crucial to the vision that the exhibition tried to construct. The exhibition was a hall of mirrors, a reification of vision itself, and the Canadians put up the biggest mirror of them all. They turned panopticism back onto itself, so that their exhibit was something to look through, as well as to look at. Canada worked the politics of visibility to national advantage.

The Canadian timber trophy objectified the operation of vision within the exhibition; the Geological Survey exhibit symbolized the extension of this operation to the outside world. The one provided height of vision; the other, depth and breadth. Logan's victory was both a scientific and a popular triumph. Not only did Europe's leading geologists (and Canada's, too, for Thomas Sterry Hunt of the Canadian Survey was on the jury) award Logan the top prize but the London Geological Society conferred on him the Wollaston Medal; further, Napoleon named him to the Legion of Honour, and Victoria knighted him. The Canadian press went into raptures and a whirl of banquets and conversaziones awaited Logan when he returned to his native country. The Canadian Institute had his portrait painted, his fellow Montrealers presented him with a silver plate, and, in the spring of 1856, John A. Macdonald personally took charge of the bill to perpetuate the Geological Survey another five years and increase its funding from £2000 to £5000.[56] The fortunes of the Geological Survey, the glory of nineteenth-century Canadian science, were inextricably bound up with the fortunes of the Canadian exhibits at the international exhibitions.

Exactly what kind of science Logan performed has been disputed. Suzanne Zeller argues that Logan emphasized its practical utility, and, indeed, Logan insisted that the exhibit was 'an Industrial one,' designed to impress not geologists but 'the public at large.'[57] Nancy Christie argues that this rhetoric was mere posturing on Logan's part, that his work had little utility, and that it was done to impress scientists at the London Geological Society.[58] Classical liberals like William McDougall decried his work as self-serving and impractical, and they demanded an end to the government's intervention. Certainly, as both scholars point out, Logan needed to exaggerate the economic aspects of his work and to publicize them at the exhibitions. Victories at these temples to practical science helped Logan prove to cavillers at home that his work had an eminently useful application. The extent to which it actually facilitated the extraction of Canadian minerals is difficult to assess, although Logan was able to cite individual sales, especially of fertilizing minerals, in subsequent survey reports.

The exhibit illustrates the difficulties in distinguishing between the practical and the scientific. The samples themselves had almost no economic significance: They had hardly been mined at all because Canada lacked the labour and capital to dig and transport them. For the moment, these

In 1855 the Canadian exhibit at the Paris Exposition was dominated by a vast timber trophy. This monumental piece of architecture symbolized the vastness of the land available for immigrants.

'inexhaustible riches of the earth' were just rocks.[59] Their value was conceptual. Logan added value to these inert lumps by adding mental labour. Once representative samples had been chipped out, carried away, identified, labelled as belonging to a particular location, a particular geological era, and offering a particular commercial use, these rocks had begun to undergo the transformation from 'wilderness' to 'artifact.' The Geological Survey provided a map of the land, thick with practical and intellectual significance, and it was just this sort of map that the *Exposition Universelle* tried to impose on the entire world. The Canadian survey extended the scientific gaze wider and deeper, across and into the earth; it helped 'spread scientific order over the earth's surface.'[60] Because Canada was so undeveloped and so invisible to metropolitan science, Logan's collection enjoyed a virtual monopoly in representing much of that land. Industry had not done its work, so the task rested with the scientist, the lonely geologist toiling in the heart of a vast, unknown continent, opening it up for future industry.

Logan helped to constitute Canada as an object for science, giving it a prominent place in the international discourse among scientists. But at the exhibition, this discourse was largely constituted by the means and language of industry and was elaborated before a lay as well as a scientific public. Moreover, as much as the exhibitions legitimated geology, geology legitimated the exhibitions. Logan described the Crystal Palace exhibition as an aggrandized geological museum: 'The Industrial Exhibition itself was nothing more than a grand and instructive display, of the same kind, in which the idea was carried beyond minerals, to all substances which nature yields, and to all the applications of which they are capable, – beyond the materials and industrial arts of one country, to those of the whole globe.'[61] Exhibitions were inventory science applied to artifacts, and geology provided a guarantee that the activity was intellectually respectable. The practical and the scientific aspects of this discourse were inseparable.

Apart from the timber tower and the medley of minerals, Canadians had only middling success. Because, Logan told Francis Hincks, 'our manufactured articles are not of that shew and description which will so much figure in the *Palais* or main building, it appeared to me prudent to request to be permitted to display them with the raw produce in the annexe.'[62] Altogether Canada sent 405 exhibits, for which the exhibitors received 93 prizes. This proportion was about half the average: 10,336 prizes for 22,243 exhibits, but, because most of the competitors were more advanced eco-

The prize fire-engine exhibited by the Perry brothers. Alfred Perry had wheeled it out to extinguish a fire that threatened the grain intended for the Crimean troops, for which he became something of a hero in Paris and London.

nomically, Canadians were satisfied with this performance. Several exhibitors obtained first-class medals, including a prize to Alexander Shaw of Toronto for the heaviest wheat in the exposition. Canadian woodworking machines took prizes for cheapness and ease of repair rather than novelty.[63] Carriages and a sewing machine failed to secure prizes; nor did leather, oil, maple sugar, meat, hemp, flax, or even woollens, which were, according to Perry, poor even for Canada. The tobacco was so bad that, on the judge's advice, it was withdrawn from competition.[64] Paul Kane sent art, but French-Canadian painters were, according to Taché, too modest to compete.[65]

In manufactures, Canada's great triumph was another fire-engine from George Perry, again the best on the grounds. A thrilling account appeared in the Montreal *Gazette*: 'A gentleman in the front exclaimed – "Now, Canada, all eyes are on you." At the given signal, Mr. Perry rushed to the breaks and shouted "Vive l'Empéreur" and the words seemed to send electricity through the pumps.'[66] The drama became real when, at the close of the exhibition, fire broke out in a Paris granary and Perry drove the machine out to help extinguish the conflagration. The story made headlines in Europe and the United States. Perry's and another Canadian engine were sold to the British Board of Trade at a loss.

Robert Romaine of Peterborough sent a steam plough subsidized by the Executive Committee, but the machine did not work: 'It propelled itself over the ground, but was a very imperfect machine, being defective in the boiler and in the steering apparatus.'[67] Still, it looked more promising than any other prototype, and William Croskill of the Beverly Iron Works bought the machine. After tinkering with it for several years, however, he abandoned it as uneconomical.[68] When the portable steam engine did arrive, it was on another design.

This, then, was the Canadian exhibit, assembled at a total cost of £17,513 2s 3d, less £1382 4s 7d profit on articles sold. Sales were sluggish all over the exhibition. The wood-working machines were sold to Englishmen, Matthew Moody's clover-cutting machine and horse rake went to Prussia, and the Austrian government spent £5 on a Canadian plough. A Hungarian ordered B.P. Paige's threshing machine and inquired after the patent. Swedes bought one of the cabinet tables, a planing machine, and a nail machine that made 80 nails per minute (one French machine could produce more than 300 per minute). Prince Albert ordered Canadian axes for the Crimea, and Prince Napoleon bought a carriage by Clovis Leduc of Montreal for £100. C.N. Tripp won an honourable mention for asphalt and an order for seven shipments for Paris streets, though this did not stave off his bankruptcy in 1858.[69] However, there was no relaxation of tariffs against Canadian products.

The exhibits made up only half the display; the other half was discursive. In Paris, Logan took charge of the artifacts and J.-C. Taché undertook public relations. The jury awarded Canada the medal of honour as much for its interpretive paraphernalia as for the articles they described. But the discourse proved more controversial than the objects themselves. The committee went to some lengths to collect documents to send to Paris, including a topographical map by engineer Thomas Keefer advertising the St Lawrence route into North America, a written guide to the geological map, a descriptive catalogue of the exhibits explaining prices, quantities, and uses,[70] and a series of essays. It also offered sizeable prizes of £160, £60, and £40 for the three best essays on Canada's resources and institutions.[71] Competition for the title of prosaist laureate was keen, attracting nineteen competitors. Three runners-up were given £25 to help them publish: the journalist and politician Hector Langevin, the geologist Elkanah Billings, and the secretary of the Bureau of Agriculture, William Hutton. The top three essays were sent on to Governor General Edmund Walker Head, who awarded first prize to John Sheridan Hogan, a journalist for the Toronto *Colonist* and future MPP; second prize to Alexander Morris, law clerk to John A. Macdonald and future MPP; and the third prize to J.-C. Taché, MPP and secretary of the Exposition Executive Committee. All were Tories. Head's preference for Hogan's essay may reflect ethnic bias on his part, for a few months later he let slip a remark about 'Anglo-Saxon superiority,' raising hackles in the francophone press.[72]

Despite all the rhetoric about cooperation between French and English Canadians, tension was never far below the surface and, in 1855, the essays served as the pretext for squabbling. Perry complained that Taché occupied himself in Paris publishing his essay, while the others, published in Canada, were not ready until November, when the exhibition closed. The Toronto press was irate: Here was Taché attending to his own rather than the nation's business. The Bowmanville *Statesman* saw in the contrast of Logan and Taché the problem of Canada in a nutshell: 'Mr. Tache is quite in keeping with the country to which he belongs, in his sluggish backwardness to aid in anything that pertains to Canadian advancement, more likely he will be found among the heirachy of France plotting against the liberties of Canada. Of Mr. Logan's antecedents we know nothing, but we are sure that he was not born and trained among the pure Cunnocks, of the land of scrubby sheep and shaggy horses.'[73]

On cue, a series of letters by Taché to the Rimouski Institut, published in the *Journal de Québec,* revealed just what he was up to in Paris: playing the *flâneur,* a literary dandy, languidly strolling along Paris boulevards.[74] 'On s'arrête, on se dandine, on se regarde, on rit ou se parle; d'autres prennent des sorbets ou des glaces assis aux portes des cafés; d'autres circulent dans les couloirs qui débordent sur les boulevards; les uns contemplent les objects contenus dans les vitrines; d'autres s'assayent sur les sièges, au fin bord du trottoir près de la route des voitures et là conversent ou s'oublient dans une méditation comme celle qui nous saisit après quelques instantes de rêverie sur le berge d'une rivière.' The francophone press rallied around Taché and accused Hogan of inaccuracy, plagiarism, and 'racism'; the Reform press accused him of inaccuracy, plagiarism, and bad style.[75] Even John A. Macdonald, try though he might to keep 'aloof from the squabbles of the committee,' had to intercede to soothe Morris and Hogan.[76] But the bad feeling persisted well into 1856, when Taché was snubbed by the same public that lionized Logan. In December 1856 he resigned rather than face 'almost certain defeat' at the polls.[77] He went on to edit an *ultramon-*

tane journal in Quebec with Hector Langevin and, after a stint as inspector of prisons, was lured by D'Arcy McGee to become secretary and later deputy minister of agriculture. In this position he directed Canadian participation in international exhibitions during the 1860s. Taché also engaged in a life-long feud with Alfred Perry. In 1885, during a dispute over Catholic asylums, he accused Perry of harbouring 'hatred for every thing connected closely or remotely, with the principles of catholicism and with French-Canadian nationality.' Hatred ran deep: On that night in 1849 when Perry led the riot that fired the Parliament Buildings, Taché had stood guard at LaFontaine's house and shot dead a rioter.[78] Beneath the 'lovingly intertwined' banners, what glares the two must have exchanged!

Only a handful of the essays saw the light of day. Those not chosen were reclaimed in secrecy, except one by Adam Lillie, rejected as illegible, which the disgruntled Congregationalist minister published. What did they say about Canada? Scanning the essays by Hogan, Morris, Taché, Langevin, Hutton, and Lillie,[79] one finds that they advertised the land rather than the people: geology, geography, and topography dominated, leaving an impression of a handful of people perched atop and amid the rocks and trees, rivers and lakes. If this was Canada, Canada was a land with no literature and no history beyond occasional statistics of school-goers and lists of laws. Langevin and Hogan tempered this grim picture. Langevin's essay dwelt on Quebec, which he painted as thickly saturated with cultural institutions. Hogan, a Toronto resident, did not conjure up Langevin's dense social world, but his essay vividly sketched social types – the gentle and lazy French Canadian; the enterprising and industrious Upper Canadian farmer who had done well by dint of simple determination. But Hogan's Canada was a congeries of individuals rather than a people. Schools were praised, but other institutions received scant measure, and politics was virtually denied: 'Their municipal system is but a small remove from the leader of the '*logging bee*' being elected builder of the bridge, and their parliament is but a higher class in the same school of practical self-government.'[80] The essays also harboured exaggeration and deception: They ignored native peoples and boasted of a fine climate and general prosperity.

Under this combined onslaught of things and words, journalists raved about the Canadian effort. In a botanical guide, Count Joubert remarked: 'Now we can form an estimate of the value of those few arpents of snow ceded to England with such culpable carelessness by the Government of Louis XV.' The Paris *Patrie* echoed regret for France's loss of 'ce magnifique fleuron de sa couronne coloniale.'[81] The French *Monde Industriel* remarked that it had known Canada was rich in resources, but 'now we see it contains an active, industrious, enterprising, and progressive population.' *The Times* of London was hyperbolic at this object lesson in nation-building: 'What can be more delightful than to be able to watch the first stages of progress in infant communities – how hardily and industriously, applying every resource of modern science and skill, they hew out wealth and independence for themselves from primaeval wilds – how they subjugate nature with a rapidity and completeness unknown in any past age of the world, and, self-governed and self-relying, tread with confidence in the face of the nations, the path of greatness to which their destiny manifestly calls them!'[82] A 'gratified' Brown Chamberlin lectured on 'the importance of having a good character in Europe. It is something to be famous for energy and progress, instead of political agitation and rebellion, and it is something to have the real resources of the country understood.' Indeed, only the year before, *The Times* had remarked: 'We seldom hear anything about [British America] unless it be something unpleasant ... As it is, we only hear of these Provinces as we do of Mount Etna, that is, when there is an Eruption.'[83]

Logan worked hard to prevent just such an eruption from occurring. He canvassed the officials to correct errors and omissions in the prize list, which, as he told Lyon Playfair privately, 'will be made matters of state in the colony & a foundation for newspaper war.'[84] Certainly the Canadian press followed events and eruptions of praise closely. Puffery and bickering filled columns. Exacerbated by the squabble over Taché, rivalry between Upper and Lower Canada was fierce. The Montreal press hoped to 'eclipse ... our puffing Upper Canadian neighbours,' and the Toronto *Leader* retorted that Lower Canada 'had overdone the thing,' sending useless junk.

The Paris exposition struck a serious chord in the French press. The Montreal *Herald* mused: 'Certainly, to the eyes of the future student of history, the war of 1855 will be found a more important event in the progress of humanity than even the Universal Exposition.'[85] The *rouge Pays*, published by A.-A. Dorion, reversed this judgment: 'Aujourd'hui les peuples acquièrent plus de richesse et plus de gloire par le génie de l'agriculture du commerce et de l'industrie, que par la génie de guerre ... L'avenir du Canada est donc aussi sérieusement engagé à Paris, dans les combats de l'exposition, que l'avenir de la Russie, de la France et de l'Angleterre, l'est dans les combats qui se livrent sous les murs de Sébastopol, pour sa defense et pour sa destruction.'[86]

Taché agreed. His experience in Paris influenced his thinking about the future of French Canada. In 1857 he published the 'most fully documented, detailed and articulate' plan of British American Confederation devised before the Quebec Conference. This document shaped the final plan, though it was less centrist, and Taché doled out advice to the Quebec Conference.[87] Most striking is his self-confident tone. Taché was convinced that Durham had been proved wrong: French Canadians were going to survive as a distinct people after all, thanks to prosperity and responsible government. He now sought independence from Upper Canada in cultural matters, and yet in politics he looked to federation, not independence. He decided that prosperity must come first, and prosperity would be more likely in a confederated British North America. Cultural matters like religion and education should be decided provincially, but trade would be a federal affair. At the Paris Exposition, Taché had learned to view Canada from the outside, as a commercial nation among others. He argued for a federal government on the British model because 'elle donne plus de force, plus de lustre et de splendeur à la nation dans ses relations avec les autres peuples.'[88] This, like the Canadian exhibition, was about packaging. Canada's external form was as important as its internal structure. The federal government would advertise the nation and negotiate commerce and tariffs, and this centralization would speed the development of commercial relations with France and other countries. The federal government, with its 'splendeur' and 'lustre,' would function like an exhibition. Taché wrote a powerful appeal to his compatriots and probably did more than any, save Cartier, to win them to Confederation. His arguments were seconded by others active in the exposition: In 1858 Alexander Morris published *Nova Britannia*, which argued that the wealth of central Canada was recognized worldwide, thanks to 'our representations at the Universal Exhibitions,' and he called for a new nationality formed of British North America. Hector Langevin had advocated Confederation since 1847.[89] (Hogan was murdered in 1859.)

Canada's success at the Paris Exposition promoted a self-consciousness that favoured nation-building. In 1861 commentators remarked that Canada had won, in London and Paris, benefits 'impossible to calculate. Our position as a member of the great confederacy of civilized nations, has been recognized and respected.'[90] Exhibitions encouraged this view of the world as a confederacy. The prince consort had spoken of the first Great Exhibition as bringing together representatives of local governments, different political parties, and different nations 'representing only *one interest!*'[91] The exhibition was a leviathan, wherein a multitude was transformed into a 'real unity' of one will, one interest: economic development.[92] It was this vision of cooperation across national or ethnic boundaries that Taché and other Canadian nationalists applied to Canada. Yet the economic benefits that had been the point of the effort were elusive. Immigration to Canada did not increase one whit, nor were the financial results more satisfying. When, in 1861, Brown Chamberlin pleaded for representation at an approaching London exhibition, John A. Macdonald told him flatly: 'The truth is we have found the expense incurred at London & Paris was so much money thrown away, & there was no use in following these dismal failures.'[93]

London, 1862

The Great Exhibition of 1862 was a lacklustre affair. Albert's death from typhoid, in December 1861, sent Queen Victoria into mourning and cast gloom over the festivities. The American Civil War horrified the world by its bloodiness and moral ambiguity, and it caused outright starvation in the cotton districts of England. Despite talk at the Society of Arts of making these great exhibitions periodical, this was the last one England held. The idealism of 1851 vanished: The *Saturday Review* saw only a vast shop that would not 'regenerate either society or trade ... It is not a school nor a university. It is an illustrated trade catalogue.'[94] In Canada, a few still dreamed, like the Toronto student who won an English prize for his poem:

Art's embracing temple shrine,
Heaven-born Peace! the work is thine;
    Earth rejoicing shouts all hail!
Being again the Olympian year,
Sheath the sword reverse the spear,
    Let, O Peace! thy work prevail.[95]

At this second British exhibition, as at the first American one, Canada, which had done well in the past, now relaxed its efforts, while the other British North American colonies tried to efface past errors. Although Macdonald might repudiate the results of past exhibitions, Canada could not entirely ignore the event, especially since the other provinces were exerting themselves to do well. In the end, Macdonald passed a paltry vote of $6000 to assemble a modest collection of natural resources and pay the freight of any manufactures sent privately.[96] The 'sadly inadequate' collection of 196 exhibits included minerals and woods, wine, flax, cloths, petroleum, photographs by William Notman, an invalid bedstead, and model railway cars. It was dwarfed by exhibits from the Antipodes, whose agents viciously heaped insult on injury by warning intending emigrants of 'the disasters that have befallen our ocean steamers; and the dangers connected with such close proximity to a turbulent people like the Americans.'[97] It seemed as if Canada would 'lose her exalted position as the first of England's Colonies, the brightest gem in the British Crown, which she has held indisputed since 1851.' The *Witness*'s correspondent reported that 'disappointment' and 'unmitigated disgust' appeared on the faces of Canadian visitors and, where other colonies showed progress since 1851, Canada had 'positively gone back.'[98] He abandoned the exhibition for a walking tour of Scotland.

It can hardly have comforted Canadians to see their sister colonies be counted among the progressive nations. The *Scottish American Journal* recalled that 'the Lower Provinces ... were very badly represented in 1851, and their show must have done them more harm than good. Nor have they taken pains to make themselves known since. They have no representatives in the press here, no agent, and no recognized place in London, where information could be obtained about them. Now the Lower Provinces have taken a stand on the Exhibition. The collections of Nova Scotia and New Brunswick, Prince Edward Island and Newfoundland, are more than creditable.'[99] Nova Scotia in particular produced a 'noble display' of minerals, including iron ore, gold, and a 30-foot column of coal. Its grains were eclipsed by Canada's and Australia's, but the vegetables were admirable (save for a squashed squash), as were Andrew Downs's stuffed birds. Just a few months previously, gold had been discovered, and Nova Scotians were convinced that news of this find would bring 'multitudes' of immigrants to their shores. But a substantial exhibit was needed to belie the slander of British papers like *Punch*, which mocked 'the absurd idea that gold is to be found in Nova Scotia.'[100] Meetings were held across Nova Scotia in support of the exhibition, including a 'large assemblage' of the province's leading ladies and gentlemen, where even the political opponents Charles Tupper and Joseph Howe joined forces to rouse public opinion. Howe recalled the shame he had felt at the last Paris exposition when, amid the 'countless treasures' of the civilized world, he 'came to two or three compartments having the names of New Brunswick and Nova Scotia painted in large letters above them, and beneath *nothing at all.* Then I bowed my head, and thanked God that nobody knew me; I was glad to be able to skulk out with the conviction that nobody knew where I came from. And when I went back to London, I said nothing about Nova Scotia that season; I passed for an Englishman, and got home as quickly as I could. Let us not exhibit ourselves in such a plight again.'[101] Nova Scotia showed one of the best colonial collections at the exhibition and won glory at a horticultural show that fall, where apples of prodigious size astonished the natives. 'We are now fully advertised,' boasted the journalist Thomas Annand.

The other British American colonies distinguished themselves as well. Samuel Leonard Tilley, premier of New Brunswick, was also present at the Nova Scotia public meeting to promise that, far from occupying the 'contemptible position she did in 1851,' New Brunswick would provide a 'noble rivalry.' That province beat Nova Scotia for grain and showed some admirable minerals, but the furniture was not very good. Prince Edward Island sent some well-made homespun cloth and leather goods, and a canoe widely admired as being the work of a one-armed Mi'kmaq, Peter Snake. One visitor concluded that the display showed 'an industrious, careful, painstaking people, not destitute of the "polished arts which humanize mankind."'[102] Newfoundland also determined to show the mother country that it had resources and enterprise aplenty, and sent a small but good collection, securing medals for minerals, skins and furs, preserved salmon, and cereals. The cod, unfortunately, rotted in the damp British climate, having been harvested too late to be properly dried. But, for all that the Atlantic provinces sent respectable collections, they also revealed shortcomings. The editor of the Halifax *Witness* confessed that 'all the lower Provinces together fail to make the impression that Canada easily produces.'[103] The Scottish-American commentator recommended that, as their attractions were insignificant taken individually, these colonies should combine, the better to impress the British public.

The westernmost British North American colonies also sent a display that prompted reflection on the wisdom of closer political union. Vancouver Island and British Columbia sent a small exhibit each, composed of minerals, fish, grains, Indian manufactures, and an enormous spar of Douglas fir. The Victoria *British Colonist* urged this first foray into the international arena with a mercantile metaphor: 'We have been a small shopkeeper in a backstreet. We have been living off of a few customers. Our stock on the shelves has been reputedly large; but owing to our want of business capacity, we have eked out a petty shopkeeper's existence on the profits of the sales of a few wares that have met a ready sale. We should long ago have moved to a front street, occupied an inviting establishment on the most popular thoroughfare, arranged our manufactured goods, or our raw materials in the most attractive manner, and been fully up to the age in business tact to supply the local and the foreign trade.'[104] Public meetings were held, and preliminary exhibitions were held on Vancouver Island

and the mainland in the fall of 1861. The mainlanders complained that islanders, and especially Governor Sir James Douglas, frowned on their efforts to make an independent exhibit and 'would be pleased rather than otherwise if the movement should prove a failure,' for this would make the mainland more subservient to Victoria. When every essay on British Columbia submitted for competition was rejected, the press complained that the rebuff stemmed from the essayists having refused to toady to Douglas.[105] The experience did not bolster autonomy, for the next time British Columbia exhibited it was as a united province within the dominion of Canada.

Events in 1862 gave Canada a sharp nudge in the direction of autonomy. In May, when the Canadian parliament vetoed a bill to make Canada pay for its militia, Britain seethed with rage against this miserly colony that refused to help pay for its own defence. In Westminster, Samuel Roebuck stormed: 'We do not care one farthing about the adherence of Canada to England. (Hear, hear.) ... I want the Canadians to understand that England would not be sorry to see her depart from us tomorrow. (Hear, hear.) They do us no good, or, at least, not more than New York.' At a dinner in London in July, Brown Chamberlin, serving as Canadian commissioner, 'begged them all to believe that though the people and Parliament of Canada might feel too poor to embody a large army, yet they would be ready to give their blood and sell their lives dearly for the old flag.' The liberal English historian Goldwin Smith answered that 'the Colonists, on the whole, are far better off than the mass of people in this country, and they have no national debt on their shoulders. To tax our people for their defence is injustice.'[106] During the 1860s Little England sentiment was at its height, and British fury over the militia bill caused great resentment in Canada.[107]

Relations were no less strained inside the exhibition buildings. Short of money and floor space, the Canadians erected a five-storey tower of rough boards, which the New Brunswick agent complacently described as an 'ugly timber trophy that the Times finds great fault with.' Every day *The Times* printed a harangue, 'A pile of slabs of wood must be unsightly, do what you will with it.'[108] Instead of a graceful 'coup d'oeil,' the nave offered only a 'hideous anarchy,' a veritable 'Chamber of Horrors.' The Earl of Granville (a Manchester liberal and no friend to colonies) and Henry Cole took Logan aside and asked him to lop off the top three stories of the trophy. Logan refused. They reiterated in writing the demand that the 'extremely unsightly' trophy, which blocked a stained glass window by an English exhibitor, be curtailed, on pain of its removal. The Canadian com-

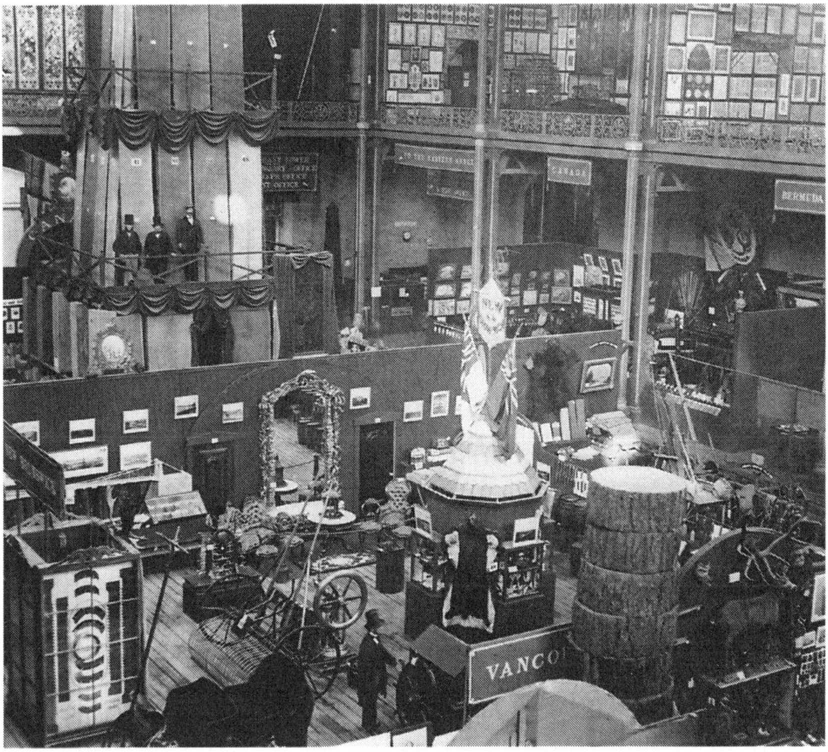

This picture shows both the New Brunswick Court and the Canadian timber trophy at the International Exhibition in London in 1862. In comparison with the trophy of 1855, this was a crude and rough structure that failed to find favour among cosmopolitan journalists.

missioners would not be moved. Though 'mortified' by the aesthetic criticism, they defended the trophy and the cause of the mistreated colonies, which had been placed in a cold, wet space at the rear where their tools rusted and their fish rotted. The Royal Commission threatened to take up the matter with the Canadian government. Nothing daunted, the Canadian Commission referred the correspondence back to Canada.[109] There was Canada, invariably referred to as an 'adolescent' nation, faced with an adolescent's worst nightmare, as mother country approached, knife in hand, threatening to castrate the towering erection that symbolized it.

British North America was handicapped because its commodities were

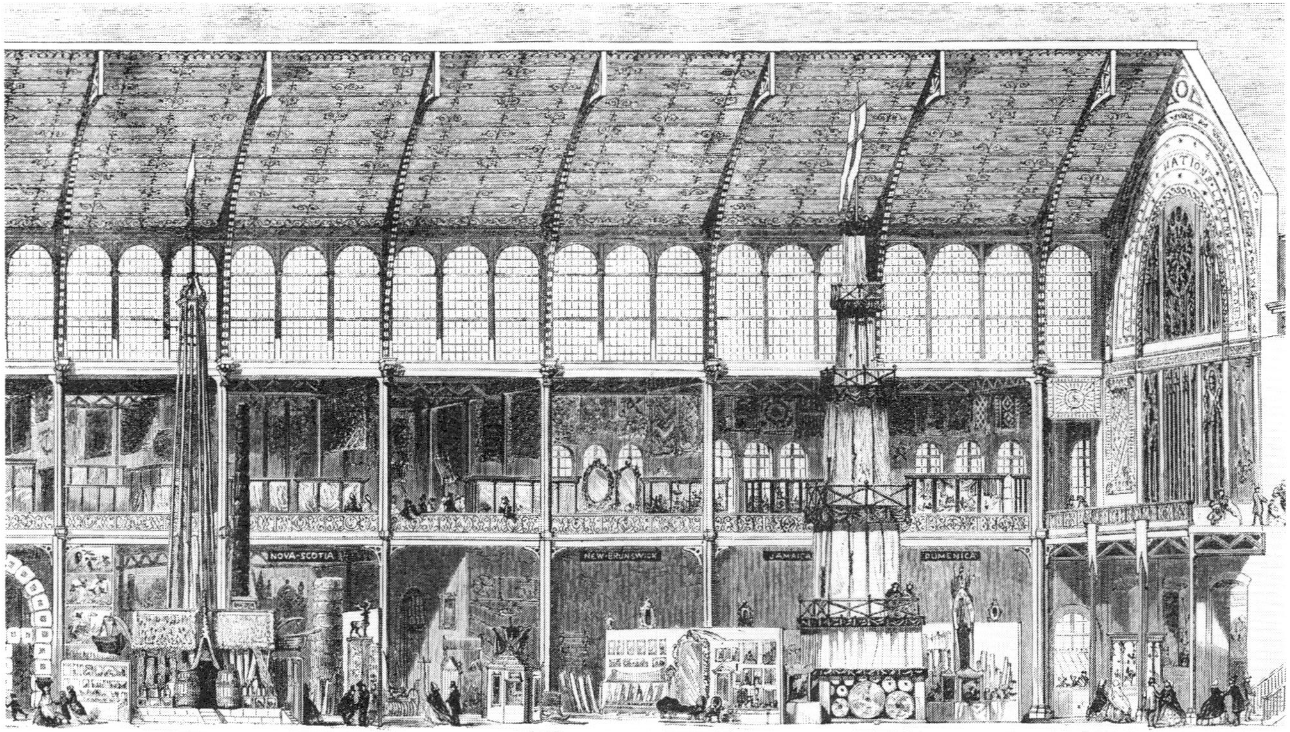

A larger view of the Colonial Annex in 1862, occupying a cold, damp corner of the exhibition. The Canadian timber trophy clearly blocked the stained glass window at the end of the annex.

not eye-catching. Rather than sending picturesque furs, sleighs, and Indian curiosities, as in the past, the Canadians sent enormous specimens of wood and minerals to impress the untrained eye as well as the savant. To those who could see past flashy exteriors to the deep principles governing the world, the colonial courts were the most spectacular part of the show: 'On foreigners particularly those Courts, through which visitors seeking only for something striking have passed heedlessly, casting only a passing glance at their manifold natural riches, have conveyed a more striking idea of the greatness of England than even the varied marvels of industrial skill and energy which are to be found in those parts of the building devoted to Great Britain proper.'[110] Not only the stained glass but even the machinery of England was, according to this view, the superstructure of imperial power, the infrastructure being the rocks, trees, fish, and grains of the colonies. The colonists dislodged the position of the viewing subject from London by raising it to encompass the empire, viewed from one corner of the empire, Canada. The jutting trophy disrupted the planes carefully laid out by the building's architect to insert its own perspective, and in so doing affronted the authority of Her Majesty's commissioners, an authority that, Michel Foucault argues, rested with the control of perspective and visibility.

Relations among the Canadian commissioners were strained as well. Asked to head the commission, a reluctant, exhibition-weary Logan agreed only if he could leave once the judging ended. His letters complained of 'multitudes of invitations to dinners, soirees, and conversaziones,' that he was 'sick and tired of all such hubbub,' and that he wished he was 'at home again attending to my geological explorations.'[111] After his departure, tempers flared among his remaining assistants – Brown Chamberlin, editor of the Montreal *Gazette*, and J.B. Hurlbert, principal of a private Ladies Academy, who collected the timber. To this twosome was added H.H. Miles, vice-principal of Bishop's College, thanks to a petition from the Eastern Townships. Miles apparently spent half his time gluing slips of paper with his name onto the commission letterhead and the other half writing a pamphlet extolling the Eastern Townships. In Montreal, Principal J.W. Dawson was livid when this self-styled 'official catalogue' described McGill as irreligious and weak in Classics and mathematics, and he ordered Chamberlin to refute the 'untrue, base and groundless' remarks.[112] Miles stole specimens of petroleum, which Chamberlin had to retrieve, all the while keeping an eye on Hurlbert, who, he feared, 'means to sell all the Upper Canada wood professing to have special instructions and pocket the proceeds.'[113] Miles and Hurlbert despised one another and almost came to blows. Fate interceded when the latter, a protégé of Isaac Buchanan, was replaced by T.C. Keefer, an engineer, after the Macdonald-Cartier government fell with the militia bill. His departure was no great loss, for, according to one exhibitor, when Lloyds asked about woods for shipbuilding, Hurlbert showed them tamarack and said it was white cedar: 'They soon found out that all the Doctor knew of these woods ... might be put in a snuff-box.'[114]

When the exhibition closed, the colonial commissioners agitated for a colonial museum to be formed from the exhibits. The Sydenham Palace housed collections left over from 1855, but the Canadian section was now moth-eaten and tarnished, 'a monument of Canadian indifference and neglect.'[115] Lord Newcastle approved the new scheme and offered storage in the exhibition building until the museum was ready. The Canadian government, however, was more interested in saving on freight by shipping things back in the spring than it was in the museum, and the project fell through.[116] When Chamberlin finally sailed for Canada, he left behind him angry, unpaid contractors who had set up the Canadian court and who went to arbitration to force the recalcitrant Canadian government to pay in full.

Dublin, 1865

If Canada did badly at London in 1862, it did worse still in Dublin in 1865, and the other British North American colonies were not notably more successful. Canada and Nova Scotia both spent about $5000, the one filling three courts, and the other only one, while a few individuals and firms in Britain sent some specimens of fish, fur, and Indian goods on behalf of Newfoundland and British Columbia.[117]

Nova Scotia's government, led by Charles Tupper, decided to make a good show. Sixty-nine exhibitors, four more than in London, sent their goods, consisting of gold, coal, iron, wood, tobacco, cereals, biscuits, fish, stuffed birds, fancy-work, dried fruit, and cheese, which suffered from the summer sun, as did some cordials: 'The heat made a great flying of corks and bursting of bottles.'[118] Manufactures included skates by Starr, scales, and stoves and cutlery of Acadia iron. The commissioners declared themselves content with their forty-one prizes, but one Nova Scotian who visited the display declared it 'poor in the extreme': Downs's stuffed birds were hardly likely to attract emigrants, nor were 'maple leaf fans and Indian knickknacks.' Grains were 'scrimpy' and the herrings had gone off. This correspondent likened the 'silly sentiment' that prompted Tupper's government to exhibit in Dublin for the sake of seeing 'Nova Scotia' written on the labels with its pursuit of Confederation: 'The *name* of a great confederacy would induce them to throw away substantial advantages.' As Nova Scotia thrashed out the issue over the next few years, the metaphor of a glittering and superficial spectacle on which the foolish fanciers of federation 'gazed and feasted their eyes' became a natural weapon of the anti-Confederates.[119]

Canada's display brought it no discredit, but the risible antics of its commissioner certainly did. The exhibit included the inevitable grains, minerals, woods, and furs, a table top of Canadian woods by the Toronto firm Jacques and Hay (one of Canada's first important exporters of furniture), edge tools from Galt, oils and paints, flax and linen, blankets and woollens, a rifle, the arms of Canada in stained glass, china, leather, and brushes. Although Belfast men praised the flax and others admired the minerals, oils, foodstuffs, and tobacco, these compliments did not translate into sales or prizes. After two months and distribution of 361 pamphlets, two pairs of boots had been sold.[120] Canadian prices were too high for European markets. Still, the display did result in sales of Crown lands worth more than the expense of exhibiting. As for prizes, Canada ended up with 49 of a total of 1475, barely more than Nova Scotia and well below the Antipodes. The agent, Arthur St George Cuff, blamed Canada's poor performance on a spiteful jury which insisted that the goods were not Canadian at all and complained: 'It is too bad to send over these samples and take orders for the goods, bring them into this country to undersell us Scotch and English manufacturers, while at the same time you come to us with a poor story and get us to lend you our money.'[121] The curator for Nova Scotia, David Honeyman, a geologist and, after 1868, curator of the provincial museum, confirmed that 'the members of Jury XII were greatly annoyed to see such fine Tweeds come from the Colonies, that they would not look at his,' and officials told him that 'they did not care whether the Colonies were represented or not.'

Many Irish frowned on mass emigration from Ireland, and this opposition may account for their hostility. Public feeling was aroused when the Canadian minister of agriculture and immigration, Thomas D'Arcy McGee, visited the exhibition and gave a controversial speech at Wexford, where he had grown up. McGee fled Ireland after an abortive uprising in 1848, took up journalism in the United States, and finally settled in Montreal, where he cornered the Irish Catholic vote. At Wexford, he urged the Irish not to emigrate, especially not to the United States, where they would be looked down upon and find their advancement blocked. Better that Irish men and women remain home, but if they did emigrate, it should be to Canada, where the Irish were welcome and could make a decent living. McGee's defence of British institutions offended Irish nationalists in Canada, the United States, and in Ireland. It was one of a series of provocations that led to his assassination in 1868.[122]

Visiting the Canadian court, McGee found Cuff inattentive to his duties. Rebuked, Cuff agreed to take a stronger stand and bribe the press: 'I must remark that the Press do not make a direct charge for notices of the kind but it has been arranged that the person who would be sent to inspect the articles and write the notice should receive compensation for his trouble. As you seem by your last communication to desire that the Canadian Court should receive a more extended notice I shall take immediate steps to carry out your wishes which is my sole desire.'[123] Many newspapers accepted libations or cash in exchange for favourable reviews. Henry Yule Hind attributed the custom to the economics of journalism: In 1861 he had advised sending a striking exhibit like a piano from Canada 'such as many a penny-a-liner would delight to enlarge upon and probably take as a text for a wordy column on "Art in the Backwoods of Canada."'[124] Canadian papers would then republish these effusions, with little regard as to their accuracy.

As his London predecessors had done, Cuff quarrelled incessantly, first with agents from the Eastern Townships, then with the Department of Agriculture and Immigration. McGee kept Cuff shamefully short of funds, so he ran up debts which, years later, creditors, including a wine merchant, were still asking the Canadian government to make good.[125] Finally he tangled with the exhibition superintendent, whom he was prosecuted for assaulting. Cuff accused this man of misconduct with women and spying, and, when called a liar, Cuff took what the (well-bribed?) Irish *Times* called 'the course most gentlemen would have taken under the Circumstances and horsewhipped Mr Iselin on the spot.'[126] Iselin's story was that Cuff barged into his office, hurled accusations, thrashed him, and threatened to drink his blood. The judge fined Cuff £2. The exhibition authorities felt bound to inquire into the allegations of misconduct. Asked for specifics, Cuff obliged with lurid innuendos of 'two persons in sexual connection in the verandah'; he had seen the Eau de Cologne woman 'surrounded by young fellows, some of whom passed hands over her bosom and pinched her thighs.' Cuff produced witnesses, but, alas, all completely failed to corroborate his story. The Irish demanded that the Canadian government dismiss Cuff, who, unmoved by McGee's Wexford speech, emigrated to the United States.[127]

Paris, 1867

The Paris Universal Exposition of 1867 was the grandest yet. The building and grounds occupied 72 acres and housed 45,000 exhibitors, three times the number of the Crystal Palace. There were special exhibitions and conferences, as well as live plants, animals, and even people, for the Imperial Commission invited artisans and labourers from around the world to illustrate work and life. Labourers were not the only people on display: Napoleon invited rulers from around the world to attend the exhibition, and dozens came. Above all, this exhibition was the most intensely rational ever held. Where previous organizers had to choose between exhibiting goods by product or by nation, in 1867 the designer, Frédéric Le Play, did both: In one extremely large building he combined concentric circles of articles 'of cognate natures,' with national displays radiating outwards from the centre. 'Admirable in theory, you passed down one of the spikes of this monster wheel, and you saw all that the country had to show; you went round an ellipse, and the relative qualities of similar productions in various lands were all presented.'[128] So magnificent was the exhibition that it was predicted that none could ever equal it, and in future only partial exhibitions should be held.[129] Rev. Henry Scadding of Toronto visited the exposition, circled the building the requisite eight times, and admired 'its beautiful exemplification of law and order in the midst of an unparalleled multiplicity.' Another Canadian, a youthful H.P. Hill, wrote to his parents, 'We only managed to do the outside circle & it *was* a hard day's work.' He was struck by the pay toilets: 'One amusing spec. at the Exposition is a large water closet establishment kept by women; to "piddle" in which costs 2 sous; t'other thing is 4 sous! Funny, is not it?'[130]

With Confederation approaching, Canadian politicians decided to erase past mistakes and mount the most impressive display yet. Taché urged its importance on Hector Langevin, now Macdonald's Quebec lieutenant: 'Je vous en supplie et conjure, veuillez veiller à ce que M. McD. ne taise pas pendant dix semaines les communications qu'il pouvait recevoir à propos de l'Exposition.'[131] Anxious to keep the *bleus* contented, Macdonald bowed to these blandishments, and in October 1866 an order-in-council voted $50,000 for the exhibit, much of it for outright purchases of resources and manufactures. D'Arcy McGee and Taché insisted that the Department of Agriculture organize the display. The boards of arts and of agriculture began the usual task of wheedling and purchasing specimens in industry and agriculture, while the Geological Survey took charge of minerals, and Crown Lands collected timber.

Surviving receipts record that the Board of Arts for Upper Canada spent $814 for a display of 500 stuffed birds (which won a silver medal), $125 for insects, $100 for educational apparatus, $205 for books and bookbinding, $300 for furniture from Jacques and Hay (which took a bronze), $233 to the Galt Machine tools company, $50 for boots and shoes made at the Kingston Penitentiary, and $175 and $280 for spectacles and false teeth.[132] Lower Canada sent a Grand Trunk sleeping car, photographs by William Notman, and paintings by Krieghoff. Grain and flour collected by the Boards of Agriculture did well, as did cheese and wine. L. Davis sent 'a small quantity of segars also a segar stand in the shape of a Gothic structure.' Napoléon Bourassa exhibited a massive mural entitled 'The Apotheosis of Christopher Columbus,' which depicted Western civilization from Homer to the Fathers of Confederation, the obvious culmination of history. Bourassa had begun the work after viewing Ingres's Apotheosis of Homer at the Exposition of 1855. Unfortunately, when he glued the cartoon onto cotton, it wrinkled badly.[133]

McGee stopped in at the exposition after seeing the British North America Act signed, but soon rushed back to Canada for fear of losing a cabinet post.[134] Taché took charge of the exhibits: He was criticized when the collection, mislaid by the Grand Trunk, was not ready on time, but in the end he won great praise and a gold medal for his ingenuity. Instead of a trophy, Taché arranged squared trunks in columns to resemble a forest, covered by a canopy of painted leaves. Against another wall, he built columns reminiscent of a Greek temple. Another gold went to Abbé Louis-Ovide Brunet, a botanist and professor of natural history at Laval University, for woods, which were bought by the Prussian government. The Geological Survey garnered the usual accolades and *La Minerve*'s sarcasm for bringing Canada ample glory but no wealth.[135]

Generally, the press voiced pride and satisfaction. Government organs linked the success with Confederation. *La Minerve* speculated: 'Un an plus tard il est probable que nous aurions pu faire beaucoup mieux avec les resources de la Conféderation.' It also remarked of the American display: 'C'est le défaut de l'organisation, de classement; l'anarchie y règne comme dans le gouvernement de l'Union.'[136] But the anti-confederates could play this game too: Honeyman was 'proud as Pharoh' at having Nova Scotia's display ready on time, and the opposition press drew a contrast between the 'admirable order' that reigned in his court and the 'confusion' and 'frightful mess' of the Canadian court, 'a large empty space full of loose boards and stones.'[137] They added that an exhibition hall would be a more creditable capital for the 'bogus Dominion' than two Ottawas, with all its sawdust and mosquitoes.

Ever the exhibitionist, Charles Tupper had organized another substantial display to represent Nova Scotia overseas, in an attempt to offset the decline in trade due to the repeal of reciprocity. The fish collection won a gold medal, and silver medals were awarded to displays of coal, gold, and other minerals and to Andrew Downs for taxidermy. There were also cereals, fresh fruit and cheese, furniture, and other manufactures. Nova Scotia's exhibit was accomplished without the proffered help of the Canadian government. New Brunswick's contribution to the Paris Exposition, however, was owed to McGee's having entered a crew from Saint John in the international regatta. When these pink-clad oarsmen astonished the world by winning, the Saint John *Journal* bragged that 'it could not but be gratifying to the people of Ontario and Quebec to know that should they feel unable to maintain the honour of the Dominion in any future aquatic contest, as we fear – with all due respect to them – too many of them would, they have only to call upon our St. John boys, and they will see them safely through.'[138] Prince Edward Island and British Columbia did not send exhibits to Paris, but Newfoundland sent a small collection of roof slate, grains, marshberry jam, wool, photographs, fish and whale oil, seals, and Inuit work.

Did the exhibits sent overseas help to construct the new dominion? Certainly they emphasized the benefits of economic integration, real or potential, as visitors frequently remarked. In 1851, after viewing the displays from British North America, *The Times* advised Canada to confederate with the Maritimes to gain access to coal; in 1855 that paper's exposition correspondent advised it to colonize the coal lands to the west. Clearly, the Geological Survey provided economic arguments for Confederation.[139] Exhibitions amplified the public impact of inventory science and encouraged the tendency to conceive of Canadian nationality in material terms. During the debates of 1864, Canadian supporters of Confederation used economic, resource-based arguments to further their views, while those opposed raised primarily political objections. The Maritime delegates were taken round the provincial exhibitions in Canada East and Canada West to gaze at the wealth of the land.

Both the economic considerations and the international arena in which the exhibitions took place encouraged Canadians to reflect on their relationship with the rest of the world. The Paris Exposition of 1855 gave Taché an opportunity to view Canada from the outside, just as a trip to England had the same effect upon George Brown a few years later.[140] Taché concluded that strength lay in unity, which would raise the nation's market strength and international profile. His eloquent appeal to his compatriots, who had so much to lose in Confederation, probably had a decisive impact in generating French-Canadian support for the union.

On the other hand, opponents of Confederation outside Canada, such as Joseph Howe, proclaimed that their colony was wealthy enough to go it alone, and international success bolstered this argument as well. Indeed, exhibitions were hardly mentioned in the Confederation debates,[141] and Macdonald himself remained unmoved by them. He saw through the journalists' conspiracy of puffery, but others did not. In 1863, for example, John Sandfield Macdonald had questioned whether the Geological Survey and its displays were of any use at all: 'They had sent specimens of Minerals and Woods to the London Exhibition, but had any practical benefit followed?' In response, however, he was severely taken to task by other politicians. Nonetheless, considered only as representations, as texts, the displays of Canada did inspire nationalist enthusiasm. Sir John A. Macdonald was a political realist with no time for frills like exhibitions; D'Arcy McGee was an idealist who spoke eloquently of a new Canadian nationality and, like Charles Tupper, believed in the power of exhibitions to construct it. All were Fathers of Confederation.

The exhibitions helped to establish a discourse about Canada which praised its economic resources and its bustling, practical populace, and claimed a place for it in the 'confederacy of nations.' Exhibitions neither originated nor monopolized the parameters of discourse about Canada, but they did constitute one of the most popular and powerful expressions of it. The eclectic, concrete collections of objects found in the exhibition were not obviously more or less representative of Canada than the political entity that now spoke from Ottawa. They helped to construct the identity of Canada as an object susceptible to, and largely consisting of, practices of representation. As Ian Hacking has pointed out, the idea of a real object is a second-order idea, parasitic upon the idea of representation, and it 'has content only when there are first-order representations.'[142] Even the obvious misrepresentations fuelled demands for more and better representations, demands that were insatiable because, as with political representation, there could never be closure through full presence. After Confederation, as Canadian immigration and press agents grew skilled at selling the country abroad, they became more conscious of the instability of this boundary between reality and representation and learned to capitalize on it.

The exhibitions did contribute to the debates over Confederation that occurred during the later 1850s, a time of prosperity and confidence. Canada's perceived successes in Europe fuelled this confidence, even if it began to seem misplaced by the 1860s. The model of friendly rivalry also provided a useful paradigm for relations within Canada. During the Confederation debates, George-Étienne Cartier used the classic rationale for exhibitions when he stated that 'he viewed the diversity of races in British North America in this way: we were of different races, not for the purpose of warring against each other, but in order to compete and emulate for the general welfare ... They were placed like great families beside each other, and their contact produced a healthy spirit of emulation.'[143] Emulation, as seen here, signified a good-natured rivalry and was usually described as 'healthy' or 'noble.' This was a sanguine idea of human nature, and the Confederation based on it was perhaps over-sanguine. It remained to be seen whether subsequent exhibitions would portray a Canada that was integrated and strong, or troubled by regional and ethnic conflict.

7

Exhibitions in Europe after Confederation and the Commodification of Canada

Chapters 7 and 8 focus on Canada at the international exhibitions in Europe and North America until the end of the century. This division of the material interrupts the chronology somewhat, but it does so to reflect the two kinds of displays that Canadians made. Exhibitions in Europe were dominated by the government because of the expense of transatlantic shipping and the relative disinterest of most Canadians. After Confederation, the Canadian government no longer purchased the displays sent overseas, but orchestrated them and spent increasingly large sums of money employing a growing army of reporters, organizers, and exhibitors to ensure that the new country was well turned out from coast to coast. The exhibitions held in North America, which were closer to home, reflected a broader range of Canadian participation and provided some sense of an expanding North American experience.

The Canada that was displayed was still not a finished product. In 1869 the Hudson's Bay Company land was added to Canada; subsequently, Manitoba, followed by British Columbia and Prince Edward Island, joined the confederation. The sea-to-sea land mass was realized, but much of this land remained, to Anglo-Saxon eyes, empty, inhabited only sparsely by native groups thought to be incapable of extracting its wealth. Canada was no more than a facade until it had been peopled by industrious immigrants. Thus, the task of selling Canada abroad was resumed in earnest. Confederation had made the country easier to advertise: Now the government could offer free land in Manitoba, rivalling that on offer by the midwestern United States. Federal and provincial governments in Canada each developed their own network of immigration agents and, during the 1870s, held annual meetings to coordinate their efforts under the aegis of the federal Department of Agriculture. Barely 10 per cent of British emigrants went to Canada during this decade, while 20 per cent went to Australia and most ended in the United States.[1] The comparison drew complaints from British Canadians, that 'Lower Canadians do not look with special favour on the introduction of a large increase to the English-speaking population,' and that 'the paralysing influence of this feeling in the movements of the Bureau of Emigration is very palpable.'[2] J.-C. Taché, secretary and later deputy minister, earned this sort of remark. He turned down requests from England for pamphlets and confided to Langevin that at every hour he fought 'cette terrible lutte' against 'les ennemis de notre race.'[3]

Gradually the department became, if not more efficient, at least more prodigal, with exhibitions becoming one of its heaviest expenditures. The high cost of transport and a generally low return on investment discouraged individual businessmen and producers from exhibiting overseas. The government's role in orchestrating and subsidizing exhibits was critical. When it abstained, Canada was virtually unrepresented, as occurred in Vienna in 1873 and in Paris in 1889. When it sunk hundreds of thousands of dollars into the display, as in London in 1886 and Paris in 1900, the Canadian display was reckoned a success. Between these two poles were the quasi-failures of Paris in 1878 and Antwerp in 1885.

What difference did success or failure make in the long run? The most obvious way of answering this question concerns Canada's material well-being: the extent to which successful exhibits attracted and unsuccessful exhibits repelled the capital, markets, and immigrants that Canada sought to attract through exhibitions. Thus, Canadians showered endless giant cheeses, gilded pyramids, and tree trunks, sheaves and jars of grain, apples, pianos, agricultural implements, and other goods on host cities year after year. This sameness led to a certain monotony in the displays (and does as much for the describing of them a century later), but by such efforts, Canadians built up substantial overseas markets in these different industries. The ever grander displays marked increasing association among resource producers and increasing concentration among manufacturers, as well as developing alliances with the state.

Also worth considering are the ways in which Canadians used exhibitions to reflect on their relationship with one another and with the world. Attempts to display Canada's assembled charms often jarred with local goals of superiority over the other provinces or of autonomy at the expense of the federal government. Yet when Canada appeared poorly, local pride was injured as surely as national pride. Whether the international exhibitions had material consequences or not, they told a story of world progress and benevolence, a story Canadians yearned to figure in with at least a bit part, if not a starring role.

There were other consequences. The fact that the Canadian government took such a prominent role in advertising the nation, that it indulged in the sort of fraudulent and exaggerated boasts that advertising requires, had an impact on its authority and on people's expectations of government. Moreover, the government's need to gather materials for display prompted it to infiltrate the private life of citizens in new ways. These consequences followed from all the international displays made by Canada in the late nineteenth century, but were particularly marked in the European exhibitions because the government's intervention there was critical.

London, 1871–1874

Lethargy in the Department of Agriculture during the 1870s contributed to lackluster Canadian participation in a series of exhibitions in London between 1871 and 1874. It was England's turn to play host, but the last experience in 1862 had been costly and humiliating because English industry was seen to be declining relative to Germany and the United States. Henry Cole and the 'Kensington Museum gang' blamed poor workmanship and shoddy mass production for the decline, hoping to correct it with small annual exhibitions of art manufactures and horticulture, to educate public taste and artisanal skill, and to 'give beauty and refinement to every description of object of utility, whether domestic or monumental.' The first exhibition, opened by the Prince of Wales twenty years after his father had opened the Crystal Palace, was a success, with a profit of £17,671.[4] Canadians took little interest in this endeavour: The government merely publicized the invitations it received. In 1873, prodded by the Colonial Office, it sent samples of Canadian wood, which arrived just as the exhibition closed. By this time, even the English had lost interest in this rational, elevating entertainment.[5] The final exhibition incurred a loss of £17,821, and no more were held. Cole complained that as each exhibition 'improved in quality and technical instructiveness, the attraction declined for the general public.'[6]

Plans were afoot to establish a permanent colonial exhibition. The Australians, enthusiastic participants in the annual shows, advocated the scheme in 1872 and, promising that this colonial display would 'familiarise the English people with a sense of the great extent, manifold opulence, and natural unity of the colonial Empire,' they obtained support from influential Britons.[7] The Canadian provinces were enthusiastic, but everything depended on Ottawa's response. In London, Canadian financier Sir John Rose attended the planning sessions, and in November 1872 he wrote to his old friend Sir John A. Macdonald advising him that 'the idea of a Colonial Court is a good one,' for political reasons: 'There is a disposition on the part of the Australian and other Colonies to take common cause wherever Colonial interests are concerned and they perhaps might feel a little sensitive if Canada were to hold aloof from this project. The influence that a combination of Colonial interests can exercise is becoming clearly apparent.' A confidential note added that Hugh Childers, a British MP and close friend of William Gladstone, was excited by the scheme and 'it could be well spent money to fall in with his plans to a moderate extent.'[8] As Rose was trying to float a loan for the Canadian Pacific Railway, he needed friends in London. Macdonald doubted that any great material benefits would ensue, but political considerations won him over. On St Patrick's Day, 1873, he sent the documents to the minister of agriculture, J.H. Pope, and gave his own view: 'I think that now the question has been mooted by the other Colonies we cannot hang back. If we do it will be made a subject of attack.'[9] But two weeks later the Pacific Scandal broke, and soon Macdonald was fighting for his government's survival. The colonial museum was delayed for more than a decade.

Vienna, 1873

In 1873 the Austro-Hungarian Empire held the largest exposition yet, with separate pavilions for different industries, but a cholera epidemic lowered attendance and the affair lost nearly $7.5 million. France, Britain, and the United States were all poorly represented.[10] The Canadian government abstained because the cost was prohibitive. The display of 1867, at $50,000, had almost doubled the immigration budget. Canadian policy favoured German-speaking emigrants, but Germany discouraged immigration agents, interpreting their propaganda as economic and political criticism.[11] No German lobby in Canada existed to demand that the government promote immigration, and there was little hope of cracking the Austro-Hungarian market, as much of what Canada had to offer – furs, timber, grain, minerals – could be obtained from Russia. A few Canadians went on their own initiative. Wanzer Sewing Machines of Hamilton won three awards; the company also attended smaller international fairs from Australia to

Russia to Chile during this period and exported thousands of machines around the world.

Macdonald did leap at the chance to send some cronies for a holiday at public expense. A sum of $5000 was voted (and $8000 was actually spent) for a series of reports on the exhibits. The chair of the commission was H.B. Witton, who worked for twenty years in Grand Trunk shops and in 1873 was elected MP for Hamilton, the first working man in parliament. He had been promised a reward for saving Hamilton for the Tories, and this tour was the form it took. Labour leaders recalled the episode with disgust a decade later: 'They had a man in Hamilton who was sent to Parliament. What did he do? The first situation that was offered him he took, and went to Vienna on some government exhibition business.'[12] Witton advised sending working men to make reports and suggested using the Boards of Arts, but he concluded that 'the selection could not safely be left to them: we could not get the right kind of men.'[13] If his objection was to their class, he objected in vain, for nearly every man sent was a wealthy Tory, including the strike-breaker Edward Gurney, who wrote a protectionist report. The report on bricks was by Henry Bulmer, a brickmaker who had compelled his employees to vote for Cartier.[14] A foreman at the Grand Trunk works in Belleville gave a technical report on the machinery, and the report on farm implements was also solid. On 4 May 1874 Macdonald, now in opposition and trying to defend his government's record, called for copies of the reports as 'interesting to the people,' but he fell silent when the Liberals threatened to produce the correspondence and expose the inflated cost. The Grand Trunk foreman, A. Davis, wrote to Alexander Mackenzie commending him for having 'well caught' Macdonald, and he itemized the sins of the other partisan, patronized, and plagiarizing reporters.[15] This exhibition marked the last attempt by Canadians to learn about new technology from this source.

Paris, 1878

In 1867 the French had decided to hold another exposition in 1878, halfway to the centennial of the French Revolution. Despite the humiliating defeat in the war against Prussia, the Third Republic proceeded with this act of optimism to show that France still ruled in matters of taste and that Paris was the permanent great exposition of the world.[16] The United States attended, as did England and the colonies.

Returning to Europe as a dominion, Canada determined to make an effort, at a total cost of $113,905.[17] However, the formal arrangements were far from satisfactory. The chief commissioner, T.C. Keefer, was a strange choice: He did not speak French, and in 1849 had described French Canadians as 'priest-ridden drones.' One French-Canadian assistant described him as 'probably the most negligent man I have known, and there have been many complaints on his account.' Keefer, a prominent civil engineer who first came to public attention with his *Philosophy of Railroads* in 1849, was being rewarded for years of devotion to Liberal interests and commerce.[18] He had two secretaries, Samuel May, from the Ontario Department of Education, and Joseph Perrault, the Quebec agronomist. Perrault was troubled by Keefer's subservience to the English, through whom all correspondence with the French had to pass. He complained to Premier Joly of Quebec (with whom he had fought against Confederation in 1864) that there were no Canadian flags or judges, and all labels were in English: 'En un mot on n'en pas su profiter de l'excellente occasion qui se presentait pour le Canada de renouer des relations commerciales avec la France ... Je ne vois que l'indépendance comme rémède à tout cela.'[19] Relations with the British commissioners were not good. France granted a block of space to England and her colonies, and England, Keefer sourly remarked, kept most of it for her own exhibitors.[20] Worse, the British did not reserve space for Canada in the machinery building, so Keefer had to round out the exhibit by means of a handbook that described Canadian machinery, as well as its healthy climate, contented Indians, and other attractions.[21]

Prince Albert made amends by inviting the dominion to erect a Grand Trophy in a vestibule that formed a principal entrance to the exhibition. The Department of Public Works designed a monumental structure, several floors high, that cost the dominion more than $14,000. Goods that Canada lacked space to show on the ground were hung from the trophy: Agricultural implements covered one side; model ships, cordage, and fish another; wooden wares the third side; and manufactures the last. Smaller trophies surrounded it, including a 556-year-old Douglas fir, plumbago and coal, with a large gilt octahedron representing all the gold mined in Canada. At each door stood a stuffed bear, tamely holding a basket for visiting cards. Flags and the arms of Canadian cities completed the decoration. Inside the trophy were stairs leading to an upper floor, which provided the best view of the galleries. Just climbing these steps was an experience in itself: 'As you ascend the stairs at every turn you are met by the glaring eyes of our wild animals, a brown or black bear, wolf or a deer, a fox or racoon, a meditative owl, a vulture, or some other of our raptorial birds peep out from the many nooks, while our domestic fowl sit here and there in quiet corners looking sedately on the great stream of humanity flowing on and ever onward up and down the great stairs of this magnificent Trophy from morning until evening, day after day, and week after week.'[22] The Prince of Wales climbed the stairs and drank some wine at the top; since the structure 'swayed threateningly,'[23] his visit showed a certain bravery. One critic denounced the stuffed beasts and Indian relics as 'fit to frighten anyone from coming to Canada. At the foundation we saw some barrels of disgusting looking meat, old cheese boxes and firkins. Some had old, torn, dirty bagging tied about them, and the scene was anything but inviting.'[24]

The government bought a few manufactures for the trophy, but could do no more without angering those who sent their own goods.[25] It did pay for a collection of natural resources, minerals, and a display of public works in models, photographs, and maps, for which it received the top award, a 'Diploma of Honour.' Samuel Keefer, Thomas's brother, won a gold medal for his design of the Niagara Suspension Bridge, the widest span in use around the world. Canadian woods and workmanship were illustrated in wooden showcases that cost the department close to $10,000, filled out with botanical specimens, for which naturalist John Macoun charged only $15 per hundred.[26] Otherwise, the government relied on representations of Canada, such as the *Handbook*, photographs, and an enormous map of the country which, at 10 miles to the inch, was 35 feet long, conveying at a glance Canada's great size; it won a gold medal. Even the Geological Survey exhibit was mainly figurative. A stratigraphical collection won its director, Alfred Selwyn, a silver medal, but the Diploma of Honour going to the survey itself was awarded in the geography class for maps and illustrations.

A total of 264 awards were divided among the 534 Canadian exhibitors. Ontario dominated the display, with 328 exhibits, compared with 107 from Quebec, 65 from the Maritimes, 29 from British Columbia, and 4 from Manitoba (all grain). The major winners were Nova Scotia with 20 prizes, Quebec with 88, and Ontario with 138, thanks to Keefer's connections with Ontario businessmen. To overcome his lack of contacts in Quebec, Keefer had asked the Montreal Board of Trade for help: 'There being no manufacturers' association here to take the matter up collectively as is the case with the Fruitgrower's Association, the Dairymen's Association, Lumberman's Association, etc. of Ontario, I beg respectfully to ask the co-operation of your Board to induce individual manufacturers, and firms of the city to enter heartily and vigorously into this subject from patriotic as from commercial considerations.'[27] The Board of Trade welcomed the initiative, but, when the Council of Arts and Manufactures failed to take an interest, it let the matter drop, blaming the 'seemingly general backwardness on the part of manufacturers and others.'[28]

Among the goods sent by individual exhibitors was a life-size lion made of paraffin wax sent by Waterman Brothers, 'the handsomest piece of work we have ever attempted to manufacture.'[29] Ontario dairymen sent a giant cheese, which one visitor liked enough to return for a second helping, although another said that its bad taste haunted her for some time.[30] Robert Malcolm of Toronto sent mailbags and 'the largest and most excellent show of saddlery,' but French law forbade him to import further saddles for sale. Still, Malcolm sold goods worth $171 and won two silver medals, which he accepted less than graciously: 'The medal sent is one of the worst I have ever seen. It is like a piece of pewter and *will require silver plating* before it can be displayed.'[31] John Labatt showed ale and porter, winning a bronze medal that adorns the Labatt bottles to this day (although the drink inside is no longer 9 per cent alcohol).[32]

Many Canadians visited the show and nearly one thousand signed the register, among them businessmen Peter Redpath and C.H. Waterous, engineer Sandford Fleming, and politician Edward Blake.[33] Many foreign visitors were ignorant as to '*where* and *what*' Canada was, such as one noble duke who thought Canada must be tropical because it produced maple sugar.[34] Keefer had a colour map of Canada put on every exhibition tag.

In October 1878, when the Liberal government was voted out of office and Macdonald resumed power, Keefer suddenly found his acts subject to hostile scrutiny. The new government refused to pay many of his expenses.[35] The fate of the unsold exhibits also caused trouble. The government humiliated Keefer by refusing to pay transportation for the exhibits to London for another projected colonial museum.[36] As for the private exhibits, sales totalling $5809.98 were made by the Canadian Commission on behalf of private exhibitors, but many were too costly or unsellable, such as English-style scales. Keefer sent home, COD, all those he could not sell for half their value. The shipping agency jacked up its prices, and exhibitors flooded the Department of Agriculture with complaints.

There was little consensus as to the outcome of the exhibition. The Ontario commissioner boasted that one French dealer ordered 1000 reapers over five years, and that demands from Europe, Australia, and South America approached half a million dollars.[37] He and another assistant, Gustave Drolet, recommended a permanent office in Paris, but Drolet thought it should replace exhibitions because 'ni l'Etat, ni l'exposant canadien n'ont retiré de benefices appréciables des expositions universelles continentales, auxquelles l'Angleterre nous a priés d'assister depuis vingt-cinq ans.'[38] The English, he claimed, reaped all the benefit, but jealously snubbed Canada at every opportunity. Drolet and Perrault broadcast their complaints, Perrault doing so in a short-lived journal that advocated commercial union with the United States.[39] Gordon Brown, brother of George and his successor as editor of the *Globe*, was named to represent Canada at a Postal Congress in Paris, but the English would not admit him to the discussions. Brown declared that, in future, he would do all he could to stop Canada from competing, except as an independent nation.[40] Brown had reported on the exhibition of 1851 for the *Globe*, but, in 1878, he paid an expatriate Montreal *rouge*, Louis-Antoine Dessaulles, to write weekly columns that praised the exposition and French republicanism.[41] Dessaulles thought that Canada benefited by projecting the image 'of a united Dominion,' and trade would follow.

Trade between France and Canada did not greatly improve because Canada was not included in the preferential tariff between England and France negotiated after Confederation. Some Canadian goods were taxed at twice what English makers paid, and others, such as Malcolm's saddles, were prohibited. Canada's great hope was that the exposition of 1878 would remedy this situation, but it did not happen until 1894, so, in that respect, the exposition was a failure. On the other hand, as Magella Quinn has shown, after 1880 French capital was increasingly made available in Quebec, adding up to $25 million in loans to the province and other investments by the end of the century.[42] Frédéric Gerbie, writing in 1884, attributed to the exposition closer relations with French capitalists, leading to the creation of a Credit foncier franco-canadien, a short-lived Franco-Canadien sugar beet company, a Franco-Canadian company to mine Quebec phosphates, and a Credit mobilier canadien, all of which Perrault helped to organize.[43] However, no great rise in commerce could come before a direct steamship line was established, which occurred in 1900. Another scholar, surveying French writing about Quebec from 1850 to 1910, finds that it peaked during the 1880s, a fact contemporaries attributed to the exposition.[44]

Colonial and Indian Exhibition, 1886

One consequence of the Paris Exposition was the Colonial and Indian Exhibition of 1886. The colonial commissioners again proposed a museum in London, but the British feared it would deteriorate into another Sydenham collection, of which it was said: 'Nothing was more dismal or humiliating than a collection of dingy and ill-arranged Colonial produce – odd lots of dirty cotton, wool, or other such materials, or bits of ores and stones – which one ought to appreciate and cannot.'[45] Still, it was pointed out, France had a splendid Algerian museum, and Holland, too, 'is a little country, but it has a Colonial Museum.' It was not enough to have an empire; one had also to put on display the trappings of empire. This speaker, a member of the Colonial Institute, intimated that these displays of power and glory were preferable to the real thing, that France 'has won in the Champs de Mars a victory perhaps more glorious than that which her great rival registered in blood at Sedan.' Some historians echo this view, arguing that British imperialism rested largely on spectacle and was so illusory and theatrical that it was remembered less as 'a political organism than as a sort of giant exhibition.'[46] What better expression of the spectacle of empire than an imperial exhibition?

During the 1880s exhibitions escaped the rigid art-manufactures framework of the past. A Fisheries Exhibition in 1883 sponsored by English fishmongers drew entries from foreign nations. Canada's display, with a working fish hatchery, was one of the highlights, although the inevitably quarrelling commissioners had to combat reports that the hatcheries failed to increase fish populations.[47] But Sandford Fleming thought the display gave 'information and satisfaction,' and it coincided with increased fish exports, at a time when the British working classes were beginning to eat canned salmon.[48] Newfoundland's government also made a great effort to be well represented, believing it a matter of 'paramount importance to the Colony.'[49] In 1884 and 1885, education and health exhibitions were held. But the 'Colinderies' or 'Kith and Kinneries' (*Punch*'s names for the Colonial and Indian) was the most important of these exhibitions.

During the 1870s and 1880s Little England sentiment gave way to a yearning for a formal statement of imperial power, as recession heightened predictions of England's industrial decline.[50] Manchester liberals had argued that formal imperial ties were expensive and useless, for trade alone would maintain common interests.[51] But English exports declined as the United States and Germany expanded, and a growing lobby, which formed itself into the Imperial Federation League in 1884, advocated formalizing relations to keep trade following the flag. Colonies stretching over half the globe offered privileged markets.

The league was purposely vague about its goals for fear of alienating colonial sentiment. Ultimately, federation implied colonial representation in a Westminster parliament or council. Colonial administrators objected to this threat to their autonomy, but often subscribed to other league planks: military pacts, penny postage, Australian federation, and a centralized system to promote emigration within the empire. Sir John A. Macdonald, whom the Imperial Federation League improbably claimed as one of their most enthusiastic members, valued the National Policy too highly to espouse the League's free trade policy.[52] Canadian high commissioner Sir Charles Tupper upset English imperialists by arguing that Canada had done quite enough for imperial defence by building the Canadian Pacific Railway and that it should recoup some of this investment through assisted emigration sponsored by Britain. Yet he was wildly enthusiastic about the Colonial Exhibition, which he urged on his old friend Macdonald.[53]

Macdonald had good reason to throw the resources of the Canadian government behind the exhibition. Canada wanted what it had always wanted: markets, capital, and immigrants. Shut out of American markets, Canada sought to increase sales in the empire.[54] The exhibition provided privileged access to these markets, as well as to the English money market. Although Manchester was no longer the workshop of the world, London was banker to the world. One-quarter of British investment at that time was spent in the colonies, where high returns on investment could be gained.[55] Canada, at the time the Colonial Exhibition was mooted in 1884, wanted a parliamentary guarantee for a CPR loan and British investment in the loan. Mammoth fruits and grains from the northwest would contradict reports in the London press that the land was so barren that only a fool would invest in a railway through it. As interest rates on Canadian loans dropped and subscription to them increased during the 1880s, Tory organs attributed this success to the 'grand displays made by Canada in the various exhibitions in Europe.'[56] The Canadian agent in Liverpool advised showing oil paintings of Canadian cities 'whose loans are quoted on the European Stock Exchanges,' for New Zealand got lower interest rates by such a tactic.[57] His advice was followed.

Above all, Canada needed immigrants. In 1880 Sir A.T. Galt became Canada's first high commissioner to England: His most important task was to oversee the small army of agents who puffed and prattled their way around the country, always visiting the agricultural exhibitions where farmers and farm labourers, the preferred class of emigrants, were sure to be found. In this manner Canada could succeed in 'easily reaching the Agriculturists of the British Isles *without saying a single word about emigration.*'[58] Subtlety was needed because many Britons opposed emigration, thinking the mass deportation of citizens a poor way of solving economic problems. It was this 'loophole for the consciences of bad rulers' that Goldwin Smith despised: Ireland, he claimed, was unable to sustain its population because it had been so badly governed, while the governors masked their misdeeds with pompous speeches about extending imperial might by planting Britons around the world.[59] Agricultural depression emptied the countryside as rural labourers flocked to the cities and overseas, causing farm wages to rise. Powerful landed interests opposed emigration and Canadian agents often met with hostility.

Canada had always needed immigrants, but there was a new urgency. The railway to the Pacific Ocean was nearing completion (the last spike was driven in November 1885), but it could only be paid for, as Galt pointed out in 1880, by taxing central and eastern Canada or by peopling the West with immigrants: 'The importance of a large and good emigration to Canada, at this moment, cannot be exaggerated. Without it the vast tracts of fertile land in Manitoba, and the North West, must remain unoccupied, and the burthen of constructing the very costly works, necessary to develope the country, and render it accessible, will become intolerable, and will press most severely upon the industry of the Dominion. Expenditure in connection with Emigration is, therefore, the best possible investment that can be made.'[60]

By 1884 immigration had slowed to a trickle and, as Donald Creighton recounts, pressure had mounted on Macdonald: 'Without a successful north-west his whole great plan of national expansion and integration was meaningless. If it began to seem really probable that the north-west would not become the homeland of a prosperous and contented population, then there could be no justification for the loan he was about to propose in aid of the Canadian Pacific, and no reason for the Canadian Pacific itself. It was as simple as that.'[61] Macdonald had 'to gamble upon belief which Canadians – and outsiders – held in the future of the Dominion ... It all depended on faith in the end.' He could inspire faith in Canada by exhibiting in Europe, just as MacNab had done thirty years earlier to justify his public works.

The Colonial Exhibition was an opportunity to address Anglo-Saxons, the preferred immigrants. Moreover, in the absence of Europe and the United States, Canada, described by Lord Lorne as 'the ripest of the Colonies,'[62] would appear to good advantage and would monopolize the machinery gallery. As the Australian colonies were not yet federated, Canada could mount an unsurpassed collective display. Finally, Canada had twenty years experience in emigration propaganda to draw on, dozens of agents in Europe and the United Kingdom, and hordes of hack writers in Canada. The government decided on a major propaganda initiative for the exhibition. It drew on administrative apparatus, the press, voluntary associations, manufacturers, artisans, farmers, labourers, artists, teachers, schoolchildren, and housewives. This massive effort had only mixed results. The loan guarantee was secured and Canada maintained its privileged access to British capital, while investment in the United States declined. The value of Canadian exports to Britain, which had been declining, rose from $41.5 million in 1886 to $44.5 million in 1887, but by 1889 it had slipped below $40 million. Immigration did not increase at all.

Although the Canadian display at the Colonial Exhibition had little impact on its audience, it had a great impact on its author – the Canadian people. The government drew on commercial methods to advertise Canada as if it were a commodity, backed by the power and authority of the state, to generate and promulgate its message. This was the act of a new kind of nation determined to obtain a new kind of citizen. Conservative political philosophers warned against the dangers of permitting citizens to think they could renegotiate the terms of their membership, but here were governments advertising their lands, and even the political tenets guiding public life, as if they were commodities like any other. This emigration propaganda was more ambitious than business advertising: advertisers tried to persuade their customers to buy a new hat or food, but emigration agents sought not only to create brand loyalty (Canada rather than Australia) but to convince the average farmer and his wife that their situation was intolerable, that it would worsen, and that they must quit the life they knew for one distant, foreign, and unknown. At agricultural fairs throughout Britain, Canadian grains were on display to convince English farmers they could not compete with Canadian growers. Gladstone had advised farmers to give up wheat and 'grow fruit and make jam,'[63] but Canadian fruit, fresh and preserved, betokened fierce competition there too.

In the late nineteenth century, commercial advertising was becoming a powerful social force, both awesome and troubling. Canada emerged as a nation just as this new force of commercialism arose. Advertisements affected Canada's material condition and prospects, as well as Canadians' sense of themselves. Canadians read copious descriptions of exhibitions published in the daily press and took pride in medals won. If Canada was not represented, it would be misrepresented by rival colonies and shameless Yankee agents who made outrageous claims about long winters and disgraced morals, insisting that 'there is more drunkenness to be witnessed in a single day in Toronto with its 60,000 inhabitants, than in Cleveland in a month.'[64] Exhibitions were a good place to combat such slanders, although they might provoke new ones: In 1886 an American paper claimed that the Northwest exhibit sent to London in 1886 was 'produced by [railway magnate] Sir Donald A. Smith under "acres of glass" at Silver Heights.'[65]

Canada's internal organization was partly designed to produce advertisements. The boards of agriculture and of arts were established during the Union to oversee exhibitions in Canada and to skim off the best for forwarding to the periodic world fairs or to immigration offices in Britain. If a single backwoods farmer came up with a single monster squash, the system was designed to discover it and place it where it could do the most good. In Winnipeg in 1876 a 'large-sized box to hold "Manitoba Products"' sat in a city office, where farmers could drop off exhibits for the eastern fairs.[66] Provinces and even counties worried that a bad local exhibition would injure their reputation. In 1861 an Ontario paper complained that Lennox County's failure to exhibit at the shows in Canada was damaging its prospects: 'Do our farmers and artisans know, that in other parts of this land, as well as in the United States and the Old Country, the impression upon the minds of the people, as to the comparative excellence of the different Counties in Canada, is made to a great extent by the reports of these industrial exhibitions, and if they look for us in these exhibitions they find us *nowhere.*'[67] Exhibitions in Canada were designed to 'enable an intelligent stranger to observe at a glance as complete a representation as possible of its entire resources and products, natural and artificial.'[68] The Canadian government brought British farmers and agronomists to Canada and trundled them around the agricultural fairs. In 1881 Professor Sheldon of the Agricultural College at Salisbury obligingly reported that 'the Canadians throw themselves with great spirit into enterprises of this kind, and these shows are a great credit and ornament to the Dominion.'[69] The principle of visibility linked the smallest township fair to the grandest world exposition.

Advertising was Canada's stake in a brilliant future, and 'exhibitionism' penetrated Canadian society to the core. One might claim that Canada was a 'promotional commodity,' a term used by Andrew Wernick to refer to commodities designed to serve as advertisements.[70] Confederation itself proved a clever advertising tactic. In 1873 an agent remarked that, before Confederation, 'it would then have been useless to address the emigrating classes of Great Britain in the name of Canada, for they would have been bewildered by territorial distinctions which they could not understand. Confederation has, however, changed all that, and the Dominion of Canada is coming home to the minds of Englishmen as an idea which they can grasp and comprehend.'[71] Tupper had made the same claim in 1860: Confederation would distinguish Canada from 'Nova Zemble and similarly favoured regions.'[72] In 1864 George Brown also argued that the point was 'to establish a government that will seek to turn the tide of European emigration into this northern half of the American continent' as well as to develop resources.[73] There were other reasons for Confederation, but this one was never far from anyone's mind.

Canada, more than most countries, existed in advertisements. The representations of its friends and enemies could have a decisive effect on its future. Historian Norman Macdonald mocks the 'pathetic faith in advertising': 'What began as a feeble attempt to deflect British emigrants from the United States, developed gradually into a settled conviction, to the neglect of other aspects of the question, that advertising was the first step towards securing settlers for Canada.'[74] Tupper certainly subscribed to this conviction. In September 1885, speaking at the Toronto Industrial Exhibition, 'he showed that all that Canada required to become a great nation was that the world should become possessed of accurate *knowledge* of her resources, government, and institutions, and no better means could be devised of spreading this *knowledge* than the projected [Colonial] exhibition, if Canada would avail herself of it to the fullest extent. Once the *knowledge* he had alluded to was given to the world, there would be an influx of capital and population to the Dominion, which would develop the country with marvellous rapidity.'[75]

This ambition probably sounded as naive to the crowd as it does to the modern reader. Tupper was undoubtedly excessively sanguine. But it would be wrong to suggest that Canadians foolishly believed they could overcome real obstacles with mere words, that there was an opposition between representations and reality. The advertisements contained overstatements, even lies, that were easily disproved, but the overall message was not one that could be called true or false. If the exaggerated accounts of good roads, jobs for the asking, and healthy climes convinced people to come, they could become self-fulfilling prophesies. The Americans had been pulling this stunt for years. By exaggerating Canadian prosperity, Canadian agents could generate the faith in Canada that Sir John A. Macdonald so badly needed, and in doing so bring that prosperity within reach. As one advertising manual published in 1887 stated, 'The usefulness and value of most things depends, not so much on their own nature as upon the number of people who can be persuaded to desire and use them.'[76] In short, truth was relative, not absolute, and could be shaped. Exhibition advertising sounds rather like the pragmatic philosophy propounded in the United States at the end of the nineteenth century.[77]

To the cynic, Canada was an ugly sprawling hulk of a country where ice and snow were relieved only by mud and mosquitoes. Exhibitions were merely advertisements, a frivolous, foolish, and ultimately marginal form of communication. To the idealist, Canada was a land of wealth and civilization, the 'germ of a great nation,' and exhibitions were one of the highest forms of civilized activity. Some remarks made by a Halifax editor in 1868, a propos of a local industrial exhibition, distinguished between barbaric man, who is 'insensible to the beauties that surround him,' and civilized man, who improves himself and grows flowers. At a certain level of improvement, 'the spirit of manly emulation or rivalry becomes active, which begets local exhibitions as a test of skill.' This step is both important and beneficial. 'The next form of friendly rivalry is district competitions,' followed by national and then international ones, at which point global intercourse and peace have been obtained and civilization is complete.[78] In other words, first there was barbarism and then there were exhibitions. Once exhibitions kicked in, they transformed civilization just as they transformed Canada. The act of representation changed the thing represented, invariably for the better.

The act of display, of formal representation, separated exhibitions from the sordid and squalid world that lay outside the representation. It was not a lie, in the way that Victorian fiction, with its idealized portrayal of society, was not a lie. The ideals that humans could imagine were true because they were God-given; it was man's nature to improve himself and the world. Every idealization was a possible prediction. Advertising capitalized on this way of thinking, for it, too, was a promise of improvement. The description of exhibitions as lofty acts of civilization was, of course, a way of advertising them. Lofty and vulgar, disseminating truthful knowledge and misleading advertising, exhibitions mediated between the two categories and weakened the distinction. A survey of the Canadian display in London in 1886 illustrates how this fusion occurred by revealing the practical problems involved in creating a quasi-mythic national identity for the marketplace.

The government threw its considerable resources into an 'exhaustive' representation of Canada in 1886. The Liberals seconded this effort: Sir Richard Cartwright approved of a $60,000 preliminary vote with the remark: 'I am not disposed to criticise very much. If this thing is done at all, I have always held that it should be done well.'[79] (John Lowe, the deputy minister of agriculture, was ready to throw these words back at him when the affair later became 'a piece of folly.')[80] Money was no object, except that the government drew the line at livestock. Though Macdonald agreed to spend over $100,000 for the display (the final tally was $312,076), his interest was political, not personal. He told Tupper baldly: 'There is no chance I think of my visiting your Exhibition. Personally I take no interest in machinery and cereals, and readily satisfy myself by the descriptions of your goings on which I find in the press. I hope that Canada has made a fair show.'[81]

The Antwerp Exhibition of 1885 provided a springboard. Antwerp was an important port city, but prospects of trade were slim, so manufacturers abstained. The Rosamond Woollen Company sent odds and ends left over at the end of the season, on the 'urgent solicitations' of the government agent. (The company sent a better exhibit to London.) The model settler's cabin sent by the CPR was admired, but the Geological Survey exhibit was so badly catalogued as to attract ridicule.[82] The whole was not very creditable to Canada, as John Lowe admitted privately, and the Liberals accused J.H. Pope of 'culpable neglect.'[83] This failure probably led to a determination that the Colonial Exhibition should be a success. Certainly the display in London prompted the Durban *Natal Mercury* to remark: 'The whole Canadian Court suggests from first to last a people terribly in earnest.'[84]

Responsible for the display were Tupper and his staff in England, and the Department of Agriculture in Canada under the minister, John Carling. Other departments – Indian Affairs, Public Works, and the Geological Survey – sent their own exhibits. To rouse lukewarm exhibitors, Tupper toured Canada in 1885, delivering 'innumerable addresses with his customary vigour and copiousness.'[85] Carling hired one agent per province to gather exhibits, except for Prince Edward Island, which got two. Positions were scarce: Lowe told one job-seeker: 'You could hardly ask for anything more hopeless. The fact is that a considerable portion of the population has applied for similar appointments.'[86] Mostly, agents cajoled manufacturers. As Lowe complained to Tupper, 'some of our most worthy establishments have, however, shown a great want of public spirit and breadth of view' and refused to exhibit.[87] Most applications came from English Canadians, though clerks were unsure whether to blame French-Canadian reticence on 'anglophobia' or 'want of enterprise'; they might have considered the smallpox epidemic that raged in Montreal in 1885.[88]

Responsibility for the exhibits was carved up as carefully as governing powers. While Ottawa collected manufactures and minerals, the provinces provided agricultural and educational exhibits, but not all cooperated happily. New Brunswick erected a separate wood trophy. In 1886, a year when secession was being agitated widely in Nova Scotia, Liberal premier W.S. Fielding threatened to withdraw that province's display if he could not keep its mineral, fruit, and education exhibits together rather than exhibiting with the dominion. Prince Edward Islanders complained that their exhibit suffered in the same way, while Vancouver Islanders objected that Canada's exhibit was too widely scattered.[89]

In collecting for the Colinderies, the government had two formidable allies: the press and voluntary associations. Long before the doors opened, newspapers rivalled each other in 'booming' the event, since sentimental puffery sold copy. Hack writers filled columns with gushing letters: Even the *Presbyterian Witness* hired the pen of P. Stewart Ross, though it warned him that 'an account of the seamy side of a London Music Hall' would not please a readership 'very staid and full of prejudices.'[90] E.P. Biggar obtained permission to print a monthly paper called the *Colonial Exhibitor*, which contained only 'advertorials':[91] fulsome descriptions of the exhibits and of greater wonders back home. The cover showed a woman pointing to an oversized Canada, its border the 45th parallel. By contrast, Tupper's official catalogue was austerely factual.[92]

Some very fine lines had to be drawn in this alliance of state and press. Biggar asked for a subsidy, as did a Halifax paper planning an exhibition supplement, but both were refused outright.[93] The press already received patronage in the form of advertisements (dozens wrote to demand the exhibition ads) and printing contracts. The government baulked at paying directly for copy and preferred to trust to a shared interest in Canada's prosperity which would lead journalists to say all the right things of their own accord. Censorship was unnecessary: The state could count on the press to share the burden of humbug, although the *Globe*, when it published an exhibition supplement, was warned by Tupper 'to avoid the introduction of contentious political matters as such in my opinion in this instance will not tend to serve any useful purpose and lie outside the scope of these objects which it is the province of the Exhn. to promote and foster.'[94]

Certainly there was humbug and hypocrisy. The official handbook expounded the health-giving qualities of the bracing Canadian climate and the joys of sleighing and skating. There was a great deal of the sorry behaviour that led Canadian emigration agents to become figures of ridicule, 'Canadian Crackers,' whom Norman Macdonald describes as 'noisy, boasting, bragging, blustering and flatulent story-tellers.'[95] The boasts and bluster were unleashed on a truly astounding scale in 1886, as tens of thousands of pamphlets, ranging from scientific treatises to elementary catechisms, were showered on visitors. They were the work of hack writers, both departmental and freelance. Two pamphlets were entirely composed of passages clipped from newspapers describing the Canadian exhibits. Text led to text in a self-perpetuating loop that long outlasted the exhibition.

Canadian agents fought a losing battle to sustain an air of authority and objectivity. They sought disinterested testimony, but could only obtain it by paying people for it. One pathetic case in point is Alexander Somerville, an influential free-trade advocate in England during the 1840s; by the 1870s, between bouts of insanity, he wrote endless unsolicited letters to newspapers puffing Canada, for which the government paid him $12 per month.[96] Agents routinely published the testimony of emigrants because, as CPR agent Alexander Begg told Tupper, 'the testimony of actual settlers, and therefore of disinterested parties relating to the Colonies in which they have settled, is accepted by people in this Country much more readily and with much more confidence than anything which may be said by those like myself in Official positions who are looked upon as interested individuals.'[97] Tupper and Begg printed handbills designed to appeal to local loyalties. In 'What Devon says of Manitoba and the Canadian North-West,' Mrs McCormick of Fairmount, Manitoba, wrote eloquently: 'I came from Glasgow, Scotland, and arrived in Winnipeg, on August 13th, 1869. I left my native land without a soul that I knew. I was a single girl then, got married the next day after I arrived in Winnipeg, spent my honeymoon on the prairie, travelling with a yoke of oxen, and one cow and a wagon – that was all our start. I feel content that I left my native land and set smoke to poverty as soon as I crossed the Atlantic.'[98] The CPR did not have a separate exhibit but lumped its goods in with Manitoba, blending private and public identities. Begg helped erect the agricultural trophy and gave a paper on emigration at the exhibition conferences – which displeased Tupper by criticizing British policy – and he got into trouble for passing himself off as a Canadian commissioner.[99] The close relationship between Begg and the Canadian Commission paralleled that between the CPR and the Canadian government.

Speech at the Colinderies did not need to be formally policed, but elsewhere the government faced detractors, such as labour groups that sent reports of joblessness and low wages to English newspapers. This adverse propaganda was combatted by great barrages of pamphlets and letters. Scientists were among the most eager propagandists for Canada. Doug Owram has described how they began, from about 1857, to argue for the agricultural potential of the North-West.[100] One western booster was John Macoun, a naturalist with the Geological Survey who wrote reports exaggerating the region's fertility. His appointment to the exhibition was urged by John Fletcher, the dominion entomologist, who described Macoun as 'a most sanguine believer in Canada, [who] has, in my opinion, done more to point out & talk up the fertility and value of our North West lands on scientific & unassailable grounds, than any other man in the Dominion.'[101] Needless to say, Macoun got the job.

Politicians and the press directed things, but the exhibition mainly represented the private sector. Here, voluntary associations came to the fore. The agricultural societies, long-time recipients of government largesse and exhibitionists par excellence, proved useless. The provincial society usually rooted out samples for the world's fairs, procured either from the provincial exhibition or by personal application to leading farmers, but local agricultural societies, almost without exception, failed to respond. On the other hand, the associations of fruit growers, dairymen, and beekeepers, which grew up around the exhibitions and also received public money, all laboured to send substantial exhibits. They were aided by the CPR, emigration agents, and the Board of Trade, which collected the agricultural specimens in Ontario. While the agricultural societies stubbornly followed their own dictates, enjoying a measure of autonomy, the larger system of agricultural encouragement created intermediaries between the state and the producing public which helped the government to achieve its objectives.

Many important collections were sent by these bodies. The Beekeepers' Association sent 40 tons of honey. Ontario dairymen inevitably sent a large collection, including 40,000 samples of cheese sold at a penny or tuppence, with total sales of $12,000. Guelph Agricultural College professor James Robertson supervised the dairy exhibition and gave lectures in favour of scientific dairying and the factory system which British newspapers printed, providing, according to Robertson, 600 miles of newspaper columns of free advertising for Canada. He complained that Tupper refused to grant Ontario space, for fear of humiliating Quebec, which sent the 'worst lot' of cheese he had ever seen: 'I am sufficient of a politician to imagine the use which they saw might be made of this fact in the local elections of that province, against the [Conservative] Quebec Government.'[102] Robertson won his spurs in 1886 and, four years later, beat out Édouard Barnard for the position of dominion dairy commissioner. Cheeses were also sent to dairy fairs around Britain, where they took top prizes. During the last quarter of the nineteenth century, Canadian cheeses did so well at international exhibitions that they became world famous and were worth a penny more than American cheeses in England. Canadian cheese exports to Britain increased during the 1880s and topped American sales in 1891.[103]

The fruit growers' associations of Canada, with Ontario and Nova Scotia taking the lead, organized a collective exhibit of wax and fresh fruit which won a medal from the Royal Horticultural Society. Local exhibitions in Canada were the venue for fruit collection.[104] Experiments in shipping were attempted, including steamers outfitted with cold storage, but only apples and pears arrived in good shape; the plums, melons, grapes, and tomatoes rotted.[105] William Saunders, longtime president of the Ontario Fruit Growers' Association and now director of the Dominion Experimental Farm that had recently been voted into existence, sent 1000 jars of preserved fruits, collected from around Canada. The *North British Agriculturist* praised both the fruit and the collaboration among fruit growers: 'Their associations, which are established in every Province, are the medium for the communication and dissemination of facts and ideas concerning the improvement of fruit culture, and the transactions of each association are at the disposal of all for the common good. This is a contrast not very favourable to us at home. We are content to grope in the dark.'[106] During the late 1880s a substantial apple trade with Britain developed, as exports rose from $78,890 in 1884 to $649,282 in 1887 and $1,277,577 by 1889. Thus, as with dairy products, international success and cooperation among the producers, and between producers and the state, developed around exhibitions.

The fruit adorned a mammoth agricultural trophy, along with cheeses, butter, meat, milk, ham, and fish. Archways, decorated with sheaves of wheat, grains, and tall North-West grasses, opened into the centre, where the inevitable timber trophy formed the apex. Agricultural implements lay around the trophy, including a prairie turf-breaker. To the east and west were life-size plaster figures of a woodman and a dairymaid; atop were Ceres, Pomona, Flora, and an allegorical figure of the dominion. Tupper called this striking display 'completely representative': 'A vivid and comprehensive picture of Canadian agriculture from ocean to ocean, in all its ramifications, was presented to the spectator in one moment, suggesting to his mind all the blessings attending the possession and the skilful use of boundless and fruitful expanses, and telling of skies of unfailing kindness.'[107] The ability to convey complex and extensive information 'at a glance' counted for much at an exhibition, as did metonymic reduction and condensation. Equally spectacular was a game trophy assembled by J. Hubbard, an avid Manitoba hunter, designed to lure the leisured classes to Canada for hunting trips. One man whom Hubbard directed to Canada, although 'handicapped' by his wife, killed several moose and deer.[108]

Canadian food was featured at a restaurant and a market, but there were endless squabbles about apples and oysters of dubious nationality. The

The Canadian agricultural trophy at the Colonial and Indian Exhibition held in London in 1886 was a riotous assemblage of foodstuffs, timber, and implements, along with allegorical figures symbolizing those items. Pink-cheeked and hale, these figures also represented the promise of a healthier, happier life in the dominion.

oysters were a ticklish matter, originally from Prince Edward Island but transplanted into English waters. Exhibition official Sir Philip Cunliffe Owen remarked gravely: 'It seems to me that it must be a very difficult thing to say for certain that an Oyster originally came from Canada.'[109] The island of Anticosti, whose owner was trying to sell shares at the exhibition, had enormous potatoes, alternately denounced as being of wax and purchased in Reading.[110] Indian manufactures and perfume smuggled in daily were sold as Canadian, and lace came from Ireland, not Montreal. But it was not easy to say what exactly was colonial. Tupper tried to bully the other colonies with a policy that only goods manufactured in the colonies could be shown: colonies exporting cocoa beans could exhibit the beans, while Canada, if it processed them, could show the cocoa itself. Otherwise, he said, the exhibition would become more English than colonial. This policy would have increased Canada's advantage over the other, less industrial colonies had they not rejected the proposal.[111]

Manufactures, especially agricultural and musical, distinguished the dominion exhibit from those of the other colonies. The Leeds *Mercury* described the implements in motion as 'wonderful and deafening,'[112] but this criticism applied equally well to the pianos and organs that attendants played all day long. Both industries provide useful case studies. John Watson sent a reaper and a mower, but his display ceded the spotlight to Harris and Sons and Massey Company, both determined 'to surpass any exhibit they had ever made before. Paint gave way to more elegant enamel, simple burnishing to bronze and silver washes, and the heads of both concerns made arrangements to attend in person.'[113] There were no field trials, but the press provided an informal judging, praising the light but strong machines or decrying their gaudiness as showing 'wretched taste.'[114] Watson's reaper did not work smoothly and, his agent W.H. Pellow added, the impossibility of buying replacement parts discouraged sales. He need not have worried: A mower sold in Scotland worked for forty years before enquiries were made about repairs in 1927.[115] Massey sold binders to Lords Lansdowne and Lorne and, Tupper reported, sold 100 binders and 150 harvesters in a single day.[116] Hart Massey was 'much pleased' with the exhibition – 'the wisest and best expenditure ever made by our Government for the benefit of Canada.'[117] Harris was less pleased, having sold only a binder and a mower by December. Asked if the companies would contribute to a permanent exhibition, Massey gave a resounding yes, but Harris replied that business 'cannot be done in any country in the world except by the same methods as prevail in Canada. These are the establishment of agencies.'[118] Massey established agencies when the exhibition closed. Implement sales to Britain rose from $3056 in the year before the exhibition to $24,910 in the year after the exhibition, and doubled again the next year.

Dozens of Canadian organs and pianos were sent to the Colonial. W. Bell & Company of Guelph described their exhibition policy.[119] Bell had exhibited widely until 1879 but then dropped out, 'it being now so easy for a manufacturer, no matter how shoddy his goods, to get a prize & there being so many of them that customers in buying never take into consideration how many, or how few medals or prizes, a maker displays.' Bell hadn't exhibited at Antwerp, yet had outsold its rival there. This time it was 'absolutely necessary to show, in order to defend our reputation & prevent some small and unworthy manufacturer from getting the award & advertising it at home in a misleading way.' With exports above two thousand annually, Bell's exhibit showed sixteen instruments.[120] Not to be outdone, Dominion Organs sent a twenty-man band. The piano players often got into trouble for swearing, fighting, and flirting. One morning the agents for Lansdowne, Bell, Heinztman, and Uxbridge, all too drunk to work, harassed Dominion and Newcombe players with shouts of 'Bastard of a Jew' until a constable dispersed the louts.[121] Shipments of organs to Britain rose from $118,129 in the year before the exhibition to $166,893 in the year after the exhibition, and continued to rise in later years. Uxbridge and Dominion Organ sold their wares at a good price. Heintzman had attended himself and sold all his pianos, but he reported that the exhibition had eaten all the profits and that an agent was needed for regular business. His records show no subsequent overseas sales.[122] Other firms, perhaps accustomed to a protected market, set their prices too high.

One ambitious exhibit was a display of Canadian art. Lord Lansdowne and Lucius O'Brien, the president of the Royal Canadian Academy, recommended a utilitarian exhibit to advertise Canada's resources and beauties rather than a 'general Exhibition of Canadian Art.' A recent visit to London had convinced O'Brien that the local variety was not up to English standards.[123] He thought the art exhibit should consist of a colour map showing settled and vacant lands, with paintings to illustrate them. 'In short my idea would be to make the room containing the Art Exhibit a key to the whole Exhibition.' Visitors could take in Canada's vast resources and geography at a glance. O'Brien wangled a free trip to the Rockies on the strength of his plan, but his larger goal was to convince the government that art was useful and should be funded.[124] His utilitarian view did not prevail because Tupper, with an eye to the competition, included paintings by Canadians studying abroad, who did not all paint scenery.[125]

Lansdowne asked the Royal Academy for a review of the Canadian art exibit, explaining 'what is wanted is a few words of kindly criticism and if possible of encouragement.'[126] Professor J.E. Hodgson, RA, obliged. He praised the boundless 'pictorial resources' of Canada and its picturesque people, but complained that impressionism had infected its painters.[127] He hoped Canada would develop its own style of painting, one true to the countryside it portrayed. R.A.M Stevenson, another reviewer, agreed, but remarked a propos the Canadian gallery: 'You can fancy yourself in a good European gallery much more easily than you can if you are in the Fine Art section of any other colony. This is considerable praise.'[128] William Brymner was talented but frenchified; Paul Peel received faint praise; Robert Harris's *Meeting of the School Trustees* was too gloomy, although, to the delight of Prince Edward Island, the painting was a popular success. The friendly criticism may have been poor consolation for lagging sales. W.D. Friend complained that even though he had his pictures framed, none sold, whereas 100 drawings were sold from a Bond Street Gallery.[129]

The Colonial Exhibition led to some greater imperial initiatives. Exhibitors banded together to create a commercial exchange, and the British government established an emigration office. It also inspired a colonial conference. The Imperial Federation League held a congress in midsummer which was attended by some Canadians, among them teacher George Parkin and Oliver Mowat.[130] The league asked the Colonial Office to hold a conference for representatives of colonial governments. The result was the first Colonial Conference in 1887. Political federation was not on the agenda, but a variety of economic and military matters were. Canada was represented by Alexander Campbell, the Lieutenant Governor of Ontario, and Sandford Fleming, who called for a telegraphic cable between Canada and Australia.[131] Macdonald and Lansdowne were 'lukewarm' about the conference,[132] which produced little but paved the way for the meeting of premiers in 1896, and other colonial conferences in later years. It was not federation, but it was a useful new departure, and it secured some of the same ends as federation.

Another child of the Colinderies was the Imperial Institute. On the urging of colonial exhibitors, the Prince of Wales promised to make the collection permanent, with attractions like music to draw the public. Prince Albert declared that the institute should be a gift to Queen Victoria on her Silver Jubilee, which fell in 1887.[133] The queen really wanted yet another monument to Albert, but by this manoeuvre her son could be sure that the public would subscribe. Tupper liked the plan and made another flying visit to Canada to solicit contributions and convince Macdonald to vote £20,000 towards its establishment. Tupper promised that this would be strictly a colonial endeavour and, anyway, Canada could not refuse a gift to the queen. Macdonald feared that it would 'become a bazaar or sale rooms kept up at the general expense for the advantage of a few manufacturers.'[134] Still, Tupper had his heart set on it and a difficult election lay ahead, in which the loss of Quebec 'over the corpse of Riel' made the tempestuous Nova Scotia vote crucial. Macdonald agreed to the £20,000.[135]

Almost immediately, the prince announced that plans had changed. The Chamber of Commerce objected to the institute being purely colonial, claiming that London businessmen had lost enough money to that cause, for tales of vast sales by Canadian furriers Holt and Renfrew had aroused jealousy.[136] Theatres objected to the encroachment into popular amusement. Thus, the institute was to be 'imperial,' not 'colonial': not only would it include British wares and be run by the British but the popular attractions were cancelled. Macdonald and Tupper had been duped. Macdonald told Tupper not to expect a penny above the £20,000.[137] Tupper wrote accusing letters to the organizers, complaining that the new plan would not benefit Canada: 'Our products and industries would be completely overshadowed by those of the mother country, and that, divested of the attractions originally proposed, the Institute would degenerate into a mere museum.'[138] Lansdowne tried to raise subscriptions for the Institute, but 'these proposals fell perfectly flat. I do not think the idea is by any means congenial to the Canadian mind.'[139] Reports of corruption among the organizers further discredited the project. For years, the institute resembled the neglected Sydenham Palace.

Canadians did take up imperial federation in the mid-1880s. Branches of the Federation League, hitherto anaemic in Canada, sprouted everywhere. In Winnipeg, for example, L.A. Boulton suggested that 'while the Colonial Exhibition is going on and public attention is centred in the development of the Empire, it is a good time to discuss the question everywhere and a branch of the Imperial Federation League might with advantage be formed here for the purpose of discussing the question in all its bearings.'[140] The league gave full credit to the exhibition for awakening public opinion:

> It may be safely said that the Exhibition of 1886 has done more to educate the people, to bring them up to the great idea of Imperial unity, more to make them understand the importance of Empire, than any amount of speeches, essays, newspaper articles, and the like ... We may bring a noble array of

> convincing figures, and dish them up with all the flowers of rhetoric, all the persuasiveness of eloquence, but the ears are closed, the minds are uninspired, all has fallen upon fallow soil ... Pictures are among the greatest educators of the human family, because the eye epitomises all the other senses; and suggests them all. It is to the eye then that we Federationists have at length come as to our final court of appeal. From the jury of ocular demonstration the last verdict has been demanded, and the answer is conclusive and unanimous. At last, at last, the dull mass is animated.[141]

The Times, normally restrained, seconded the observation: 'It is no exaggeration to say that the Colonial and Indian Exhibition has done more to give body and substance to the federal conception than all the discussion in the world could have done ... The idea of Federation now holds the field to the utter exclusion and repudiation of everything that savours of disruption or separation.'[142] Federation did not occur, but the appeal to the public bolstered popular imperialism. The general public had never been quite so anti-imperialist as its government, and, with the extension of the suffrage, imperialist rhetoric plumped out the speeches of politicians.

The function of exhibitions was changing. In 1851 they acted by placing the building blocks of reason – objects of sense impression – directly before the spectator so as to create a public sphere grounded on a shared objectivity. As Ian Hacking has remarked, popular governments need objectivity, which 'is part of what makes the practice of communal rather than authoritarian decision possible.'[143] Objects were arranged in space so as to create rational associations. In 1886 objects were arranged in space so as to create emotional associations and to make them more vivid and compelling than other forms of persuasion could. The emphasis was on giving 'body and substance' to existing ideas – emigration, imperialism – to make them more pleasing. Eric Hobsbawm remarks, 'The intellectual study of politics and society was transformed by the recognition that whatever held human collectivities together it was not the rational calculation of their individual members.'[144] The Great Exhibition was intended to provide a rational empirical basis for proving Britain's economic superiority, but the Colonial Exhibition signals a shift towards an irrational Jingoism.[145] The London *Standard* noted the change, remarking of the exhibition: 'It was pleasant enough, but where did the instruction come in?'[146] Leading British men of science like T.H. Huxley demanded that the Imperial Institute be a place of technical instruction, for lack of which England had 'dropped astern in the race' of nations, but when the institute finally opened in 1893 it was a glorified museum that relied on popular concerts and social events to attract crowds, while its scientific work consisted in analyses and reports for business rather than education or research.[147]

In Canada during the early 1880s, there was no significant imperialist lobby. In 1887 one formed.[148] It was largely in response to the fight for commercial union waged by Goldwin Smith and a noisy minority, but the exhibition played a part. In May 1886 the conservative English imperialist, Edward Thring, wrote to George Parkin: 'I have been to the Colonial Exhibition. It is the most magnificent of the Exhibitions, and undoubtedly will be a great Historical fact, mark an epoch, and a new departure in the history of the English Empire. I go so far as to think that, even if that wicked weak man [Gladstone], who is now trying to ruin England succeeds in plunging us into temporary ruin, the force of the great fact and the awakened life of the English-speaking races and colonies will eventually prevail. The Exhibition is a wonderful sight.' Parkin decided to attend: 'I want to take some hand in this great federation idea – and such a chance as this to meet the men interested in it from all parts of the world may not be within my reach for a long time again ... Your words about the Exhibition would themselves draw me England-wards.' He returned to Canada full of zeal for imperialism and became a standard-bearer for the idea.[149]

Exhibitions never fully lived up to their promises. Exhibiting was a discursive activity: part speech, part simple economic activity. Canadians who painted and polished their goods to send to the show were engaging in economic activity that was also an advertisement for their work. The Canadian government tried to orchestrate these practices so as to create large-scale representations of 'Canada' as a nation. It hoped to effect large-scale practical results, to create and channel a flood of immigrants to the country, as well as to secure overseas markets and capital. Things did not work out that way. Businessmen did sometimes benefit, as in Massey's case. But the large-scale representations did not act causally on history quite the way they were supposed to. Most of their effects were themselves discursive. Representation spawned representation, discourse led to discourse. Those, like journalists, who trucked wholly in discourse declared exhibitions successful because so much verbiage was generated. Canada gained bulk and substance at the verbal level, if not in population. Others, like Macdonald, who had a more practical view of history, remained unimpressed.

In the long run, though, discourse did prove historically causal. Much of the argument over imperial federation focused on the most sturdy basis for it: sentiment or political and economic ties. Goldwin Smith asked, once the colonies had rejected defence and tariffs: 'What remains but improvement of postal communication and a Colonial Exhibition, neither of which surely calls for a political combination unprecedented in history.'[150] The exhibition, by that standard, *was* the empire. 'Empire,' like 'Canada,' was in large part a representation, a way of viewing lands and peoples from the perspective of a larger purpose. No one can see an empire. It requires a feat of imagination to read the concept of empire into objects like grains, cocoa beans, and spices. But a few years after the exhibition, in 1899 and again in 1914, thousands, then millions of people chose to die for the concept of empire. The Colonial Exhibition had its part in this achievement.

Paris, 1889 and 1900

The Colonial Exhibition was almost the last great Canadian exhibition in Europe. In 1889 yet another Universal Exposition was held in Paris, but the monarchical states of Europe refrained from joining in this celebration of the French Revolution. Britain stayed home, and so did Canada.[151] Louis-Antoine Dessaulles, an enthusiast of French expositions and republicanism alike, was there as an exhibitor and he railed against Canada's abstention, blaming it on the church: 'Le Canada a fait la sottise énorme de rester à l'écart, et il le doit surtout à ses Evêques ... le Canada avait beaucoup a gagner à venir à l'exposition, la plus grandiose qui ait jamais été tenue et sa petite bouderie épiscopale n'a fait de mal qu'à lui.'[152] J.A. Chapleau, ex-premier of Quebec and now the federal secretary of state, also strongly urged that Canada participate. Unable to secure consent, he instead commissioned a report by Jules Helbronner on the social sciences aspect of the exposition.[153]

Several Canadian exhibitors and visitors made the trip, most memorably the Massey Company, which won first prize for its Toronto Light Binder at Noisel, near Paris. A Harris machine was also entered, but dropped out at the last minute. The Toronto binder worked far better than any other in the field, performing its work flawlessly and quickly. Finishing hours before any other machine, the Massey driver ostentatiously cleaned up a plot of land that had defeated rivals. This showmanship and mechanical superiority was recognized with the highest award, an Objet d'Art, of which only five were distributed.[154] This success, coupled with numerous victories in Australia, helped crack open the international market for the company. Joseph Perrault, attending the exposition on behalf of the Montreal Chambre de Commerce, acted as a juror, facilitating Massey's victory. When he returned, the Chambre de Commerce held a banquet in honour of Perrault and the Massey Company together. The Montreal *Gazette* was amused by this spectacle of French Canadians honouring a Toronto firm: 'Though speaking different tongues, they were all the same good Canadians.'[155] Perrault and Hart Massey both urged the Canadian government to give official publicity to their Paris feat, insisting that only official recognition would enable the Massey Company to compete with American firms.[156] In 1889 the government did not act on this suggestion, but a few years later it did so.[157]

Canada rejected a number of invitations. A Vienna Agricultural Exposition the following year was ruled out because Austria prohibited immigration propaganda.[158] New Zealand and South Africa were also turned down about this time, as were a Horticultural Exhibition in Hamburg in 1897 and a Wine and Electricity Exhibition in Bordeaux in 1895. Canada abandoned a plan to exhibit in Antwerp in 1894 because of the large charge for space. As late as 1889 it was still disposing of items left over from the Antwerp Exhibition of 1885, broken stoves and Hiram Walker whisky among them.[159] Ottawa unbent enough to send Canadian photos to a photography exposition at the Imperial Institute in 1895.[160] This display of interest in the institute was rare for the Canadian government, which felt it had done more than enough with its initial grant and was further annoyed when the institute negotiated directly with the provincial governments.[161] When the institute opened in 1893, newspapers across Canada voiced dismay at the sorry condition of the Canadian department. The institute functioned largely as a social club until 1902, when the Board of Trade reorganized it to make it more useful.[162] The Canadian view was that it could not attract that part of 'the enquiring public as the Department desires to run up against, either in respect to trade or emigration,' but the presence of the other colonies necessitated some show from Canada.'[163] Canada did take a modest interest in smaller exhibitions in London and around the United Kingdom.

The next European Exposition Canada attended was held at Paris in 1900. With fifty million visitors and 83,000 exhibitors, this last exhibition of the century was also the largest. But, according to its historian, the optimism and self-confidence of past expositions was gone. Journalists remarked that 'the exposition in 1889 was happy, rational, and French; in 1900, it was inconclusive, confusing, and, where strikingly excellent, not French. In 1889, praise from the intellectuals was general; in 1900 praise of the idealistic principles on which the exposition was based was passé.'[164] Voice was given to this new bewilderment that exhibitions now aroused among the thoughtful by Henry Adams, who 'haunted' the 1900 World's Fair, 'aching to absorb knowledge, and helpless to find it.' He found only 'inert facts' and prostrated himself before the electrical dynamos, 'his historical neck broken by the sudden irruption of forces totally new.' The world was changing so greatly that history could no longer educate; the past could not shed light on the present. The exhibitions, long a proud assertion of human knowledge and human power over history, were becoming symbols of ignorance and impotence. One recent work of intellectual history argues that philosophers have still not gone beyond Adams's criticism of knowledge.[165]

Canadian enthusiasm for these gigantic extravaganzas was also ebbing, and this was the last exhibition where a general effort to represent the country was made. Some Canadians objected to participation at all, given France's continued persecution of the unfortunate Colonel Dreyfus.[166] As in 1878, Canada exhibited as a British colony, with no more status *vis-à-vis* the French authorities than any private exhibitor, and dominion officials found the experience frustrating. Any ragtag sovereign state like Liberia, 'habité par des nègres affranchis!' received better treatment than the great self-governing nation of Canada, complained Maurice de la Fargue, Paris correspondent for the Montreal *Patrie*.[167]

Space was short. The British accused the French of skimping foreign empires to make their own seem greater.[168] With only 60,500 square feet for all the colonies and India, the British offered 8000 square feet to British North America, the same to Australia, 5000 to British Africa, and 3500 to Crown colonies.[169] One colony after another refused to exhibit on these terms, and Donald Smith, now Lord Strathcona, high commissioner for Canada, threatened that 'if Canada was not enabled to make an exhibit worthy of her position and her resources, it might be a question for consideration whether we should exhibit at all.'[170] By not dropping out, however, Canada inherited space from those that did, and the dominion ended with nearly half the available space, or 28,121 square feet.

There were other problems. Strathcona was named to the Royal Committee organizing the British display only after much effort, and similar complaints were needed before the Canadian commissioner, J. Israël Tarte, Laurier's minister of public works, managed to have his name inserted in the official catalogue. There was a long, drawn-out dispute about installation. The minister of agriculture, Sidney Fisher, thought it not 'wise or profitable for Canada to show her products in manufactures alongside those of the Mother Country.'[171] However, the colonial pavilion was shoddy and defective. Canadians baulked at paying their share for it ($71,000), and grumbled about preferential treatment for India in the matter of cold storage.[172]

The final cost of the exhibition to the federal government was $315,000. Despite the space limitations, Canadian industry and resources were represented, with Quebec most fully displayed. There were pavilions full of buggies, bicycles, and binders at Vincennes, the agricultural wing of the exposition. Implement makers eagerly sent machines, but the commission found that, as usual, 'inducements had to be offered to desirable exhibitors, and in some cases much persuasion was found necessary.'[173] Hundreds of farmers and the Experimental Farms contributed enormous, eye-catching sheaves of grain to advertise the Northwest. The Canadian display in fruit and dairy products was spectacular, thanks to the use of cold storage. The visitors' book was filled with praise, such as that of the consul general of Turkey, Persia, Arabia, and Siam: 'The Canadian display of fruit at Paris Exhibition is the finest I have seen.' Considerable sales were made, including 1700 packages of apples to an Egyptian merchant, the first consignment to that land.[174] Still, the United States beat Canada for the fruit prize. The inevitable trees and cheese were rolled out for display and consumption, but columns of paper were an innovation. A display of gold, copper, coal, asbestos, and building stones emanated from the survey and from mining companies, which had been conspicuously absent in earlier shows. The handbook that the Canadian commissioners gave out to all and sundry piled statistic upon statistic to show Canada's growth since the first Great Exhibition. No longer were amateur essayists called upon; government statisticians provided the figures.[175]

There was even a display of Canadian painting and sculpture, though it was a sore point. The French did not accept colonial entries in the fine arts pavilion, and Britain refused to share its space there. Sidney Fisher offered space in the Canadian pavilion, which the Royal Canadian Academy refused haughtily,[176] but a number of Canadian artists took up the invitation, among them Marc-Aurèle de Foy Suzor-Côté, who had studied and painted in Paris since 1892 and had already exhibited at the Paris Salon.[177] Statues by Louis-Philippe Hébert were prominent, including one of Alexander Mackenzie and another of Queen Victoria, at whose feet a young woman representing Canada deposited a crown of maple leaves. Pride of place was given to a massive painting of the dispersal of the Acadians.

Hundreds of Canadians visited Paris that summer to see the exposition and other sights, study, do business, or attend the congresses that accompanied the exhibition.[178] French Canadians were particularly well received: journalist Robertine Barry claimed that 'le Canada en France est un mot presque magique. C'est le "Sésame ouvre-toi", qui nous assure partout une cordiel réception.'[179] She was one of five columnists covering the Paris

Exposition for Tarte's journal, *La Patrie*, including Tarte himself and Joseph Perrault, who had, as usual, found a place on the commission. At innumerable banquets, Tarte gave countless speeches recounting Quebec's survival as a French nation, some of which were not well received by the Tory press in England and Canada.[180] One London journal cried 'Treason' when Tarte said that the French tricolour flew over 3 million French in Quebec. Gustave Drolet leaped into the controversy, recalling all the British snubs of 1878, and he praised Tarte's efforts to assert Canadian autonomy: 'L'exposition universelle de Paris était un théatre tout indiqué, un terrain naturel pour essayer les premiers pas d'une jeune nation. M. Tarte tint les lisières. L'expérience réussit.'[181] Tarte took the position that Quebeckers were loyal subjects of Britain, but were determined to maintain their French nationality – as would any French emigrant. This was too much for the Toronto *Mail and Empire*, which declared that, to be one nation, Canada must be homogenous. Complaints by Tory MPs that the exposition cost Canada too much and promised too little benefit were understood by French Canadians to signal hostility to French immigration into Canada.

If European displays brought out the best in Canada at one level – the best fruits, grains, and manufactures – they also brought out the worst. They provoked reflection on Canada's relations with other countries, a subject on which Canadians did not agree. If collaboration on economic goals had been uppermost during the 1850s and 1860s, by 1900 a new concept of nationhood, as requiring one homogenous nationality, was aired in the press and in the House of Commons. Tensions were exacerbated by the Boer War, which English Canadians supported and French Canadians reviled, including the Canadian commissioners to Paris, Senator Raoul Dandurand and Israël Tarte, who told the French plainly that French Canada shared its dislike of this British imperialist war.[182]

The French exposition was popular in Quebec, but, in the long run, it seemed more trouble than it was worth, as did the costly business of exhibiting generally. After 1900 the Department of Agriculture moved over to smaller exhibitions that were more fully integrated. At Liege in 1905, a juror remarked: 'The method of Canada of displaying her products collectively and under one heading with attractive panels citing interesting facts and statistics about them, and of classifying exhibits by group and sections, is a welcome change to the prevailing methods of individual exhibits.'[183] The same exhibits were carted around from one show to the next. Canada could not attend all of them, so selection was based on what sort of immigrants Canada wanted. Ireland, for example, created a dilemma: Irish immigrants were welcome, but in Ireland, the announcement of an exhibition led to nationalist demonstrations against the out-migration. If Canada did not attend, speculated an agent in 1907, 'politically it could be turned against the Govt by politicians who could point to Belgium where we encouraged the French; at Milan the Dagos, etc. all being good people and the kind wanted on our land; but poor Ireland, with its down-trodden people was not even recognized by the Government of Canada in the way of participating at the great Dublin Exhibition – or in other words, the Irish not wanted.'[184] France continued to hold world expositions as well as its own imperial displays: Between 1877 and 1931, thirty-six exhibitions were held, most of them in France and Britain.[185] At the Wembley United Empire Exhibition of 1924–5, Canada hung back, because of the expense, until the Prince of Wales personally solicited its attendance, and then the Canadian government became outraged when Scandinavian and American timber was used for the buildings.[186] Canada's display at Wembley was perhaps most remarkable for the life-sized carving of the Prince of Wales in full Indian regalia, done entirely in butter. Christie Brown, Hiram Walker, and Massey-Harris sent exhibits to Wembley, as did the Group of Seven.[187]

The twentieth century has seen the development of a whole technology of propaganda that makes Victorian efforts seem amateurish. This technology has its own historians. As to the nineteenth-century overseas Canadian exhibits studied here, it is difficult to evaluate their historical significance. On the whole, they seem to have failed to achieve tangible results: increased trade or immigration. They were not much more successful judged simply as texts or representations. They projected Canadian unity, but infighting among provinces at home and among commissioners abroad consistently sabotaged this image. Only a small percentage of Canadians saw them, while the Europeans who paraded through these innumerable Canadian courts were largely untouched by them. Perhaps their impact is best measured by the effect they had on the Canadian bureaucracy, expanding its breadth as well as the depth of its penetration into businesses, homes, and schools. And yet, despite their best efforts, many Canadians could not be persuaded to exhibit, and agents lamented the apathy year after year. The public then blamed the Canadian government for its failure to elicit a good representation.

Whether or not these international displays contributed substantially to an emerging national identity, in the objective sense of bolstering its economy or in the subjective sense of supplying images associated with the idea of Canada, they did foster an expectation that national identity existed. Erving Goffman and Kenneth Burke have argued that identity is dramaturgical and is formed in large part by the act of role-playing.[188] The shifting content of this identity has some interest, but the structure of its authorship is more problematic. Canadians had to assume a role on this world stage before a public of millions, driven by economic necessity and the rules of the exhibition itself. Utilitarianism dictated the content of this self-advertisement: economically important artifacts, with just so much of the nation's art and culture as would complement them. Less immediately obvious was the question of authorship. Who would speak for Canada at these grand affairs, and how would they go about it? The government's intervention was necessitated by the organizers' decision to correspond only with foreign governments, but the ultimate extent of the Canadian government's involvement was far from inevitable. However reluctant Macdonald might be personally to orchestrate exhibitions, circumstances, public opinion, and the insistence of his colleagues all conspired to increase his government's involvement. The exhibitions were not mere trade fairs, but places for the assertion of national identity – and this function was deemed too important to leave to private individuals. Canadians wanted to send an authoritative message to the world, and this goal required that there be an authority to express it. Failure was more than a technical problem of ways and means, but amounted to an abdication of identity, which was unthinkable. Thus, by requiring the formulation of an identity for Canada, the exhibitions helped to create a demand among Canadians for such an identity. They also focused Canadians' attention on the government as its source. This expectation continues to plague Canadian governments to this day.

The exhibits looked spectacular, lots of people saw them (20,000 per day in Paris in 1900), and they doubtless made a general impression on Europeans. Canadians could not have escaped reading about them in the daily papers. They merit a place in any study of nineteenth-century culture, alongside theatre, and, like theatre, they bolstered in a general way the mythologies of the day: that Canada was a progressive country with considerable resources, that England had a far-flung empire, and that France and French Canada maintained a bond of friendship. One could make this sort of claim for theatre, without demanding evidence of its direct economic and political impact. Still, exhibitions were supposed to be more than a text – they turned the world into text in the hopes of scripting the actions of the characters. If, as Aristotle suggested, theatre has a cathartic effect, substituting emotion for action, the exhibition was supposed to have the opposite effect: to inspire action and turn mere spectators into performers, by convincing them to buy Canadian apples or to set off for

Manitoba. Canadian exhibitions in Europe by and large failed to play this role. Perhaps people are not easily persuaded by state-organized propaganda, at least not in the form of exhibitions. Their organizers may have overestimated human suggestibility. Alternatively, the reason may lie with the display itself. Perhaps the organizers were to blame for failing to represent Canada adequately, or perhaps Canada was itself to blame for a lack of alluring qualities, which even the boldest lies could not conceal. Yet Canada was not fairly tested. Perrault exclaimed, 'Quel spectacle de progrès dont notre exposition de Paris n'est nécessairement qu'un pâle reflet.'[189] At the European exhibitions, Canada was represented; at those in the United States, by contrast, participation was so much greater that Canada – its people and its artifacts – was actually present on the showgrounds.

8

Exhibitions in America after Confederation and the Commodification of Everything

The North American exhibitions were shrines to the majesty of the United States, but Canada sought a small slice of reflected glory. Canada's exhibits were tiny compared with those of the United States, but they showed great strength in some strategic categories in which Canadians had competed against the Americans for years. Again, government led the way, rousing 'apathetic' exhibitors as it advertised itself as much to Canadians as to the world. But it had less control over the total representation than at European exhibitions. For a start, these shows were too big. They embraced every aspect of life, many of them outside the state's purview, ranging from the Congress of Religions to the midway. The proximity of these American fairs encouraged greater public interest as well as independent action among Canadians. Exhibitors openly defied the government's attempts to control the display: one Saint John man who had been forbidden by the chief commissioner to exhibit did so anyway; a native troupe performed dances at Chicago which had been banned in Canada; and the province of Manitoba exhibited on its own, outside the fairgrounds. Canada was there, in all its myriad forms and contradictions, both as exhibitors and as spectators, as tens of thousands attended. The Philadelphia Exhibition and the Chicago World's Fair sprawl across history as unwieldy, untidy objects of knowledge, robust signifiers of the vitality of a popular culture that was both allied with and defiant before the state and the marketplace.

This chapter explores these exhibitions from a Canadian perspective, but the drama of the emerging Canadian identity takes second place to the drama of the cultural logic of exhibitions within a broad North American context. The exhibitions can be shown to have had some impact on economic, political, and cultural events inside Canada and these events will be chronicled. But these events were epiphenomenal to the history of the exhibition. The Canadian government's loss of control over the exhibits reflects changes in the process of exhibiting as it inexorably overturned and subverted the meanings that early organizers and theorists had tried to impose on the act of display. Indeed, the Canadian exhibits provide an excellent lens through which to observe this train of events. Canadians throughout the century consistently produced some of the most typical, if not the most spectacular exhibits – that is, the most thoroughly representative of the principles that underlay the exhibition. The geological collection in 1855 seemed to epitomize the best hopes for the conjunction of high science, popular science, and economic science. At the Colonial Exhibition of 1886, the Canadian exhibit exemplified the convergence of political and economic propaganda, the blurring of state and business within the common interest of advertisement. And, at the Philadelphia Exhibition of 1876, the outstanding collection of Canadian educational artifacts carried to its highest logical development the principle of the object lesson. In reflecting on the significance of the exhibitions, I will show that the Enlightenment attempt to create a more practical immediate form of knowledge led in fact to the reification and abstraction of knowledge. The final section of this chapter, devoted to the Chicago World's Fair, describes it as a place where popular culture and the process of commodification had taken control.

The structure of Part Two on international exhibitions mimics the section on agricultural exhibitions in Canada: Both begin with serious and earnest origins, move on to discuss the politics of knowledge and its relation to political forms, and finish with the triumph of popular culture. Although a strictly chronological account, beginning with the Crystal Palace in 1851 and finishing with Paris in 1900, would more clearly show this progression, Canadian geography prompts a division between the European and the American international exhibits. Because Canada's participation in exhibitions overseas was so thoroughly mediated by the government and because only wealthy Canadians could afford to attend, the gradual triumph by popular culture cannot be examined through the prism of Canadian participation at these European affairs. Part Three brings together political and cultural concerns by examining both the agricultural and the international exhibitions from the perspective of sexual and racial identities.

Philadelphia, 1876

Only an international exhibition was grand enough to mark the American Centennial and reconcile the strained country.[1] Circumstances were not auspicious, for a severe economic depression darkened the mid-1870s, and high tariffs caused foreign manufacturers to stay home. But it was still quite a show, attracting more than 30,000 exhibits and 10 million visitors (30,000 of them Canadian, by Canadian calculations). Dotted across Philadelphia's Fairmount Park were dozens of pavilions, restaurants, beer gardens, popcorn stands, sculptures, and memorials, including the torch of the unfinished Statue of Liberty. Across the street a raucous collection of beer halls, hucksters, and sideshows served as a midway. Power was supplied by a Corliss steam engine, to all accounts an awesome sight: 'Silent and irresistible, it affects the imagination as realizing the fabled powers of genii and afrit in arabian tales, and, like them, it is subject to subtle control.'[2] Americans also showed typewriters, Edison's quadruplex telegraph, and Bell's prototype telephone. This telephone impressed Lord Kelvin and also Hamilton entrepreneur H.C. Baker, who ordered the first commercial switchboard in Canada.[3] The fair confirmed the Yankee stereotype as expressed by *The Times*: 'The American invents as the Greek sculptured and the Italian painted: it is genius.'[4]

One innovation was in the system of judging. Exhibitions were supposed to formulate and publicize right conclusions about commodities, but they were failures in this regard. Typically, jurors merely grabbed prizes for their own country. Organizer N.M. Beckworth decided that the judges should be consumers rather than puffing producers, and medals should be for absolute, not comparative, merit. Every worthy object would receive a bronze medal.[5] He also required written verdicts, for medals spoke only a 'primitive and feeble' allegorical language. Knowledge of value had to be present as text, and the exposition left behind volumes of reports crammed with details of practical mechanics or farming through to political economy.[6] The Enlightenment desire persisted, to reveal the good and unmask the bad, to subject even the hyperreality of advertising to the discipline of truth, and to distil from the experience the knowledge needed to compose an encyclopedia of the material world.

Not only artifacts but also people were juxtaposed and judged, according to a 'space by race' design that ordered peoples according to hierarchies of 'progressiveness.' A Smithsonian exhibit portrayed Indians as a dying race which, one visitor remarked, obviously merited extinction.[7] To the west, Colonel Custer died trying to extinguish them. This design reveals the weakness of the knowledge-as-juxtaposition program. Simple juxtaposition conveys no meaning: There must be a subtext to explain it. The subtext in 1876 was that the white races were the most energetic and progressive. Even commodities were only inferior or superior under a description, which, as Beckwith said, had to be made verbally explicit. The absence of any social context for a piece of technology meant superiority was decided on technical grounds. Thus, labour-saving machinery, so necessary in the New World, could be called superior to the labour-intensive methods used in India, even though these methods might be more suited to its economic and social fabric.[8] In an exhibition, comparing civilizations seemed as easy as comparing ploughs, and the bias towards spectacular technology prejudged the conclusions. As Leo Marx has remarked, machines were, and are, so visibly impressive that 'to see a powerful efficient machine in the landscape is to know the superiority of the present to the past.'[9]

Cultural relativism clashed with the purpose of the world's fairs – to encourage trade among nations. Trading countries must agree on prices or values. Value rests on social determinants, implicating all of a nation's social and cultural life. Exhibitions facilitated commercial exchange, which meant breaking down cultural distinctiveness. Many likened them to Babel, but the more perceptive realized that, in the end, everyone spoke the same language. Thus, one Canadian, Henry Lacroix, concluded that the cunning of progress harmonized all the diverse languages into one chorus, so that 'the confusion of tongues in Babel times, is not possible now.' Two decades later a visitor to the Chicago World's Fair described it as a Babel, made orderly by 'the harmony imposed by commercialism and profit-making.'[10] The Tower of Babel had crumbled, but this generation began the work of building it again, believing that commerce was a universal tongue.

Canadians harboured much dislike of vulgar and violent American ways, but, at a world's fair, America set the standard. Its material culture was undeniably spectacular, and it aroused Canadian jealousy and emulation. Every exhibit was compared with its American equivalent, and the many obvious gaps between the two countries forced Canadians to insist all the more stridently that they were progressive. The usual boasts were that individual provinces (usually Ontario) won more laurels than individual states, and that Canadian progress was proceeding at a faster pace. Certainly, Canadians insisted, their display was more like the American than the 'exhibition of tawdry aboriginal finery, chiefly the spoils of war, and rude native contrivances indicative of barbarism such as would have been exhibited a hundred years ago, and of which striking illustrations were afforded in the case of Egypt, China, &c.'[11]

Alexander Mackenzie's Liberal government, in power from 1874 to 1878, took exhibitions seriously. It sent displays to Philadelphia, and later New South Wales and Paris, in a vain effort to buy the goodwill of manufacturers, which, Ben Forster has shown, had become a powerful political lobby by the 1870s.[12] These manufacturers demanded tariffs, which the Conservative Party offered to introduce. The Liberals instead promised to open foreign markets to Canadians, using exhibitions to do so, and the money voted ($100,000) would be spent directly on the voters, in paying for the transport and safeguarding of exhibits. At the time of the invitation, in October 1874, trade talks between Canada and the United States held out the prospect of reciprocity, in which event, a commissioner advised, a lot of space had better be reserved for Canadians.[13] The talks failed, and Canadian products faced prohibitive tariffs.[14] This wall created anti-American feeling and 'marked apathy and indifference,' which the government valiantly tried to overcome. The Ontario Board published an article in the *Monetary Times* warning that if Canada showed badly or not at all, 'it would be alleged in every paper in the United States, and very many in other parts of the world, that Canada did not exhibit because afraid of the contrast between her productions and those of the Republic.'[15] Foreign markets would be lost and immigration agents would never live it down. Federal commissioners also insisted that if Canada 'was seen at all it must be to take rank as an important American power.'[16] To bolster interest, Ottawa offered to distribute purely Canadian awards, alongside Beckwith's bronzes. Even if snubbed by the judges, Canadians would have something to put on their advertisements. The British helped with the judging and the scheme succeeded, although Dominion Organs refused the bronze Canadian medal because it did better in the American competition, and Ottawa Mills complained because it received only an American medal.[17] Still, the medals were a poor substitute for either the American market or a protectionist tariff.

The federal government was also embarrassed by its choice of commissioners. The minister of agriculture, Luc Letellier St Just, headed the commission. His first task was to secure for Canada the right to independent representation, which the British Commission conceded.[18] Joseph Perrault was secretary, but proved so unpopular that other senior commissioners were added: Senator E.G. Penny, editor of the Montreal *Herald*, and David McDougall of Berlin. These men disgraced themselves by spending enormous sums on entertainment – including great quantities of liquor and ice cream – which the Tories gladly publicized.[19] The *Mail* composed a little ditty:

> We can fancy Mr. PERRAULT exclaiming
> For I am of a numerous house
> And with many kinsmen gay;

Where long and largely we carouse,
And who shall say me nay.

Each week a birthday coming on
We drank, defying trouble;
And sometimes two would meet in one,
And then we drank it double.[20]

Perrault offended the Quebec board, which almost refused to work with him; he also offended the British, by complaining his office was too small; and he offended all of Ontario by placing its machines 'in the background' behind Quebec's furs and fabrics.[21] The Geological Survey's attendant, B.J. Harrington, called him 'the slimiest of snakes.'[22] The term 'commissioner' became one of abuse: At a meeting of the Montreal Board of Trade in August to discuss sending a display to an exhibition in Sydney, Australia, miller W.W. Ogilvie remarked (in Penny's presence): 'They should, however, select some other name for them than "Commissioners," as that name was not popular at present; it was good, practical business men they wanted.'[23]

By contrast, the provincial collection boards were voluntary associations of leading businessmen (a system the Tories later replaced with paid agents). Plans for a preliminary dominion exhibition in Montreal fell through because of the legal quarrels over the Crystal Palace. New Brunswick held a provincial exhibition, but then sent everything on, for fear of offending exhibitors.[24] Slowly the entries came in. Of the 150 carloads sent to Philadelphia before the opening (excluding livestock, fruit, and dairy, which came later), 100 came from Ontario, 30 from Quebec, and the other provinces made up the rest.

The Advisory Board in Ontario was a model of activity. It convened prominent industrialists such as Massey, Watson, and Wanzer, who sent out pamphlets arguing that a good display would win the American public to reciprocity, and they themselves exhibited in strength. Wanzer, for example, sent five sewing machines and won an award that bolstered an export business already worth $157,337, half of it with the United States. In general manufactures and machinery, Ontario sent more exhibits than the other provinces together: 290 in total, as compared with 133 from Quebec, 142 from the Maritimes, and 3 from British Columbia. This was a creditable collection, although the judge of machine tools remarked that despite a 'great and noble effort ... the true tool spirit is not yet so fully developed as it undoubtedly will be,' and a steam fire-engine failed the practical trials.[25] In agricultural implements, Ontario sent sixty-eight entries, compared with thirteen from New Brunswick and just three from Quebec. A British implements expert remarked: 'The strength of the Exhibition was in the American and Canadian sections; the latter making a most creditable display, astonishing the world, not only in this, but in other departments, by the steady solid progress that was visible.'[26]

The other provinces also distinguished themselves in various ways. The Nova Scotia Board concentrated on economic minerals, with coal and Bay of Fundy granite leading the honours. Halifax business was well represented, with prizes going for Starr skates and de Wolfe carriages, as were the city's women with an extensive display of fancy-work. The New Brunswick Board concentrated on woods after its preliminary exhibition, organized by the Mechanics and Manufacturers Association, failed to attract industrialists. However, the Albert Manufacturing Company not only exhibited plaster of paris but also supplied it for the pavilions.[27] British Columbia was served by an Advisory Board and by George Mercer Dawson of the Geological Survey, who collected minerals, grains, fish, and timber, including a Douglas fir 8 feet in diameter, and a gilded pyramid representing the $37 million worth of gold mined in the province since 1858. The Geological Survey itself under Alfred Selwyn won prizes, but none so spectacular as Logan's, and a hostile critic declared this and the Paris display failures.[28] Manitoba crops were devoured by hoppers in 1875, and the Manitoba Board received only two applications to exhibit. Nonetheless, it assembled several carloads of grains, roots, minerals, furs, and a collection of seeds, wild hops, and tea that took a prize.[29] Squabbles undermined Quebec's exhibit: Impatient Ottawa commissioners named a Quebec Board in August 1875, whereupon the Quebec commissioner of agriculture, outraged at this interference, refused to foot its bills. The Council of Agriculture was unable to wheedle grains from agricultural societies and bought them from a Montreal seedsman.[30] Still, Bulmer, profiting from his Vienna jaunt, won a prize for his bricks, and the Canadian Meat and Produce Company of Sherbrooke won for canned ox-tongue, game pie, chicken, duck, veal, venison, beef, pork, and sausages.

A log house advertising Canadian lumber was another successful exhibit. Canadian lumbermen decided to collaborate so as to beat the Yankees and prevent infighting. One Ottawa merchant remarked, 'It would never do for Quebec and Ontario to compete in sawn lumber when the Ottawa interests are so closely allied together.' Each province provided sections of the cabin, although Quebec had to buy timber from Ontario to obtain beams large enough.[31] Overhead, the highest flag in the park fluttered from an enormous New Brunswick spruce. According to the judges (among them exhibitor James Skead), 'Canada had the largest, most varied, and most attractive collection of timber and lumber in the Exhibition.'[32] Several of these lumbermen received orders, including Skead himself, but one year later, in 1877, the square timber trade began to decline for want of fit timber.[33]

Canadian cheese was becoming an important export, worth $674,486 by 1870 and $3,886,226 by 1875, but Canadian dairymen determined to improve these figures.[34] Ignoring the summer heat that discouraged American dairymen, the Canadians sent the most cheeses (29 tons compared with the 26 tons of American cheese and 500 lb from other countries), and the best. The average American cheese score was 76.82; the Canadian, 87.36. The highest score awarded any American cheese was 96 points, whereas J. Ballantyne, MPP, obtained a perfect 100 for a cheddar cheese with a 'rich buttery flavour.'[35] All told, Canada won forty-nine cheese prizes, four more than the Americans.

The Ontario Fruit Growers' Association had beaten American exhibitors at a fruit exhibition in Boston in 1873 and determined to beat them again. In 1876, despite drought, it sent 1480 plates of apples (of 120 varieties), as well as many pears, plums, grapes, peaches, crabapples, and berries. California apples were larger, but the Canadian product could not be beaten for taste.[36] There was, growers claimed, no better way to advertise Canada's fine climate than these rows of sweet, red apples and succulent peaches. The Nova Scotia Fruit Growers' Association and the Montreal Agricultural and Horticultural Society also sent prize-winning fruits.

Canada's greatest victory in 1876 was in livestock. Canadians won the sweepstake prizes for the best long-woolled sheep and the best boar on the grounds, and they would have won the sweepstake for best bull if the judges hadn't hastily cancelled the contest. Americans grumbled that 'Canada had the largest bulls, but they seemed to be coarser than the American, and were in some cases so large, fat and unwieldy as to be scarcely able to waddle around.'[37] James and William Russell, of Richmond Hill, showed the best Shorthorn cow on the grounds, 'Isabella,' who was 'hailed as the greatest cow of her breed anywhere.'[38] In horses, Canada accounted for one-third of the entries and 'the palm was worthily borne off by the Canadian exhibitors.' Canadians had the best coach horses, agricultural horses, and, as for draught horses, the Canadian teams 'seem as if they would carry off half a small town if they were only made fast to it.'[39] Horses were an important export to the United States and Great Britain. Canada dominated the long-woolled sheep prizes, as it had done at American agricultural fairs for many years; the judges remarked that the term 'Canada wool' had become synonymous with long wools. About this time, however, tastes changed and the demand for coarse wool dropped off, devastating the Canadian woollen industry.[40] Coming in the middle of a commercial depression, this victory helped stabilize Canadian livestock exports to American and foreign markets: Horse exports tripled between 1875 and 1878, sheep exports rose substantially, and cattle exports rose by nearly one-third, although sales to the United States declined.[41] Decades of agricultural exhibitions paid off.

Canada also did well in the sporting competitions. At the regatta, sculler Ned Hanlan won his first international victory.[42] The New Brunswick rowers who had won at Paris in 1867 were soundly defeated by a Halifax crew, which looked set to win the championship, with only England left to beat. In Halifax, crowds gathered around the telegraph office, betting was heavy, and 'anxiety was manifest on every face.' When the announcement came, it was met with incredulity and then fierce anger: The English crew had rowed down on the Halifax crew, which turned to miss them, but collided. Halifax was first across the line, but the English crew cried foul and won the prize.[43] The Halifax crew was given a hero's welcome home just the same. Tar-barrels burned in the harbour, while 'THAT REFEREE was burned, hung, shot, etc. in effigy all over the city.' Anger turned quickly to grief, for just a few weeks later Obed Smith, the bow oar, drowned.

In fine arts, some 156 Canadian works took their humble place among the 2971 paintings and 675 sculptures on display. There were a lot of nudes, including some shocking French sculptures: 'No such array of feminine pulchritude in marble and paint had ever been seen in Victorian America before.' Abbé L. Provancher, recounting his adventures in *Le naturaliste canadien*, was appalled. Reality, he insisted, should be veiled.[44] He was, however, captivated by the typewriter. Canadians sent landscapes and flowers, among them Daniel Fowler of Amherst Island who won a bronze for his *Hollyhocks*. American judges still deferred to European art, awarding England, France, and Germany prizes on more than 10 per cent of their entries, themselves less than 5 per cent, and Canada less than 1 per cent. With gentle irony, Fowler credited the award with ending his self-doubts: 'But now all was settled for good and all; the most impartial and most competent tribunal that could be established had decreed that I possessed 'Artistic Excellence' enough to merit world-wide commendation, and my position had been made for me.'[45]

No sales of art assuaged dashed hopes of glory. The Americans were too bowled over by the nudes to covet the Canadian works. J.S. Ingram probably summarized the prevailing view of Canadian paintings: 'Those illustrative of Canadian scenery and life were of course the most valuable contributions.'[46] Frederick Verner showed landscapes and visited the exhibition himself; thereafter he abandoned painting scenes of the West, which he had never actually seen.[47] Although the Montreal Art Association submitted Fowler's piece (it paid $118 for it when he won a prize in 1870), Ontario led in art, as the Ontario Society of Artists combined with the board to organize a collective display. The artists were almost uniformly English: French Canadians and natives were more painted than painting, as in the 'spirited' work by G. Scott, *Habitants attacked by Wolves.* William Notman upheld Quebec's honour with a collection of photographs and the lucrative position of official exposition photographer.

One success story of 1876 was the Ontario educational exhibit. Its organizers pitted themselves against American state boards and won such recognition that, for years, Ontario boasted of 'the best education system in the world.'[48] The exhibit cost close to $10,000 and included 2000 articles, spread over an expanse of 110 by 25 feet. At the centre were photographs of the public buildings of Ontario, maps, the English Great Seals, and two suits of armour. Atop the glass cases were 'ethnological' busts of scientists, writers, and monarchs, from Shakespeare to Herschel to Prince Albert. Inside were 'Maps, Charts, Globes, Diagrams, Models, Object Lessons, and a most extensive variety of School Apparatus.' Samuel May, the attendant, was often asked if these articles were used in the schools and he answered: 'The Object Lessons, Maps, Charts, and Globes are in pretty general use; but many of the more expensive kinds of Apparatus, or more difficult instruments, are rarely used.'[49] The collection was vast and far more impressive than the usual run of educational exhibits consisting of students' workbooks.

Robert Stamp has suggested that this success proved the downfall of Ontario education. Whereas the Americans began to reform their educational system in 1876, Ontario was mired in a 'conservatism resulting from international praise' that retarded reform. Moreover, the display was fraudulent because, despite May's claims, Ontario schools were poorly equipped. One visitor was astonished at such a 'magnificent' exhibit in Philadelphia and in Paris, but in the schools themselves in 1880: 'In forty-four no globe is there; in thirty-nine no map of Canada itself exhibits; in alone two are there charts like the charts of Paris and Philadelphia, of the body, and of plants, and the teachers say they find no time to employ. For object-lessons, in all the schools did I discover not one. Philosophical apparatus, most of the teachers do explain they have never heard with ... As now appears by me, *your exhibit was nothing but a show extracted from a museum.*'[50]

The educational display showed potential rather than actual conditions.

The Ontario Education Exhibit at the Philadelphia Exhibition in 1876 was a world's fair in miniature. Models of schools and written schoolwork were interspersed with educational artifacts representing natural history, history, geography, and science, and illustrating the steady progress of humanity.

This was not good enough: There was supposed to be some relationship between what one saw on display and what was really 'out there.' This claim was difficult to test, and was mostly assumed by judges and spectators. Ontario profited from this assumption even as it broke faith. Embarrassments like this one eroded the claim that exhibitions generated reliable knowledge. If exhibits were untrue, they were hardly educational and improving. The erosion of this truth claim was one reason why exhibitions became more frivolous over time.

In 1876, however, the Ontario exhibit eclipsed its rivals, to the embarrassment of Quebec. George Brown had supported Confederation so that Ontario could forge ahead by shaking off Quebec's influence in matters like education. His predictions seemed fulfilled in this sphere, where state and church had worked out an uneasy alliance. Most distressing, the Quebec Advisory Board had made a special effort in education. Siméon LeSage, its secretary, issued circulars asking schools and colleges for models and photographs of buildings and statistics, Laval to send its scientific apparatus, and the Urselines to loan their magnificent paintings.[51] LeSage confided his hopes to a friend: 'Nous ne pouvons pas espérer d'éclipser les Américains dans l'industrie, mais je les défi bien de figurer avantageusement à côté de notre province avec leurs établissemens d'éducation & de charité ... Le sorte des armes a voulu que nous perdissions la suprematie materielle de l'Amérique du Nord mais nous pouvons nous estimer encore assez heureuse & assez honoré si nous établissons que le sceptre de l'intelligence & de la charité chrétienne n'est pas tombé de nos mains.' In February he cheered: 'Vous verrez que la petite Province de Québec ne fera pas pitié à Philadelphia. Hurrah for Quebec!'[52] The mid-1870s was a time of optimism in Quebec, when nationalists like Curé Labelle predicted that French Canadians would overrun the North American continent some day.[53] But the public-works style exhibit of school buildings sent by Quebec impressed nobody. When Perrault realized this, he wrote to the superintendent of education, Gédéon Ouimet, soliciting documents, but it was too late.[54] Ouimet visited the exhibition and departed a convert. 'If it is possible to retrieve our lost credit it should be done: for, hereafter, to remain behind would be to acknowledge our inferiority. Educational expositions have assumed so much importance in our day that a country which desires to keep a good name must take part in them ... It rests upon us to prove that our nationality and our religion do not prevent us, as certain fanatics are constantly proclaiming, from being the friends of progress and learning, and that we are able to march forward at an equal pace with the English provinces, whose success at Philadelphia has done so much to honour the Canadian name.'[55]

Battle was joined. Ouimet sent a collection to Paris in 1878 where, as he said, 'we endeavoured to show to that France, which knows us so little and which we love so much, to that Paris, the Capital of the civilized world, that education was widely spread through the Province of Quebec, and that we were not exactly savage Iroquois.' The principal of the Montreal Commercial Academy, Urgel-Eugene Archambault, went along to neutralize Samuel May, the Ontario educationalist, who, Archambault feared, would attempt to 'faire ressortir le système d'Ontario au détriment de celui de la Province de Québec.'[56] Both Ontario and Quebec won medals.

The rivalry continued to the end of the century, though Ouimet's enthusiasm was dampened after 1880, when the three branches of Quebec education – the religious orders, the lay Catholic, and the lay Protestant boards of education – competed at a provincial exhibition.[57] Ouimet forgot to invite the Protestants, who were '*defeated* foot, horse & artillery,' while the Catholic orders were bested but refused to admit it, especially after a prize awarded them in fine arts was retracted on technical grounds.[58] All Montreal buzzed with the affair when Abbé Martineau preached a mass at Notre Dame attributing victory to the teaching orders because 'Dieu est le maître des sciences ... il est clair que les congrégations religieuses sont plus près de Dieu que les maîtres laiques: donc leur succès fera toujours le désespoir de leurs persécuteurs.' André Labarrère-Paulé remarks: 'Cette affirmation est grave. Poussée jusqu'à ces dernières conséquences, elle peut éliminer les laiques de l'enseignement.'[59] As the exhibition revealed, Ouimet's lay teaching system was embattled. In 1893, at the Chicago fair, Catholics won prizes and praise, which *ultramontanes* used as a defence of Catholic schools in Quebec and Manitoba.[60]

Educational exhibits became an important measure of national standing. In 1876 Ouimet concluded that 'il est certain que, dorénavent, la partie scolaire des expositions internationales sera le critérium, la pierre de touche de l'état social et de l'activité industrielle de chaque peuple, et, par conséquent, prendre part à ces expositions devint un devoir national pour tous les pays qui peuvent le faire dignement.' The New South Wales agent in Philadelphia remarked: 'I am sure there is nothing which so fully shows the extraordinary progress of Canada as the educational display of Ontario.'[61] In 1850 Canada had tried to impress the world with its material prosperity, but now 'civilization' was measured by state support of science and education. This emphasis signals a change in the exhibitions. The first, in 1851, was devoted to the 'Works of Industry of All Nations,' with no provision for education or science, as distinct from industry; by the 1870s this omission seemed a 'deplorable lacuna.'[62] Exhibitions were born of the urge to bring knowledge and practice into closer connection. Now organizers began to separate them again, claiming that they had discovered education, when in fact they had discovered education as a distinct or reified category. In 1851 the whole exhibition was educational, but, by the 1870s, it had become a bazaar (as contemporaries complained), with one section distinctly labelled educational. The Ontario exhibit typified this change.

The Ontario educational display fulfilled the highest aspirations of visual education. Here, the 'intelligent enquirer' could take in at a glance the outlines of the 'entire structure of our Educational System, and could understand its practical working.' It taught effortlessly, and it taught about Canada, as well as about education, for its 'methodical classification' provided a 'tabular synopsis of the resources of Canada in all branches of Natural, or Industrial, production. One would say it was a book that one opens, where the matters are arranged chapter by chapter, following a logical chain that goes from the simple to the composite in such a way that, having reached the end, the reader has his memory stored and his mind edified without effort as without confusion.'[63]

This exhibit was not just an advertisement for Canada, but also a restatement of the act of exhibiting. It was as encyclopedic as the fair itself. It re-enacted in miniature the international exposition, itself a microcosm or 'world in miniature.'[64] Both the exhibition and the educational exhibit gathered everything exhibitable into a systematic display, transforming 'things' into 'objects of knowledge.' In 1876 the educational exhibit brought together collections of miscellaneous goods – minerals, flora, fauna, globes, busts, mechanical productions – but now classified them first as educational objects and only second according to their nature or properties. They carried the logic of the exhibition one step further. And the more 'complete' the collection – the more the world was reconstituted under the heading 'education' – the more successful was the exhibit. The Ontario exhibit, the most 'complete' on the grounds, most nearly fulfilled the exhibition impulse towards encyclopedism, to create a total representation or discourse of the world.

The exhibit embodied the Enlightenment conception of knowledge as rational associations occurring in space, but now it added the nineteenth-century dimension of evolutionary time. The display gave an overview of mental progress from childhood to adulthood, from the first playful interaction with animate objects to the highest literary and philosophical works. This sequence recapitulated the intellectual advancement of the species, from savagery to advanced civilization, conveyed in the anthropological exhibits. It was during the 1870s that anthropologist A.H.L.F. Pitt Rivers began to arrange artifacts, such as primitive and modern weaponry, to show progress towards complexity 'in the form of small steps linked to one another in an irreversible and unbridgeable sequence.'[65] Intellectual development imitated biological development, as described by Ernst Heinrich Haeckel in 1867 with the phrase 'ontogeny recapitulates phylogeny': The growth of the individual parallelled the growth of the species.

Exhibitions were a place for people to ponder material goods, commodities, art, and crude natural resources. They turned the world into discourse, making every item within their walls intelligible and educational in terms of philosophical arrangement and classification. The exhibition would then become a 'generalizable mode of functioning,' teaching visitors how to interpret the 'real' world, But, in the end, rather than the exhibition remaking the world in its own image, the world remade the exhibition. Outside the education court, it was business as usual, a marketplace, not a classroom. The exhibition was becoming a trade and fun fair, as businessmen and visitors won out over the bureaucrats and educators.

To some extent, this departure signalled a failure in the project of generating practical knowledge. Instead of combining with practice, knowledge tended to homogeneity, so that only artifacts constituted as object-lessons could be adequately defined as objects of knowledge. These objects included many commonplace things as well as the expanding intermediary range of tools that existed only to educate: globes, maps, and busts of great men and pretty princesses. Dr May remarked of the Ontario display: 'The whole exhibit was so arranged as to show that the development of the intellectual and physical faculties, and the acquisition of knowledge and science is more easily acquired by the use of models, maps, apparatus, &c., than by any other method of teaching.'[66] The intervention of all this apparatus, these layers of representations, contrasts with the prior attempt to close the gap between student and thing studied. Over a century earlier, Rousseau had expressed impatience with all such paraphernalia: 'You want to teach geography to this child, and you go and get globes, cosmic spheres, and maps for him. So many devices! Why all these representations? Why do you not begin by showing him the object itself, so that he will at least know what you are talking to him about?'[67] Exhibitions stood for the Enlightenment ideal of a direct, practical knowledge, but they became less referential and more self-referential. Once it was decided that the representation of the world, at an exhibition, was more instructive than the world itself, the 'descent into discourse' was inevitable.[68] Education was not a transaction between the individual and his or her environment, but required the intervention of intellectuals who would manipulate the world and then present it to the public in suitably predigested form.

Exhibitions began as a vanguard for popular education, but, by 1876, education had acquired a momentum of its own. Its own more direct and effective techniques of education, techniques that relied on human intervention and management, made exhibitions seem frivolous. Exhibitions could only amplify and reflect popular knowledge, reinforcing what already existed in people's minds. Rather than subjecting public opinion to elite direction, exhibitions amplified all those aspects of opinion – its frivolity, its emotiveness – they were meant to mitigate. Rather than raising the crowd up to the level of the philosopher, exhibitions lowered the philosopher to the level of the crowd.

Intervening Exhibitions

When the Centennial Exposition finally closed in Philadelphia, no one knew what to do with the Canadian exhibits. Most were faded or damaged, the market was glutted, and the tariffs prohibitive. It seemed a godsend when Australia solicited them for an exhibition in New South Wales in April 1877. To send the whole collection would cost only $25,000.[69] Manufacturers, loath to trundle their shop-worn goods home again, took to the idea, and they sent delegations that convinced Mackenzie and Sir Richard Cartwright to pay the freight and name a commissioner, the aging and impecunious John Young. He was a failed grain merchant who, John Lowe confided to Morris, 'has rendered some public service in his day; but not as much as he thinks.'[70] Roderick Cameron, a Canadian shipper based in New York, ferried the goods to Australia. A billiard table proved an expensive mistake, but reapers, ploughs, sewing machines, and a host of smaller wares were sold. Filling orders regularly was another matter, and only regular steamer service would make freight cheap. Cameron claimed 'considerable imports' resulted, but an agent with the Australian Steam Navigation Company told Lowe that 'from the miserable returns our Canadian manufactures received from the last Exhibition here I fear very much, the inducements to risk further will be small indeed.'[71] The government did not consider it successful enough to be worth repeating, for it merely named commissioners to the next Sydney Exhibition, in 1879, while a few Canadians exhibited privately. An exhibition in Melbourne followed soon after, and again Canada sent only Cameron, in an attempt to bribe him into establishing a branch of his shipping company in Montreal.[72] A decade later, there was still no steam line, and Canada's sales to Australia were still 'dismally slight' and declining.[73]

The Australian exhibition was a sop to public opinion, briefly infatuated with the idea of a large Antipodean trade. Shut out of the United States, merchants cast about for alternative markets to save the struggling manufacturers and farmers. Goldwin Smith heaped scorn on these enthusiasts and urged Canadians to pursue continental free trade, but protectionists feared that such a move would cost Canada its political autonomy.[74] At the exhibitions of the 1870s, Canada learned the lesson that the United States was flourishing under protection, and that no other foreign markets provided an adequate demand for its manufactures. Doubtless this lesson bolstered support for Macdonald's triumphant return to power on a protectionist platform in 1878.

The Canadian government turned down an invitation to New Orleans in 1884, although some Canadians sent manufactures, fruit, and livestock.[75] Many Canadians visited this exhibition, as well as another in Boston, where a score of Canadians from Quebec and Ontario sent boots, sleighs, wire, wagons, a fire extinguisher, and fine art.[76] Some visited an exhibition in Cincinnati in 1888 which Ottawa declined to attend. Ontario sent a hastily assembled mineral exhibit, weighing 10 tons, in the care of David Boyle.[77]

An alternative to the impenetrable American market was the West Indies. Ottawa did send an exhibit to the Jamaica Exhibition of 1891 in hopes of edging out American goods there.[78] Though the three-month fair attracted over 300,000 visitors, it lost money and inaugurated a depression. Canadians made the largest display of any foreign nation, and their 257 exhibits won 205 prizes (as compared with Britain's 140 and a paltry 79 to the United States). The Department of Agriculture published trade statistics to encourage manufacturers and named agents to rustle up exhibits, especially in cheese, flour, beer, fish, woollens, furniture, and machines, but all was in vain. By 1895 Canadian trade had dropped from 1.5 per cent to 1 per cent of Jamaica's exports.

Chicago, 1893

The World's Columbian Exposition was an enormous, sprawling event, not quite the largest fair ever held (Paris in 1889 attracted a few more than the 27.5 million who came to Chicago in 1893), but grand enough to set the style for years to come.[79] Despite a financial crisis and years of depression, Americans were at their most confident as they reflected on the continent's growth since Columbus's voyage 401 years earlier. At the opening, orators 'surrounded by the stupendous results of American enterprise and activity' welcomed visiting nations 'to illustrate with us the growth and progress of human endeavour in the direction of a higher civilization.' Observers declared it the greatest event since Bunker Hill: 'It has been in concrete; Greek art and philosophy, Roman heroism; and the Sermon on the Mount.'[80]

In 1893, Chicago, 'that fire-tried, uncultured, greasy, pork-making, matter-of-fact, windy city, Chicago,'[81] was the hub of the livestock trade, the nation's abattoir. Brash millionaires and bordellos flourished. For the World's Fair, a White City of neo-classical structures was erected along the lake front, done in plaster resembling marble and entrancing visitors as the 'fairest dream of civilization.' Canadian tourists and columnists gushed extravagantly: 'The grossness must die out of the heart of any man or woman as they enter the precincts of this White City – it is so nearly a realization of our ideal city, that Sacred City wherein we all hope some day to dwell.'[82] More than any other exposition, perhaps any cultural event, this one best accommodated all the extremes of late-nineteenth-century life: It was sublime, it was ridiculous, and it was profitable.

The exposition began with a battle over where to hold it, Chicago or New York, with most Canadians voting for Chicago (Quebec City Hall was an exception).[83] The corporation set up to manage the exposition sent invitations around the world, attracting 70,000 exhibitors, two-thirds of them from outside the United States.[84] As the States' nearest neighbour, Canada could not watch the preparations indifferently. Newspapers urged a good show from the beginning, and the Canadian government early took up the cause and pledged support for it, with the backing of the Liberal Party.[85] A sum of $125,000 was voted to canvass for exhibits, transport them, and pay for caretaking. The provinces took responsibility for resources and agriculture, while Ottawa tapped the manufacturers. By May 1893 the vote was exhausted and expenses were running at $250 per day. In a final estimate, the dominion government spent $250,000, the provinces $200,000, and exhibitors another $250,000, not including the value of the exhibits, invoiced at $763,510.[86]

The Conservative government had a vital interest in the fair. Led by Sir John A. Macdonald raising the loyalty cry, the Tories had won only 48.9 per cent of the popular vote in 1891, and Macdonald's subsequent death had further weakened the party. Robert Craig Brown argues that, while the 1880s had seen 'sporadic' efforts to expand trade abroad, in 1891 the work began in earnest.[87] The McKinley tariff of 1890, which aimed at keeping Canadian agricultural products out of the United States, had injured Canada's farmers, who blamed the National Policy for taxing their raw materials and blocking them from foreign markets. Organized into granges to express their dissatisfaction, farmers still made up the majority of Canadian voters.[88] The exposition provided an opportunity to advertise Canada's resources, agriculture, and manufactures to foreign markets, if not to Americans. But perhaps even more important than selling Canada abroad was selling Canada at home, proving to Canadians who would visit the World's Fair, or read about it in their local papers, that the Canadian economy was flourishing. This meant proving that the National Policy had indeed created a healthy manufacturing sector and a local market of workers to buy wheat, meat, and cheese from Canadian farmers.

But this promotion was only half the task. Farmers knew that, for example, the binder twine they used was made in Canada, and perhaps they knew that no twine was made in Canada before the National Policy imposed a 25 per cent tariff. Their complaint was that cordage and twine makers had formed a combine, headed by David Morrice's Consumer Cordage.[89] Liberals were demanding that binder twine be put on the free list, for it cost farmers $400,000 in extra taxes. Tory MP Nicholas Flood Davin, torn between party loyalty and the demands of his Saskatchewan constituents, admitted that the tariff must be reduced.[90] The government needed to prove that Canadian manufactures could compete with American products in quality and price, and the World's Fair offered a way to accomplish this goal – the only way short of actual market competition. The fair, once a harbinger of freer trade, was now to serve as a surrogate for it.

This scheme might have improved Tory fortunes, but Canadian businessmen had no desire to compete with the Americans either for real or for pretend. With no prospect of sales, responses trickled in. Ottawa named agents to persuade, cajole, or coerce if necessary. The Quebec agent reported: 'I find no enthousiasm on the part of intending exhibitors,' and there was 'apathy' in the Maritimes.[91] Throughout the eastern provinces, the 'best' manufacturers refused to exhibit;[92] one of the worst was undoubtedly F.H. Coombs, of Saint John, a dealer in miscellaneous items he styled as 'inventions.' When Commissioner William Saunders tried to reject the goods, telling him that the World's Fair was not a saleshop, Coombs responded sharply that the fair was indeed a saleshop, the Canadian exhibit itself 'little else than an advertisement,' and that he pressed his claim to space 'as a Right not as a favor.'[93]

Meanwhile, Saunders's agents were begging Rosamond Woollens to exhibit, wheedling a display from Waterous Engine Works, and pleading with Brown Brothers to send books.[94] By January 1893 the cajoling intensified. Saunders and his successor J.S. Larke (Saunders resigned because of ill health) warned a Nova Scotia engine firm that abstention would indicate that 'we know our goods are inferior to those which will be shown by other countries: and this will tend very greatly to strengthen the opposition to the protective policy of the Canadian Government in the minds of those who are opposed to it through prejudice or ignorance.'[95] A Hamilton wire company was urged to exhibit even at personal sacrifice,

> in the interest of the National Policy generally. The Exhibition will go on whatever individual Canadians may do. There will be shown there a display of Canadian manufacture. If we have not the best goods that are made in Canada, the manufactures of Canada will be judged by those that are sent. One purpose we have in view is the instruction of our own Canadian people who will go there by thousands. The attacks made upon the National Policy require that our own people should be educated as to what is being done in the Country. If the manufacturers do not take some steps by which can be shown there the highest quality that is made in Canada that education will be erroneous, and all the agitation which is sure to take place during the coming year and next session of Parliament, will have an evil effect upon the Policy as a whole.[96]

Set in the Mid-west, the fair would speak directly to western farmers, a Truro buggy maker was told.[97] David Morrice, sulking because he had lost twine contracts to the prison system, refused to exhibit until Prime Minister Thompson had a private word with him. Even then, Morrice insisted that his goods be displayed only, and not entered for competition. Larke disobeyed him, entered his goods for competition, and Morrice won a prize.[98]

The Canadian government encouraged exhibiting in the interests of Canadian business, but it was a partisan effort, with Tory agents soliciting Tory manufacturers. This tactic alone may have discouraged many from entering their goods. When the Kennel Club chose a Grit to supervise the dog exhibit, Conservatives were in an uproar.[99] The government was converted from its earlier indifference to exhibitions out of concern for its own survival. In 1876 the Mackenzie government exhibited to sell Canadian goods abroad; in 1886 the Macdonald government exhibited to sell Canada abroad as a place to live and invest in; but in 1893 the Tory government exhibited to sell Canada's government to Canadians at home. Confronted with an opportunity to expand relations with nations around the world, the Canadian government could not raise its sights beyond its immediate interests. These were the last few years of a government that had been too long in power and looked only to its own survival.

For the first time, governments hired full-time press agents. Rather than leaving coverage to the newspapers, they supplied pages of puerile puffery to advertise the exhibitors in the home market. Ontario was quick off the mark, leading Tory faithful to complain that they had to read about the fair in the *Globe.* Mowat's government voted $1000 for expenses to their agent, who used the grant to entertain American journalists in return for glowing reviews, just as Cuff had done in Dublin earlier. Larke complained that American papers demanded 75 cents a line, which the minister of agriculture refused to pay, though he did arrange for weekly letters to be sent to the leading Tory organs by two press agents, who proved quarelsome and bibulous.[100] In the end, Liberal and Conservative papers were amply provided with accounts of Canada's splendid exhibits, a testimony to the extensive machinery of propaganda at the government's disposal.

Exhibiting was a costly business with uncertain benefits, but Canadian businessmen were remarkably indifferent to the Chicago fair. A Furniture Manufacturers' Association took a slight interest, mostly in the furniture contract for the Canadian Pavilion.[101] Most Canadian and American piano-makers boycotted the exposition, though Dominion Organ applied to send 78,000 lb of organs. Newcombe and Goderich Organs also sent instruments that received good use (officials twice asked Larke to stop their incessant concerts). The Canadian Manufacturers' Association was hostile to the fair and left the industrial exhibit in the 'competent hands' of the government, but roused itself to recommend a sculpture by Hamilton McCarthy depicting a colossal Canadian maiden 'in all the glory and beauty of form and feature of her race.'[102] Government and business seemed to have changed positions. The *Canadian Manufacturer* discouraged readers from attending: 'There are but few reasons, and they very weak ones, why Canadian manufacturers should put themselves to much trouble or expense to make displays at the Chicago World's Fair.'[103] Foreign markets were not reason enough. Perhaps the 'flight from competition'[104] and the turn inwards towards combines and monopoly capitalism impaired Canadian enterprise.

Some put on a good display. Christie Brown Company sent 9600 lb of biscuits worth $965. Brewers Sleeman, Carling, and John Labatt sent exhibits, Carling's weighing in at 4173 lb. Labatt, fresh from honours at Jamaica, sent a mammoth pyramid of empty bottles, with the beer itself kept in cold storage for the trials. (This decision was unfortunate, for the cold storage building burnt down in a horrific blaze that killed seventeen firemen and, in addition to the beer, destroyed all the Canadian fruit.) Distillers like Hiram Walker were represented, though their wares soon disappeared, doled out by commissioners to visitors or enjoyed by themselves.[105] Larke pointed out that, whereas one-third of Canadian manufactures won prizes in 1876, the proportion was now 87 per cent, even though the ratio of awards had dropped from 42 per cent to 36 per cent of all entries.[106]

Still, the display was smaller than the collection sent to Philadelphia. An American survey described Canadian manufactures as 'subservient to her agriculture, lumber industries, and fisheries.'[107] Emily Cummings penned a scathing review for the *Globe*, a not impartial source: 'An uglier, commoner-looking erection than the line of wooden arches, painted white, with its two pepper-castor-looking towers, which form the portal, is not seen in the whole building.'[108] Inside were bare and dirty aisles, the floor mat made of linoleum leavings from the Danish exhibit next door. One visitor was heard to say, 'Well, of all the poverty-stricken looking places! It looks as if Canada was bankrupt.' Canada spent only about $2000 ornamenting the display, a fraction of what England or even New South Wales spent, and its initial design was vetoed by the Chicago authorities as being too niggardly.[109]

The implement makers made an important display. These men sought the farmer's patronage at local fairs throughout Canada and shipped their implements around the world, engaging in trial after trial of reaping, mowing, binding, and threshing with rivals like John Deere and McCormick. The acknowledged giant of implements was Massey-Harris, the largest maker under the British flag. Massey and Harris had combined in 1891 and swallowed more companies the next year. Export sales accounted for 40 per cent of their production by 1900, and they produced 15 per cent of all manufactures exported from Canada. In 1892 Massey-Harris sales amounted to $3,600,000, but by 1895 they had dropped to $2,821,000.[110] Massey's historian, Merrill Denison, blames the financial crisis, but the Chicago fair had a part in this decline.

Massey-Harris demanded an enormous space for its display and threatened to exhibit in the British department if thwarted. Larke made over 4000 feet to the combine, which sent sixty implements, occupying nine train cars. The machines were highly finished: the self-binder, for example, was made of sycamore, oak, and ash, embellished with silver plating and burnished brass, the wind screen was made of silk, and both screen and conveyer belts were hand embroidered. Massey-Harris was as much a master of advertising as at manufacturing by the 1890s, and its stable of writers made the most of the occasion. Articles appeared in *Massey's Illustrated,* a monthly rural paper, and pictures of the exhibit were placed in newspapers across Canada. The Chicago *Farm Implement News* described it as 'the largest, the fullest and the finest exhibit in the whole building.'[111] The Tory *Empire* drew the requisite conclusion: 'Pointing, as an exhibition of this class does, directly to the progress of manufactures and indirectly to the prosperity of the agriculturists of a country, it is undoubtedly an excellent criterion by which to estimate a nation's prosperity and attainments in industrial arts.'[112] Many other implement makers were also represented at Chicago, among them Cockshutt, John Abell, and John Watson.

The Canadians anticipated brilliant success at Chicago, based on past performance and export sales. They were disappointed. First, the authorities decided, at the last minute, to permit trials of the machines in the open field. As the implements sent were got up for show, only companies near Chicago could send field-worthy specimens in time, and the rest refused to take part. National prejudice also caused trouble. At the beginning of the fair, some foreign agents withdrew their exhibits from competition, complaining that they were under-represented on the juries. The boycott was abandoned, but bad feeling remained. On the implements jury, judges from Russia and Germany were highly impressed by the Canadian machines, but the American Awards Committee disallowed their awards on the grounds that the Canadian machines were copies of American ones. The foreign judges were deemed incompetent because of their unfamiliarity with American industry. This assessment was not wholly inaccurate. The snub was particularly injurious because prizes were not competitive, but awarded to all exhibits showing any merit.

This decision was a blow to the Canadian government as much to the manufacturers. Massey-Harris turned to Ottawa for redress, complaining in November 1893 of 'the influence that this matter of awards is having on our business in Europe especially in Great Britain.'[113] Larke appealed the awards, citing partiality, incompetence, and fraud. He wrote to Sir John Thompson suggesting the threat of a lawsuit, to help 'our friends ... if we fail to receive awards in this branch when we have succeeded so well in other branches, it will be used by your political opponents with disadvantage to the policy of the Government. It will be said that the Government is lending the support of the tariff to the manufacturers of inferior implements, thus compelling the farmers of Canada to pay a high price for inferior goods.'[114] The government did not sue, but it did publish a sessional paper recounting the indignities suffered at Chicago, which Massey-Harris then circulated.[115] At least it was an argument against successful rivals at home, whom Cockshutt Company thought should be stripped of their prizes.[116] Implements manufacturers paid dearly for the defeat: In 1894 the tariff was reduced from 35 per cent to 20 per cent.

Not everyone complained about the infamy of 'Yankee justice.' Hugh Johnston was pleased to get a medal for his patent ploughshare and scornfully dismissed the government's charges of unfairness:

> You blame the Exposition Co., and blame me because I did not go to you and several other Tom, Dick and Harry, while you and your favorite combine friends were really to blame, and badly to blame for not having more awards had you been men of less self and more true pluck ... I care not for your tariff, nor do I believe with you in your paltry excuses as to the existing jealousy of the U.S.A., and I believe they are of too generous a nature, a live and let live disposition, but like any other creature of genuine spirit, they will not put up with any would-be smart tricks, and this generosity was proved in all other Departments of the Exposition, and honor dealt out to whom honor was due and Canada got her share without a grudge that I know of, and this is the way to win confidence and respect, and therein lieth the secret of their extensive population to-day, and if our people had the brains to invent and the heart to assist in experiment and manufacture, they should not then fear any nation.[117]

His letterhead made pointed references to 'Honorable Judges' and 'Only True Awards won for Canada on Agricultural Implements.'

The reasons alleged for the injustice were, as Johnston noted, jealousy of Canadian successes in other departments, most notoriously cheese. At a competition in June for cheeses made in 1892, Canada won 85 per cent of the prizes. Larke commented that this triumph created such jealousy that other awards would be harder to win.[118] These successes reflected badly on industrial failures, and yet they, too, followed years of state encouragement of agriculture, through the agricultural societies, the experimental farms, and the dairy, stock, and fruit associations, all of which received state moneys and material aid at the agricultural shows.

Canada's great triumph was in cheese. Private cheese makers made a strong display: 'Cheese King' David Macpherson, who owned more than eighty factories in Ontario and Quebec, sent six cheeses of 1000 lb each, all judged first rate. Quebec and Ontario shared the honours, with Quebec beating Ontario in the June exhibit for cheeses of 1892, thanks to special efforts by assistant dairy commissioner J.-C. Chapais and the Société d'industrie laitière.[119] But top billing went to the dominion dairy commissioner, James Robertson. He had assistant J.A. Ruddick prepare a mammoth cheese, the 'Canadian Mite,' which weighed in at 22,000 lb, having used over 200,000 lb of milk (the amount 10,000 cows produced in a day), and measured 28 feet in circumference and 6 feet in height.[120] A Canadian Pacific train was fitted out for the giant cheese, which toured through the country with much fanfare. In Chicago, when it was installed, the floor collapsed under its weight. Although the outside of the cheese suffered from the heat, judges bored two feet into it and found it solid throughout and of 'remarkably good' flavour and 'extraordinarily fine' body. The cheese was sold to English interests, who paraded it through the streets on a horse-drawn cart with 'Free Land in Manitoba' painted on its side. The Mite was one of the wonders of the fair and attracted visitors all day long. One grey-haired man surveyed it, 'spat vigorously, and said: – "Gosh! Ef the skippers ever get into that thar cheese they'll grow as big as rabbits."'[121] The Mite created a mental association between Canada and cheese, and accolades won by the private makers proved Canadian cheese was fine cheese: 'The senses of sight, taste, and latterly of smell will have forever connected Canada and cheese.'[122] One columnist wrote that people in Chicago thought Canadians lived 'on cheese exclusively' and that restaurants automatically served them cheese.[123] A Woodstock man declared: 'The Centennial set up Canada in the cheese business, with a capital of millions, and now the stock has gone up above par.'[124]

Not all approved of Robertson's cheese. A correspondent in the *Globe* thought the Mite an 'extravagant folly,' and that the money would be better spent on the poor.[125] Robertson replied that his scheme would increase Canadian wealth by half a million dollars annually.[126] Like Sir William Logan, Robertson was a master of propaganda. Also like Logan, he quietly despaired of 'the banquets with which my life has been a weariness – 3 o'clock in the morning and only a few more toasts to propose.'[127] Canadian butter was less successful: over-salted and greasy. An American dairyman remarked of the two exhibits: 'The peacock is not all covered with gaudy plumage. He has some very ugly feet which are not only naked, but, anon, obtrude themselves in a way very mortifying to his pride.'[128] Butter became Robertson's preoccupation during the 1890s.

The Fruit Growers' Associations took charge of the fruit exhibitions, with the help of William Saunders at the Experimental Farm. He arranged for bottled fruits to be shown and he toured the West for samples in 1893. Canada made an imposing display of fresh and preserved fruit until the cold storage building burnt down; then many tables sat bare until the fall fruit ripened. Ontario and Quebec were well represented, but fruit from British Columbia and Nova Scotia were damaged by frost en route to Chicago.[129] The federal and provincial experimental farms made a substantial contribution to these exhibits, helping Canada achieve awards. Wherever private enterprise failed, the government was well equipped with bureaucrat-farmers to fill any gaps.

Canadian breeders left Chicago disappointed. The total amount won by Canadian animals was $21,461 – a goodly sum, even in comparison with the American winnings of $65,050. But this amount did not compare well with Philadelphia, Larke conceded. In 1876 Canada showed 10 per cent of the animals and won 31 per cent of the medals, whereas, at Chicago, Canada showed 31 per cent of livestock and won only 23 per cent of the awards.[130] The numbers were limited – to 100 horses, 200 cattle, 450 sheep, and 300 swine – but they were the finest animals found at Canadian fairs, so the defeat was telling. Larke blamed it on Americans having bought up the best British and Canadian beasts, but privately he complained to the Agricultural Association that the animals sent should not have been shown outside a county fairgrounds. Canadians sold almost no cattle because, Larke told Lowe, they lost and the animals were too old.[131]

The humiliation was only relative. Canadian horses won only one-ninth of American prizes, but Canadian sheep won $7503 to the Americans' $7826. Canadian swine won almost as many prizes as the Americans (64 to 67), but half the amount of prize money: The Canadians forced the Chicago officials to separate three breeds that they had lumped together and in which Canadian breeders excelled. The fact that Canadian cattle won only one-quarter of American prizes seemed a more serious problem. It reflected greater reliance among Canadian breeders on their own bloodlines, rather than British imports. Canadian breeders complained that the American judges did not like the small teats on Canadian Ayrshires, looked for pure red Shorthorns, and preferred size to evenness.[132] It is possible that the Canadian exhibitions rewarded second-rate stock on the basis of pedigree and reputation, as critics like William Weld had long complained. After all, the exhibition could only reward the best animals to be found on the grounds, rather than intrinsic excellence. It also rewarded the largest and oldest herds, at the expense of newer and more profitable breeds like Holsteins. Another reason for the defeat may lie with the divvying up of the Canadian livestock exhibit. Ontario sent most of the animals and took most of the prizes, though Quebec won more prizes for Ayrshires, and together the two provinces took $1885, compared with the United States'

$150 for Ayrshires. Canada's performance in Shorthorns was also more than creditable: The Russells of Richmond Hill surpassed their success of 1876 by winning best Shorthorn herd and best herd of beef cattle of any breed, and other prizes. Thus, in the two breeds in which Canada had specialized for most of the century, Canadians won the top prizes. The experience did not much affect the export trade, which remained steady, while sales of animal products rose. Sales of live cattle to the United States, already down to a fraction of what they had been, dwindled further from $21,327 in 1891–2 to $1076 in 1895–6.

One hundred and seventeen Canadian paintings graced the arts pavilion, and five Canadians won awards there: F.C.V. Ede, Sarah Holden, Robert Harris, John Fraser, and G.A. Reid. Economic concerns weighed heavily in this section also, for the selection committee rejected, as discouraging to immigrants, a painting by naturalist Ernest Thompson Seton entitled *Awaited in Vain*, which showed a lost woodsman being devoured by wolves. The Royal Canadian Academy objected that the 'repulsive' work would terrify children and cause 'accidents' to pregnant women. The author obediently 'removed the more repulsive details and introduced brushwood to screen the rest.' He pointed out it was set in the Pyrenees, although a Chicago man offered him money to rename it 'A Scene Near Montreal' and show it privately. The picture was finally shown, to the disgust of some visitors.[133]

Minerals were displayed by the Geological Survey and by mining companies from across the country. Ontario's exhibit featured copper, nickel, and petroleum; Quebec's, asbestos, mica, phosphates, and plumbago; Nova Scotia's, copper, coal, and gold; New Brunswick's, granite and gypsum; the Territories sent coal and building stones; and British Columbia sent coal, silver, and the traditional gilded pyramid. BC coal won few prizes, it was noted, because few of the mines had begun economic production. Well represented in minerals, fisheries, forests, agriculture, and fauna, British Columbia claimed to rank third among Canada's provinces, below only Ontario and Quebec in importance.[134]

The Maritimes also exhibited their resources, especially agriculture and fisheries. (Newfoundland, devastated by a fire in St John's the previous year, did not exhibit.) On the whole, however, Ontario dominated the fair. In machines and general manufactures, Ontario took forty-two prizes, Quebec twenty-one, the Maritimes eight, and western Canada two (for horseshoes and a buffalo robe). In mining, Ontario won forty-two prizes, Quebec twelve, the Maritimes eight, and western Canada six. Perhaps if the fair had been in New York, eastern Canada might have sent more. But

Triumph of the Wolves, or *Awaited in Vain*, by E.T. Seton, shown at the Philadelphia Exhibition of 1876. The wolf's relish for his meal was as disconcerting as the imputation that the painting illustrated conditions in modern Canada.

Ontario had already taken the lead in industrial development: In 1891, the province had nearly twice as many factories producing over $50,000 annually as Quebec could claim, and together the two provinces accounted for 78 per cent of all such factories.[135] The exposition fairly represented the unequal development of Canada's economy.

The exhibit had little obvious impact on this development. Immigration continued to decline until 1896. A scant $46,000 worth of Canadian wares were sold; others were given or thrown away; and the rest, worth $403,197, were shipped home.[136] The fair coincided with a commercial crisis in Chicago, and sales were generally disappointing. Canadian exports of manufactures rose slightly from $7 million in 1892 to above $7.5 million in 1895 and above $9 million the next year. R.T. Naylor remarks that, 'after the mid-1890s while Canadian-manufactured experts declined relatively, certain industries remained strong.' These included machinery, tools, agricultural implements, pianos and organs, and sewing machines, all industries that tended to exhibit. Elsewhere he records that, in 1894, Canadian and British capital 'began its headlong rush to the Pacific coast.'[137] The display in Chicago may have contributed to this trend.

The World's Fair of 1893 was driven by economic forces, everyone agreed, but to say as much was to do it scant justice. It was no longer quite so evident as it had been that the acts of advertising and staring, selling, and purchasing, even when practised on such a grandiose scale, had any other significance – at least any that intellectuals (as distinct from penny-a-line pundits) could expound on with dignity. The buildings might seem 'a dream of the Heavenly City,' but the fair itself was 'little more than a giant advertising scheme.'[138] Tolstoy described it pithily: 'The Chicago Exhibition, like all exhibitions, is a striking example of imprudence and hypocrisy: everything is done for profit and amusement – from boredom – but noble aims of the people are ascribed to it. Orgies are better.'[139]

People were becoming more literal-minded. They wanted to see directly at work the social processes that journalists had been crying up all along. It was no longer enough to have peaceful intercourse as a by-product of buying and selling; there were now congresses for direct communication. It was no longer enough to see cultural diversity in the exhibits and the occasional oddly dressed spectator; there were now encampments of colourful peoples from around the world, in a fun-fair, carnival atmosphere. The congresses were the loftiest part of the World's Fair, and the Midway Plaisance was the most popular. Neither attracted much direct intervention by the Canadian government, so a survey of these activities reveals, not policy, but popular cultural expression.

The congresses held to accompany the fair spanned a range of topics from business and banking to temperance and religion. T.C Keefer spoke on Canadian waterways before a Water Commerce Congress, while B.E. Walker, manager of the Bank of Commerce, read a paper on 'Banking in Canada.'[140] There was much useful exchange of information and collaboration on reforms such as uniform weights and measures. There was also much moralizing as to the benevolent effects of business on society. These congresses had a hallowed air and they revived the spirit of international collaboration – qualities that must have been useful at a time when businessmen were forming combines and describing these activities as a moral evolution rising above distrust and competition.[141]

'Not things but men!' was the slogan of the congresses. The Toronto *Globe* described them as the heart of the fair, bringing 'together all the leading thinkers and actors of the day, who are really the vital force of the world's progress.' The congresses dignified the fair by suggesting that humans were in control of the historical processes visible on the grounds. Yet there was some unease with the marriage between business and ideals. In 1851 the idealism had been imminent in the exhibition itself as a congruence of objects and visitors, with only a few technical lectures offered, but by 1893, the material and the moral had bifurcated. In 1851 the display of material wealth seemed a testimony to God's beneficience. By 1893, as criticism of materialism had developed, the religious elements seemed tacked on, as evidenced by a World's Parliament of Religions and the belated decision to close the fair on Sundays.

The Parliament of Religions is said to have initiated modern religious relativism. The Archbishop of Canterbury would not condone it, objecting that parliaments presume equality of membership, but religions are not equal; nor did he approve the public discussion of sacred things.[142] The Vatican sent a delegate to a Catholic Congress, held in conjunction with the Parliament of Religions, and liberal bishops spoke at the Parliament, but the pope later denounced both the congress and American liberal Catholicism.[143] In Canada, *ultramontane* journalist Jules-Paul Tardivel did not wait for the pope's verdict before condemning its relativism, and the *Manitoba Free Press* decried the 'parliament of errors' for the same reason.[144] The Toronto *Mail*, by contrast, thought the project a 'noble enterprize,' but objected to its being 'on exhibition for the consideration of pleasure-seekers and sight-seers.'[145]

Two Canadians, George Grant and John William Dawson, principals of Queen's and McGill Universities respectively, lectured before the parliament. The aging Dawson aired his aging views that religion and science were compatible, while the broad-minded Presbyterian Grant pleaded for greater religious tolerance, which Canada, by dint of its cosmopolitan population, had begun to initiate: 'It is easy for a people in such an environment to understand ... that no church or nation has a monopoly of the truth or of the spirit of the living God.'[146] It was precisely this message of tolerance that Allen Pringle took back home with him. Pringle was a free-thinking beekeeper, in Chicago to oversee the apiarists' display.[147] He was most impressed by religious men from the East whom, he said, 'made the best exhibit.' Writing to a Napanee newspaper, Pringle argued that the true religion was the religion of humanity, not that of dogma, and that it must include even those who denied God, or it would not be universal as a religion should. Two Anglican divines promptly responded that 'this craze for universal toleration is not Christian ... Universal toleration means utter indifference to religion.'[148] Discussions occurred everywhere. A Sunday School Committee in Toronto proclaimed its definition of the parliament's lesson: 'Religion is that which holds people together, that in short which makes society possible.'[149] Anything causing division into sect, class, nation, even sex, was irreligious. This view that religion served society, rather than the orthodox Christian view that society served religion, was the message one Anglican professor, Alfred Momerie, had delivered at Chicago: 'The love of man is the essence of religion. Religion may be lacking in metaphysical completeness; it may be lacking in original consistency; it may be lacking in esthetical development; it may be lacking in almost everything, yet if lacking in brotherly love it would be mockery and a sham.'[150]

Yet relativism could cloak dogmatism. Rev. A.H. Ball, preaching to the Toronto Zion Congregational Church, predicted that the display of 'the product and power of Christianity' would overcome the dogmatism of everyone else and 'have a far reaching effect in preparing the uncivilized peoples of the earth for the reception of the truths of the Gospel.'[151] Agnes Maule Machar applauded the study of comparative religions on the grounds that Christian missionaries schooled in heathen beliefs would be more effective.[152] This, too, was Principal Grant's final position: All religions served humanity, but only Christianity completely fulfilled all human needs.[153]

The congress was meant to bring religion back to the fair. Those who did not see the hand of God in the material progress represented at the fair (and there were many who did – one Methodist told a Brantford flock that the 'exposition at Philadelphia owed its very life to the *exposition* of truth upon this continent') saw, instead, a history propelled by commerce.

Owen Chadwick identifies secularization as a growing distance between the details of life and God's intervention.[154] The architects of the exhibition – where all the connections between commerce, industry, and national life were made visible – might tacitly agree with the astronomer Pierre Simon de Laplace, who, when asked where God fit into his cosmology, answered, 'Je n'ai pas besoin de cette hypothèse.' Confronted by a growing irrelevance, devout Christians tried to combat this trend by developing a social mission, which many speakers at the congress advocated. However, as Ramsay Cook has pointed out, this sort of activity itself led to secularization.[155]

The last gasp of Christianity within the exposition proper lay in the proposal to close the doors on Sunday. Organizers, politicians, reformers, and congregations were of two minds about Sunday opening, as they had been since the 1850s. The Crystal Palace had been closed on the Sabbath, but the Palais d'industrie had remained open to let the working classes attend. By the 1890s the sanctity of the Sabbath was debated with fierce intensity across Canada. In 1893 Toronto had a municipal referendum on the issue of Sunday streetcars, and the newspapers were filled to overflowing with arguments for and against.[156] Having organized themselves to fight this campaign, religious groups levelled their guns against the desecration of the Sabbath at the World's Fair.

To receive a federal grant, fair organizers had pledged to close on Sundays, but afterwards changed their minds. The matter went to court and was appealed, as the fair remained open on Sundays. Incensed Canadian divines, Sunday School associations, and parishioners wrote to Ottawa demanding that the government keep Canadian exhibits closed on Sundays and put pressure on fair officials to close the entire grounds.[157] One rural pastor, G.A. Gifford of Fergus, warned that the fate of Babylon and Rome would fall on those who so outraged nature: Sunday opening at Chicago would 'thrust a dagger into the very heart of our Christian civilization, through which the life blood will ooze out, until it becomes a pestilence and the nation a by word and a reproach.'[158] Canadian officials sympathized with the sentiment, but, because the display was scattered through many buildings, they could do little. Because caretakers were not paid to work on Sundays, thieves plundered the exhibits, undeterred by covering cloths and signs saying 'Closed.' In the end, the fair was closed on Sundays because of low attendance: Few wanted to come when so many exhibits were closed. This decision came in time for a Congress of Weekly Rest in Chicago in 1893 (Canadian John Charleton was a vice-president), where many speeches reflected on America's decline from its 'proud position' as a 'Christian, Sabbath-keeping nation' into one suffering from dissipation and social unrest.[159]

It had not been moral arguments, but popular demand, that decided the Sabbath matter. This outcome was one sign of the triumph of popular culture which the World's Fair heralded. The exhibitions had always been a contested cultural space, where elite control was gradually relinquished to the public that came to see the amusements. The editor of the Toronto *Mail* commented that 'the great mass of people who go to the Fair do not care much for grand buildings or art exhibits or the advance of science, or the history of the past. They want a circus ... They give a glance at the buildings, a hasty run through the pictures, and they may dawdle a little while among the manufacturing exhibits, but their real enthusiasm is displayed in the crowded side-shows.' The display was 'very instructing and amusing no doubt, but not so exalting to the soul and instructive to the mind as everything connected with the Columbian Exposition was intended to be.'[160] The organizers of the Chicago fair created a special place for the amusements: the Midway Plaisance. Faith Fenton thought this separation made the rest of the fair a place of 'repose and purity' where women could walk about undisturbed for once.[161] But it also strengthened the impact of the sideshows, making them a cultural event in their own right, with a certain legitimacy. The midway's obvious popularity forced commentators to come to terms with the scene. Agricultural exhibitions had already begun to pay for sideshows, but after 1893 they unabashedly set up their own midways.[162] Toronto got a midway in 1898. Exhibitions had tried to bring popular taste to the level of cultured elites; instead, matters had gone the other way. The exhibition now catered to the masses, and the classes had to accommodate themselves willy-nilly.

Many of the midway attractions were familiar to Canadian audiences. Buffalo Bill's Wild West Show had visited Canada in 1885, complete with 150 Mexicans, cowboys, and natives, including Sitting Bull.[163] Other more exotic displays stretched middle-class tolerance, most notoriously the Turkish belly dancers. Kit Coleman visited the display of flesh, apprehensive of 'Visions of Lady Managers, with morality sticks,' but found the dancing a little tame.[164] The dancers visited the Toronto Industrial in September, purged of offensive gestures. Coleman spent a lot of time on the midway, enjoying exhibits like the monster Ferris wheel and the Egyptian street, which, she insisted, was 'Cairo indeed and no fake.'[165]

Cairo street was one of the many recreations of exotic lifestyles on the midway and near the fair site. As well as the Egyptians and the Sudanese, there were Indonesians, Dahomans, Malays, Samoans, Fijians, Japanese,

Chinese, and Irish communities, and encampments of Amerindians from the United States and Canada.[166] Newfoundland's contribution to the exposition was fifty-eight Inuit from the Labrador coast. Coleman was not impressed by their encampment, finding it dirty, smelly, and idle. As she describes her visit, it hardly lives up to ideals of cultural exchange, but does suggest that the Inuit were not passive before the curious stares of the tourists: 'They are a vicious people, and resent being looked at. One of the women gave an old lady standing near me a whack on the back that nearly sent her into a fit, while a horrid little boy with a sore head scooped up some mud and threw it at me, after which he spat at me, for which I promptly boxed his ears.' There were also Kwagiulth from the Pacific Northwest who shocked tourists and officials by performing a gory Tamanawas dance. Chapter 10 examines this episode more closely.

Natives were both exhibitors and exhibits at the World's Columbian Exposition, but they were also, like thousands of other Canadians, spectators. At first, visitors stayed away, fearing high prices and empty booths. But prices soon fell, as did train tickets reduced to a penny a mile, and the flood began. There were various routes to the fair: One Franco-Ontarian paddled in an old punt to get his first sight of electricity. Many newspapers reported local visitors, as did the Bothwell *Times*: 'Alonzo Badgely and others who visited the World's Fair, returned last week and report a fine time.'[167] Some never made it at all: One train en route to the Fair with Canadians aboard collided with another train, strewing the ground with bodies. There was the inevitable story of white slavers luring an innocent Quebec woman with the offer of a job: The case was dismissed from court when the men agreed to pay her fare homeward. One Toronto man spent his own wages and even the earnings of his family on repeated visits to Chicago, forcing his wife to sue for alimony: '"He's nothing but a crazy show-struck man, that's what he is," said Mrs Boeckh emphatically. "All he thinks about is actresses and circuses, and that's the cause of our trouble."'[168] A carpenter from Caledon Township, David Smith, went to the fair and carried about a notebook in which he recorded facts that struck him: California displayed a tower with 13,873 oranges; the Bethlehem Iron Company showed a 125-ton steam hammer, the largest in the world, its measurements neatly jotted down.[169] Canadian Scots joined in the Caledonian Games, while French Canadians gathered for a congress. Many came for Irish Day, when the mayor of Dublin and the British commissioners traded insults. The Order of Foresters met in Chicago, and their chief, Oronhyatekhon, received an award with the remark, 'Indians are very fond of medals.'[170]

Many gathered around the Canadian Pavilion to celebrate Dominion Day. This was a solemn event, punctuated with speeches by dignitaries. Chicago's mayor, Carter Harrison, told a story he thought would please Canadians: A Siamese prince angered his king by building a castle, and he tried to avert the royal wrath by saying the castle was built for the king's pleasure. The king gave the order to continue, saying he would claim the castle when it was ready. Harrison explained the moral: 'Canada is a growing nation, and the United States is watching and waiting until she is ready for a new acquisition to her territory.'[171] The audience hissed the mayor, who, taken aback, quickly changed the subject. Larke leaped to the platform and gave an impassioned 'able and manly speech' in defence of Canadian nationalism which even the Liberal newspapers reprinted. For all their indignation, Canadians were as horrified as any when Harrison was murdered shortly before the fair closed.

The verdict on Canada's performance was, finally, a mixed one. Favourable press reviews far outweighed critical ones, but they were written by government-hired hacks and probably had little popular resonance. Visitors polled by agents and newspapers seemed well pleased, but Canada's performance was uneven and occasionally humiliating. A small fraction of visitors who signed the British Columbia guest books made derogatory comments (which were scribbled out), and various British Columbia residents expressed disappointment, but most visitors from Canada and the United States praised the Canadian exhibit as 'unequalled,' 'dandy,' 'delightful,' and 'out of sight' (a phrase that appears so often one begins to suspect a *double entendre*).[172] An American who wrote 'Splendid – join the Union,' was answered 'Never – not by a jugful.' One source of embarrassment was a less than imposing Canadian Pavilion that contained offices, a newsroom, and lounges for entertaining, done up in native woods. The minister of agriculture inspected the pavilion and found it characterless, while the premier of Manitoba declared it 'just about fit to make a good stable for a Manitoba farmer.'[173]

Manitoba staged a rebellion in 1893. Premier Thomas Greenway found the space Canada offered him contemptibly small and decided to exhibit outside the showgrounds, in a log cabin that doubled as an inn. The opposition press immediately dubbed it the Manitoba Sideshow, situated among the fakirs near the midway: 'The visitors to the sideshow are principally those who have a taste for freaks.' A cartoon showed Greenway as an Indian beating a tomtom to attract visitors, reported to number between 900 (government sources) and a dozen (sceptics) per hour.[174] There were grain and flour trophies, woollens, pottery, soap, lager, and Indian artifacts. Plans for a display of cattle were abandoned when a selection committee found none worth sending.[175] The education exhibit bore a sign announcing that 'Manitoba has one system of schools!' Commissioner Joseph Tassé was infuriated by this attack on the Catholic separate schools and urged: 'Soyons avant tout Canadiens.'[176] The Manitoba exhibit encouraged Greenway's critics to praise the Canadian display as exhibiting cooperation across political and sectional lines,[177] but of itself it did little for Canadian unity.

At least Manitoba took an interest in the World's Fair, unlike New Brunswick, which declared the exhibition not worth the bother. A special effort had to be made with Quebec because, one agent told Saunders, 'it is barely possible that trouble may arise from the seeming neglect of Quebec in respect to the World's Fair.'[178] This effort meant wooing the manufacturers and finding posts for them on juries and commissions: Minister of Agriculture A.-R. Angers told Larke that it was 'important politically' to find a position for the president of the Montreal Chambre de Commerce.[179]

Ontario made a great effort to advertise itself, but its commissioner, Nicholas Awrey, MPP, got on no better with the federal commissioners than Oliver Mowat did with the federal government. Premier Mowat led the provincial-rights movement and made serious encroachments on Ottawa's jurisdiction.[180] All problem files dealing with the World's Fair were placed at the beginning of the documents in the Department of Agriculture's archives, and among them is the 'Notes of differences of understanding' drawn up by Saunders, Larke, and Awrey in October 1894.[181] For example, the federal government was supposed to pay installation costs, but baulked at costly ornamental enclosures on which the word 'Ontario' figured prominently and the word 'Canada' not at all. In July 1894 W.D. Dimock, the Nova Scotia agent, remarked ruefully: 'I think Awrey must have been wise when he commenced to attack our part of the show. I thought it would have been against him but he has been elected by the biggest majority he ever had.'[182]

Jealousy, not unity, was the message the Canadian exhibit conveyed. Most Canadians were too caught up in their own problems and politics to subscribe to the fiction of a grand national display. Canada, in 1893, suffered from economic depression, lacklustre leadership, and religious and regional divisions. The display at Chicago gave a fair representation of all these qualities, as well as the achievements of agriculturists, manufacturers, workers, and educators. Shortcomings did not threaten the National Policy tariff, for the manufacturers already had enough political power to prevent that threat, as they knew when they turned down invitations to compete. The Chicago performance did not bolster support for the Tories, who were voted out of office in 1896. The World's Fair occasioned an explosion of nationalist rhetoric in Canada, but did nothing to substantiate this rhetoric.

Jealousy also provides an answer to one question that a history of Canada at the international exhibitions must ask: Why did Canada not have one of its own? The United States had three major exhibitions before the end of the century, and a half-dozen smaller ones like New Orleans, or the California Mid-winter Fair to which many Canadian exhibitors sent their goods when the Chicago fair closed. Australia and New Zealand, even Chile, had some, and it was a rare country that did not make the attempt. Visionaries like Alfred Perry had been calling for an exhibition in Canada ever since the 1850s, when Canada had seemed to do so well in Paris and London. After Confederation, fair promoters and newspaper editors raised the idea from time to time, but it never attracted much support until the end of the century. By the late 1880s, Montrealers were warming to the idea of an international exhibition to celebrate the city's 250th anniversary in 1892. The Antiquarian and Numismatic Society suggested that an exhibition would bring 'IMMENSE COMMERCIAL BENEFITS' and, in these troubled times, 'that the very circumstance of our different provinces being brought together in a friendly rivalry, in a way which would do credit to the whole country, would do much to cement their union, increase the pride and confidence of Canadians in the future of their country, and foster a self-reliant patriotism, founded on an intelligent understanding of the advance already made and the possibilities of the future.'[183] The announcement that Chicago would have a World's Fair in 1892 put paid to that scheme. Instead, the society marked the anniversary by a display of relics and manuscripts at the Montreal Exhibition of 1892.[184]

In 1895 and 1896, delegations from Montreal waited on the government to propose an international exhibition be held in their city, with the government kicking in half a million dollars. Their aim was to impress foreign visitors and Canadians alike, and to create a 'friendly rivalry' among the 'sister provinces.' The government, with Sir Charles Tupper at its head, was 'favourably disposed towards the idea': With an election imminent, any scheme to scatter about large sums of money in a feel-good project was welcome. Unfortunately, not 'friendly rivalry,' but 'a spirit of jealousy' among the sister provinces, scotched the scheme. Toronto had plans of its own to invite foreign nations to the Industrial Exhibition in 1897, and it asked only for official endorsement and a grant of $50,000. In the House of Commons, MPs for both cities advanced their claims: Montreal was the 'greatest and grandest city in the Dominion,' and could put together the best industrial exhibition. Toronto already managed its annual show 'super-excellently well' and the collection there would be 'more representative' of Canadian agriculture, and 'largely' representative of Canadian industry. In a snappy exchange, a Quebec member insisted:

> the Toronto exhibition, although possessing great importance, would be only an exhibition of agricultural produce.
>
> Mr. COATSWORTH. No, not at all.
>
> Mr. MACLEAN (York). We shall have a Midway Plaisance.
>
> Mr. FOSTER. An Egyptian mummy.

A western MP spoke up for Toronto, while the Maritimes would have preferred Montreal. In the end, though, it was clear that international exhibitions could be held in countries with only one major city, such as England or France, or they could be held in countries with many rival cities, such as the United States. But they could not be held in Canada, where either Toronto or Montreal would receive preference and the other a snub. The slight could cost the government the support of an entire province. Year after year, decade after decade, exhibitions never quite succeeded in turning local jealousy to friendly rivalry, as their organizers so sanguinely hoped that they would.

In theory, exhibitions were supposed to replace hostility with peaceful cooperation by educating society to be more thoroughly under the control of rational economic forces. By the end of the century, these early hopes had become patently absurd. War, not peace, was the more natural outcome of economic forces, according to the new theorists of empire. The surface appearance of a peaceful cosmopolitan gathering at the fairgrounds was belied by underlying national jealousies. These jealousies and economic rivalries led to biased judging and encouraged a chauvinistic jingoism among the crowd. Moreover, the economic forces that ruled the world and the fair were themselves not rational but all too irrational, dependent on the whims of the developing mass market. The crowds that swarmed to exhibitions, when not cheering gory scenes at the midway, were sure to be found admiring spangled reapers or gigantic cheeses, and ignoring less gaudy but more worthy items. Exhibitions had, if anything, intensified this process by pandering to the public's taste for the spectacular. The educative thrust of the exhibition was blunted as it turned aimlessly in upon itself. If the exhibition taught any lesson at all, it was that knowledge could not surmount opposing social and political interests, but would always become implicated in their struggles.

PART THREE

Exhibitions and Identities

9

Women and the Political Economy of Exhibitions

The Victorian lady was expected to confine her activities to the domestic sphere, rearing children and running the household. She was discouraged from voting, from labouring for wages, even from speaking publicly. Many women fought these restrictions, on many different fields of battle. Temperance and reform activities, mock parliaments, and other forms of political activity by Victorian women have already been studied by historians, but another important battlefield was the exhibition. Most exhibitions had a special Ladies Department, where women's expertise was recognized and honoured. Rows of preserves, freshly baked loaves of bread, delicately embroidered fabrics, and hand-sewn quilts testified that the exhibitors were skilled in the domestic arts. By advertising domesticity, the Ladies Department seemed to bolster a traditional concept of women's culture. In fact, exhibitions mediated between public and private realms by publicly advertising women's handiwork. The very act of exhibiting opened up new opportunities for communication and action within the public sphere. Women began to use exhibitions to demand a broader public role for their sex, including the right to speak, to labour, and even to vote. Exhibitions provided women with a back door into the public sphere.

In tracing Canadian women's participation in nineteenth-century exhibitions, this account builds on several feminist reworkings of Jürgen Habermas's theory of the public sphere, notably those by Joan Landes and Mary Ryan.[1] It explores how the political sphere and the wider public sphere overlapped. The nineteenth-century public sphere, as Habermas points out, rested on three foundations: economics, rationality, and publicity.[2] The economic roots of political activity had been explained by John Locke, who, by inserting the concept of property, refined Hobbes's argument that men join society to protect their lives: Men formed political relations to protect the fruits of their labour.[3] This theory indicated why, for most of the nineteenth century, people without property could not vote. Propertied women, usually widows or spinsters, had the right to vote, but were disenfranchised by the mid-nineteenth century. The vote was gradually extended to all men because the concept of property expanded to include labour power, which all working men possessed and had a right to protect one way or another, as trade unions forcefully reminded politicians and industrialists.

Women were not included in this extension of suffrage. There were several reasons for this, but foremost was the Victorian notion that women were not primarily economic beings. In fact, however, many wives, as well as spinsters and widows, owned property and worked for wages, or contributed to the family economy with paid and unpaid labour at home. But the Victorians underestimated the number of working women, and they pursued as an ideal the division of society into separate spheres, public and private. In their view, the woman was the queen of the hearth, nurturing her family and running the household, while the man earned the living and participated in political life. The march of civilization was supposed to eradicate the last vestiges of labouring women. In short, classical liberalism posited that all economic agents should vote, but, for reasons external to politics, it was largely blind to the notion of women as economic agents.[4]

Two other foundations of the political sphere were reason and publicity. Habermas argued that a liberal public sphere existed when rational beings engaged in public debate about matters of common concern. As far as the Victorians were concerned, here were two more reasons to exclude women, who were neither rational nor public beings. Women were thought to be more emotional than rational, and it was a standing joke that, given the vote, women would elect the most handsome candidate. Moreover, women were supposed to shun publicity, which was thought to be unwomanly, an indication of sexual disorder, and pernicious to woman's first duty of being a wife and mother. This injunction against public speaking prevented women from proving that they, too, reasoned and laboured.

Exhibitions served the three ideals of the public sphere: property, reason, and publicity. At an exhibition, farmers, artisans, and manufacturers gathered to advertise the fruit of their labours and to study improvements. Women did not occupy an important place at the earliest exhibitions, which were designed to encourage the sort of economic activity that men properly did: grain growing, cattle breeding, ploughing, and, later, artisanal and mechanical work. But women's work gradually intruded. Domestic and fancy-work began to appear before mid-century, ranging from woollens and linens to embroidery and hair pictures. Women's activities were brought before the public eye: Women could read descriptions of their work and even their names in the local newspaper. Having broken the injunction against publicity, women could then manipulate the publicity that the exhibition provided to advertise their rational and economic activity. Particularly at the international exhibitions, women's organizations tried to advertise the full range of women's work inside and outside the domestic sphere.

By showing that women laboured, politically engaged women (such as journalists, but also middle-class women who did not work) could demand the right to vote, in keeping with nineteenth-century political theory. There were other arguments for enfranchising women – pragmatic arguments concerning social reform, and humanistic arguments appealing to natural justice – but to the Victorian liberal, none was so persuasive as the argument from property, an argument coeval with liberalism itself. Politics, to the Victorians, was not an autonomous sphere of agonism, but a representative activity. It was representative not only in the sense of delegating authority but also in a mimetic sense. Politics represented society, its interests and its activities, in legal form, and women had to prove that their activities made them fully social and, thus, political beings. Michael Oakeshott restated the nineteenth-century position when he argued that the only valid reason for the enfranchisement of women was 'that in most important respects they were already enfranchised.'[5] Exhibitions and politics both mirrored society, and, by achieving material representation at the one, women strengthened their claim to representation in the other.

The first section of this chapter draws on writings from France, England, and the United States which influenced Canadians and which appeared in Canadian publications to outline, in general terms, a transatlantic discourse about the social function of women and exhibitions. Paradoxically, women were discouraged from making public displays, but were themselves a kind of spectacle for the male gaze. The second section describes women's participation in agricultural and industrial exhibitions in Ontario and Quebec as an early foray into the public sphere. The final section examines what Canadian women sent to world's fairs from 1851 to 1900, where exhibiting was not merely a gendered but also a political activity.

The earliest exhibitions gave women an opportunity to act in public and to overcome the opposition between publicity and femininity. This opposition was at root a paradox. Victorian ideas of sexual relations owed much to Jean-Jacques Rousseau, whom even Conservative newspapers quoted approvingly on the need for female modesty.[6] Rousseau blamed female vanity and exhibitionism for many social evils, and he praised a Spartan model which refused any publicity to women: Praise was itself slander and could only make a woman notorious. Examining the *patriotes* of the 1830s, Allan Greer finds that 'self-display was repugnant to good women because it signified sexual immorality, just as surely as female confinement to private pursuits indicated chastity.' Political self-assertion, he adds, was proof of sexual disorder.[7] In public, a woman could only 'appear,' as an object to be looked at. Even late in the century, many associated female shop clerks with prostitution, for 'the public nature of the store, its sumptuous atmosphere, and its low wages combined ... to make the transition from counter to bordello all too easy.'[8] Women could only address the public indirectly, as one English novelist, Mrs Hemans, explained in a passage reprinted in a Canadian newspaper: 'She can never with consistency, appear in the forum or the pulpit – in the senate or at the polls – still, without disparagement of her sexual character, or infringement upon those hallowed feelings which the delicacy and loveliness of her nature have cast around her, she may devote her leisure to the pallet and the pen, and send forth the emanations of her soul, to enlighten and to bless.'[9]

At mid-century, the Canadian press saved its worst epithets for 'public women': Lucy Stone, Mrs Bloomer, and the 'whole tribe of unwomanly women who disgrace themselves, their progenitors, and associates, by lecturing noisy audiences night after night, are the prime movers and the proper representatives of these disorganising developments of modern folly and wickedness.'[10] Women who tried to improve conditions for their sex had first to fight for a public hearing. Over the century, as women spoke at religious assemblies and at anti-slavery, temperance, and women's rights conventions, the sight of a woman at the platform became tolerable, though, as late as 1914, Andrew Macphail denounced suffragists as 'the show-women, the m'as-tu vus.'[11]

Yet, paradoxically, women were themselves a kind of spectacle. Rousseau believed that woman's good character rested on her reputation. 'When a man acts well, he depends only on himself and can brave public judgment; but when a woman acts well, she has accomplished only half of her task, and what is thought of her is no less important to her than what she actually is ... Opinion is the grave of virtue among men and its throne among women.'[12] An American lecturer explained this theory to audiences a century later: 'Nature has implanted in the heart of every woman a desire to appear well in the eyes of others; this desire should never be contravened unless it oversteps the bounds of propriety, but should be indulged so far as your means will justify. It is associated in her mind with the feeling of self-respect, which is one of the best safeguards of a virtuous character.'[13] Moral agency rested with the male audience, and women were, despite themselves, always on display.

Men had an 'inalienable right' to stare at women. 'When a woman walks the streets she leaves her virtuous indignation countenance at home; she knows well enough that the street is a picture gallery, where pretty faces framed in pretty bonnets are meant to be seen, and everybody has a right to see them.' When one girl complained of a man staring obtrusively, her father insisted she avert her gaze so that her observer 'may stare his eyes out without annoying you in the least.'[14] The power and privilege of vision was vested in men. One woman mischievously used this line of reasoning to argue that education could not impair virtue: 'Are the Turkish women, veiled and secluded as they are kept, more virtuous than those of European nations? Are we the worse for going out for a walk, when we are told to "put our virtuous indignation in our pockets," and submit to be regarded as pictures on exhibition?'[15]

Women and exhibitions shared function as well as form: Both exerted a refining influence. Individual women were expected to elevate the men in their lives, and women as a social force were to elevate society by exercising domestic virtues. The Montreal *Gazette* in 1822 attested to woman's influence 'in moulding and improving the general manners, dispositions, and conduct of the other sex, by society and example.' The *Canada Farmer* in 1870 declared that 'while she has the noble task of moulding men's characters, and instilling into their souls all that is purest and best in human nature, she need not descend from her lofty sphere into the dusty arena of politics.' Appearance remained paramount, for 'woman's influence depends largely upon her power of charming. A frowsy woman may possibly be morally very excellent, but her influence will be far less than that of a fascinating, sinful sister.'[16]

The same rhetoric was bestowed on exhibitions. The Great Exhibition of 1851 was expected to bring happiness, peace, prosperity, friendship, and brotherhood, and Canadians joined in the enthusiastic predictions. C.D. Day likened its spiritual qualities to high art and hoped it might inaugurate an age of peace, while Rev. Samuel Nelles explained how its 'infusion of mind and artistic life' into material life would lead to moral progress. The Toronto *Globe* saw in the vibrant Mechanics' Institute festival of 1851 proof that material progress 'tends, or should tend, to the comfort of the human family, to the absorption of selfishness, and to the adumbration of that unity of interest, that harmony and benevolence of design which Nature exhibits.'[17]

Exhibitions were a mirror of society, but of society distilled to all that was best in it and reflected back onto the minds of men and women so as to raise their sights and sentiments to loftier grounds. Woman, as the 'treasurer of the spiritual wealth of the race,'[18] was also a kind of mirror of social virtue, which she transmitted to her children. To use Kenneth Burke's language, women and exhibitions were each a 'constitution' for humanity and promised to remake the world in their own image, creating a better world: They were 'the very substance of the present' and 'an exhortation about what might be.'[19] But to declare something *substantially* true of the present is, as Burke points out, to remark that it is not in fact true at all. The improving influences of women and exhibitions were not in evidence in everyday life, where workers and husbands seemed all too often sullen and drunk or worse, and world peace remained elusive.

Women, like art, represented the ideal. Art and womanhood were 'closely identified,' as the *Illustrated World's Fair* remarked in 1892, 'for art would be nothing without woman and woman owes much of her idealization and emancipation to art.'[20] Woman ennobled as art ennobled – by its transparency, according to John Ruskin. Ruskin believed that art should not simply imitate life, but the more closely it 'exhibited' nature the better it was, and the true aim of art was 'representation of some natural fact as truly as possible.'[21] This idealization left women little scope for personal development. Feminists such as Anna Dickinson in 1870 insisted: 'I would make every woman understand that she was born for herself and not for another ... Woman was *made for herself* – to round out herself. Let her live to the full and make a complete woman of her in every respect.' Anna Garlin Spencer remarked at the World's Fair of 1893 that woman 'began to learn that she was a person and not merely a passive conveyer of personality from generation to generation ... not merely a purveyor to the individuality of man.'[22] Because they represented women's existence in material terms, exhibitions helped to establish the fact that women were not simply transparent vessels pointed towards a future utopia, but had some genuine presence or substance in the present-day world.

Canadian Exhibitions

The early agricultural exhibitions were designed to encourage market activity and were dominated by the manly arts of ploughing, grain growing, and stock breeding. Occasionally, propertied women appeared on the prize lists, usually widows. In 1820, 450 prizes were awarded in parish exhibitions throughout the Quebec District, and two went to widows, both for cows. In 1837 Widow Vincent had the best bull in Missisquoi.[23] In Prince Edward Island the livestock of a Mrs Blake dominated fat stock shows year after year during the 1860s.[24] The number of females winning agricultural prizes did increase over the century, as more women became farmers and breeders (over 10,000 according to the census of 1891). At the Eastern Townships Exhibition of 1892, three women won ten per cent of cattle prizes.[25] Eliza M. Jones took her magnificent Jersey cattle to provincial exhibitions in Ontario and Quebec and won prizes worth hundreds of dollars. Women won prizes in bee keeping, market gardening, fruit growing, and dairying, and in the 1870s they won model farm competitions in Quebec. In 1888 women won about 10 per cent of Horticultural Society prizes in that province. Nuns won many prizes: In the 1890s the Sisters of Charity in Montmagny regularly won for grains, forage crops, butter, cows, pigs, and horses.[26] Nonetheless, in traditional agricultural categories, men continued to take most of the prizes.

Domestic work rarely appeared at early exhibitions, though there were exceptions. The Frontenac County Agricultural Society awarded prizes for flannel, linen, socks, and cheese, all traditional female productions, during the 1820s.[27] Women's work began to appear for several reasons. First, it was an important part of the domestic economy. An emigrants' guide of 1849 advised: 'A settler's wife should be active, putting *her* hand to every household work – she must become skilled in sugar-boiling, candle and soap making, and the baking of bread, the manufacture of leaven, salting, and curing of meat, and fish, knitting of all kinds, spinning, dyeing, and making into cloth and clothes her wool and flax, for there are no tailors or mantua makers in the bush – she must also manage poultry, butter, and cheese.'[28] Another text warned that crime, folly, and intemperance resulted from idleness: 'If we abandon household manufactures, we place our females in a dangerous situation.'[29] In Quebec, Firmin H. Proulx discouraged prizes for domestic work and butter as having little impact upon agricultural progress, but, confronted with rural poverty, he concluded that exhibitions of woollens and linens would encourage women to make rather than buy these household objects. In 1876 he advised women to compete for honorific prizes and, in 1885, praised the woollens shown by wives, 'qui tiennent plutôt à enrichir la ferme que de faire des dettes chez le marchand, en achetant des étoffes et des laines importées des pays étrangers.'[30]

Another reason for the incorporation of women's work was the tradition of charity bazaars. In Montreal a bazaar held in 1831 raised £273, which was divided among Catholic and Protestant hospitals and other good causes.[31] Here was fancy-work of all sorts, including quilts, the last word in extravagance of time, labour, and embellishment. The articles exhibited were for sale, but the purpose was philanthropy, not market activity. Why not show them at exhibitions as well? No remunerative price could have been obtained for items, like quilts, on which a lifetime's work had been lavished, in defiance of market considerations. Their husbands' work, found in the implements shed or the cattle stalls, enabled genteel women to devote themselves to domestic and fancy-work. One speechifier contrasted European women who laboured in fields and mines with those represented at Nova Scotia's first industrial show in 1854: 'This Exhibition shows that the females of our land are at liberty to spend their time and display their ingenuity and taste in pursuits better adapted to their sex.'[32] Men's and women's work, side by side, operated as a binary logic to reinforce the ideology of separate spheres.

Finally, women were invited to compete for pragmatic reasons, to swell the number of paying visitors. A Lambton Society stated: 'It is a fact that the majority of paying visitors go purposely to see the ladies' handiwork and manufactures.' In Kent, in 1873, 'as usual the Ladies Department was well represented, both by themselves and their handyworks. And, had the Exhibition Building been deprived of the attractions of either, the social and pecuniary loss to the Society would have been very great.' Directors credited women with the success of the Western Fair, the first large private exhibition in Ontario: 'We have no hesitation in admitting that the splendid display in the ladies' department formed one of the chief attractions of the show, and by increasing the attendance, enabled the Directors to pay all demands in full, and left a surplus of $837.'[33]

By mid-century most township, county, and provincial exhibitions in Ontario and Quebec had Ladies Departments. Exhibits were useful or fancy, ranging from straw bonnets and woollen socks to lace and bark pictures. Women also exhibited in the art department, usually as amateurs. There were fads: One year everyone sent macramé; another year, painting on china; still another log-cabin quilts; and then, Honiton lace. In the 1880s the Toronto *Mail*, with the slogan 'Art is a new gospel,' waged a campaign to reform this section in the name of William Morris and the aesthetic movement, which applied design to household objects. 'Thank Heaven,' a reporter wrote in 1882, Canadian women had heeded the art schools, or perhaps the visiting Oscar Wilde, and a 'flood tide of beauty' swept out the old 'dull, dreary show of patchwork, cushions covered with specimens of deformed dogs and impossible flowers, antimacassars of prehistoric date, bed-spreads, *et hoc genus omne.*' The *Globe* agreed that exhibits had once been, 'whether regarded from an aristocratic or an economic point of view – discouraging,' but had become a 'true index to the advancement of artistic adornment of the people.' This verdict suggests that, at least at the larger urban shows, exhibitors were not farmers' wives but ladies of the leisured class, with a sprinkling of professionals. The Strickland sisters, for example, won prizes for needlework at exhibitions throughout Ontario, and even at Philadelphia and Paris in the 1870s. Mary Dignam, another successful exhibitor at home and abroad, was an art instructor and president of the Women's Art Association, founded in Toronto in 1892.[34] Even in county exhibitions there were complaints, as early as 1855, that, rather than catering to farmers' wives and daughters, the Ladies Department was 'too exclusively into the hands of those ladies who have had the good fortune to receive a fashionable and genteel education.'[35] William Weld remarked that women preferred township shows because the rowdy Provincial was 'a sorry place to take a lady to.'[36]

The Ladies Department was usually organized and judged by women and reviewed by female reporters, for men avowed a manly ignorance of such matters. Occasionally they stooped to mock a rag mat with a sentimental picture or a beaded footstool: 'No one who has ever used a footstool, but has felt a sense of something lacking.'[37] If men judged fancy-work, there was bound to be trouble, as in one case where a disappointed exhibitor poured out her troubles to a reporter: 'I took my dollar and paid my entrance for the quilt, and then what do I find? Why the judges were a lot of men as didn't know a patch-work-log-cabin from a cambric spread, and there they were a going around, without ever looking once or putting their hands on the quilts and they passed mine, which Mary Ann said she never saw a finer, not even when Tildy Hankins was married to John Bowers.'[38] Female judges in Montmagny County, in Quebec, attended the banquet, a traditional male preserve, and, in 1884, during the toasts, they heard a Tory MP justify female suffrage on the basis of public display of reasoned deliberation: 'Puisque nous les avons acceptées comme juges ici, c'est bien le moins qu'elles commencent par être électeurs.'[39]

Only a modicum of gain and glory could be won in the Ladies Department. Many women exhibited in their husband's names, which were duly published in the newspaper. The prizes were small: If a prize bull was worth $8, a prize bonnet might bring 50 cents or $1 at most. In 1878 domestic work accounted for 7.5 per cent of money awarded at county and township exhibitions in Ontario and 2.5 per cent of money awarded at the Provincial Exhibition in Toronto.[40]

Women were also exhibited *to* at the fairs. Early exhibitions catered to male consumers, but some exhibitors courted women. At Montreal, in 1853, women could sprinkle their hankies at a fountain of eau-de-cologne. During the 1860s sewing machines drew crowds of women. From the 1870s on most exhibitors were retailers rather than manufacturers, and the exhibition became a celebration of consumption rather than of production. By the end of the century it was 'conventional wisdom' that women did 85 per cent of all shopping,[41] and, as a result, more and more exhibits – ranging from corsets piled up in pyramids to sculptures made of Johnston's fluid beef – were tailored towards female consumers. By 1893 the Toronto Industrial Exhibition boasted a Ladies' Day, which it marked with a display of electrically powered appliances.

Women were prominent in the entertainment industry that developed around exhibitions. In 1885 the *Farmer's Advocate* published a cartoon of the 'Modern Agricultural Exhibition' in which lady riders raced around the grounds, lady trapeze artists flew overhead, and lady acrobats contorted themselves in time to music produced by a ladies' band.[42] Sideshows had always accompanied fairs – visitors to the Provincial in Three Rivers in 1856, for example, could see 'Bear Woman' and a little girl from Yamachiche who was 3 feet tall and 3 feet around[43] – but later exhibitions added amusements to the program so as to draw crowds. Some people objected to these exhibitions and complained, as one American exhibition-goer did: 'By a systematic application of stimulants we render our young ladies ambitious for public applause on the race course, where the low jests of the vulgar mingle with the huzzahs of the crowd, rather than to become good housewives and produce articles which give beauty and comfort to home.' One visitor to the Toronto Industrial Exhibition in 1892 exclaimed with distaste: 'They were out of their sphere, and, worse still, they were out of their clothes.'[44]

Middle-class moral earnestness also had a place on the showgrounds. The Woman's Christian Temperance Union put up refreshment booths to crowd out liquor vendors and, 'when there was a lull in business, young ladies sang Christian Endeavor hymns.' Total WCTU profits in Ontario in 1892 reached $1795.50, and 55,475 leaflets were given out.[45] WCTUs also petitioned officials to close down the liquor shanties that lined the streets leading to the gates. In 1881 their worst fears were realized when a woman working at the restaurant on the Montreal showgrounds was seized by a band of men one night as she left the gates, raped, then bartered for a round of drinks.[46]

Agricultural journalist William Weld did not think much of the 'Modern Agricultural Exhibition,' which was, he believed, degraded by the sight of women making a spectacle of themselves with gymnastics, dare-devil racing, and musical performances.

If exhibitions encouraged 'degraded' showgirls to display their bodies, they also permitted respectable women, who ordinarily shunned public display, to achieve a greater public presence. The fetishism of the exhibition worked in women's favour. Women displayed domestic, ladylike productions, not their persons, just as a woman writer could express herself on paper but not in the flesh. Women occasionally sent manufactures. In 1852 Mrs M. Andrews won a prize at the Toronto Provincial Exhibition for 'very pretty specimens of wire netting.' 'Orders for this kind of work, left at Mrs. Dunlop's, will be attended to,' noted a helpful reporter.[47] Yet women rarely won industrial prizes before the turn of the century, despite the fact that they entered the industrial workforce in rising numbers. Women had always worked: They farmed, mended, laundered, sewed, kept animals, or rented rooms. By 1900 more women needed to work,[48] and their work became more visible in factories, offices, and the census. Women made up 14 per cent of the industrial workforce in 1871 and 28 per cent in 1891, when 11 per cent of women earned wages, one-third of them in industry.[49] As observers remarked, usually with dismay, women were working outside the home in increasing numbers.

The exhibition, a reflection of material culture, failed to illustrate this change in women's activities. The work of secretaries, clerks, teachers, servants, or factory workers did not lend itself to display, and few women owned their own business, a prerequisite of exhibiting. In 1891 Canada had more than 3000 professional female milliners, who often advertised their wares at exhibitions, but the Ontario Provincial banned 'fancy imported work or work done by professional lacemakers, fancy goods storekeepers, milliners, shirtmakers, etc.'[50] Exhibitions were meant to display the work of farm wives or leisured ladies. Women began to object to this discrimination. In 1893 Emily Cummings demanded a women's pavilion at the Toronto Exhibition to show the full range of women's work. In 1898 Jean Grant declared the needlework and toiletries on display 'a meagre and insufficient index of what industries our women are engaged in,' and she also wanted 'a building of our own.'[51]

International Exhibitions

In voicing their complaints, Cummings and Grant were influenced by the example of the Woman's Pavilion at the Chicago World's Fair of 1893. Although local exhibitions in Canada created a tradition of female exhibiting, they involved only the haphazard participation of a few women who exhibited in their own names, usually in suitably feminine endeavours. At world's fairs, in contrast, female representation was considered seriously by women who were determined to carve out a place for women in the public sphere. This goal, in turn, led women like Cummings and Grant to object to restrictions at the local shows. Women participated in the early international exhibitions only to a limited extent, but in subsequent decades they converged on them, holding congresses and erecting women's pavilions. Canadian women entered into this struggle with enthusiasm.

The early world's fairs were designed to advertise and promote industrial progress. Canadian politicians and businessmen, determined to publicize the resources of the new country to create markets, rarely encouraged female exhibitors.[52] Until Confederation, therefore, Canadian women were largely unrepresented at the world's fairs, except for a handful of traditional domestic pieces. They did not exhibit in Paxton's Crystal Palace in 1851, although chairs made of black walnut embroidered by 'the ladies of Montreal,' were presented to Queen Victoria. When organizers in Prince Edward Island could not attract specimens of industry other than bonnets, shawls, and other 'trifling' work, they initially decided not to exhibit in London at all.[53] However, when Nova Scotia organizers found the rugs and bonnets to be the best part of a preliminary exhibition held in Halifax, they proudly sent them to the Crystal Palace as proof that refinement existed in the colony.[54]

When New York had a world's fair in 1853, British North American women took advantage of proximity to send a variety of goods. Canadian women won one-quarter of Canada's prizes, for a variety of fancy-work. Berlin carpets (needlework on perforated canvas) were sent by 'the Ladies' of Toronto and Hamilton; the latter priced theirs at $800, to raise money for a church. Women from the Atlantic provinces also sent exhibits, including butter and cheese from Prince Edward Island and a volume of poetry by Mrs Peace of St John's, Newfoundland.[55] The New York fair coincided with a convention of female suffragists, occasioning ridicule in the Canadian press.[56]

British North American women sent small, traditional exhibits to the international exhibitions held in Europe between 1855 and 1867. In Paris, in 1855, women accounted for 38 of Canada's 405 entries, and won prizes for fancy-work, straw goods, wax fruit, and stuffed birds.[57] In 1862 the Canadian government, bankrupted by public works, sent only a small display, including mounted leaves from two women. Among the other exhibitors from British North American colonies, Mrs D.B. Stevens of New Brunswick won a medal for a 'What-not, made from cones,' and 'A Lady' from Newfoundland sent snake root and poplar blossoms. Women accounted for six of Nova Scotia's thirty prizes, but were conspicuously absent from British Columbia's exhibit.[58] Canadian women sent little to Dublin in 1865 or Paris in 1867, winning three honourable mentions, for an embroidered shawl in Dublin and a tablecloth and spun flax in Paris. One of Nova Scotia's successful exhibitors in Dublin in 1865 was Mary Thomas, a Mi'kmaq woman.[59]

After Confederation, Canadian women, following the lead of women in the United States, began to demand and, finally, to obtain, greater prominence at the international exhibitions, beginning with the Philadelphia Centennial Exhibition of 1876. The United States Centennial Commission tried to rally public support for this project by naming a Women's Centennial Executive Committee to raise funds and coordinate a woman's exhibit.[60] The committee soon had representatives across the United States and, when crowded out of the Main Hall, it decided to erect a separate pavilion for exhibits by women. It asked the Canadian government to invite Canadian women and to send information on philanthropy. The invitation was sent to provincial advisory boards, and the Ontario board swiftly voted to encourage lady exhibitors and to accept fancy-work.[61] In Quebec, Siméon LeSage invited schools, convents, and colleges to send models, photographs, and statistics, and he anticipated glory in this department. But Quebec educators fumed when, on visiting the exposition, they found the convents represented at the Woman's Pavilion, 'in the midst of needlework!' Even women deprecated the display: One writer contrasted the wood and cork models of convents with photographs of American charitable women, who ran things for themselves.[62] Ontario's exhibit, with a complete collection of educational artifacts, won greater honour. At the Paris Exposition of 1878, however, an improved Quebec educational exhibit did impress the jurors.

The Woman's Pavilion at Philadelphia, devoted to 'the advancement of women through her work,' was a qualified success. The *New Century for Women*, a weekly paper published in the pavilion, declared: 'The woman's sphere is as much of the world as she can move in; her place whatever she can reach and hold.' But its writers were soon busy defending the pavilion, as suffragists like Lucy Stone, their exhibit rejected by the conservative Women's Committee, contemptuously denounced it for bolstering the 'aggrandizement of the very power which degrades them.'[63] One Canadian visitor described the pavilion as a 'temple to vanity' and looked in vain for 'the new avenues of labor that had been opened for [woman's] advancement' and 'the practical workings of that education which would make them better wives and mothers.'[64] Another correspondent, published in a New Brunswick paper, found that 'everything is either childish and elementary or else suggestive of old-maidish debility,' and that women in industrial pursuits remained unrepresented, anonymous, and ill-paid.[65] The *New Century for Women*, which did not support female suffrage, observed that much women's work could not be separated from male labour in other pavilions and that the collection was a first attempt to show women's participation in the workforce. 'If it has permitted the entrance of many petty prettinesses,' it continued 'it is for the good of the senders thereof – that they may measure these trifles with the dignified and substantial exhibits of others. And as even the wax-fruit and embroidery stand for the living of these women, their means of honest support, we fail to see the objection to it as an exhibit of industry, save as it sets forth in unmistakeable speech, by how few methods hitherto has a woman been permitted to make a living at all.'[66] Even this display was too much for some: One visitor saw only spinster attendants and women who wished they were men, while another argued that women should not exhibit at all, for 'women who regard domestic duties as their profession are little inclined to participate in exhibitions and shine with their accomplishments.'[67] In 1876, then, women tried to turn the Centennial Exhibition to practical purposes, but failed because of the paucity of the display and the hostility of spectators. Nonetheless, the potential for political expression had been established.

The exhibits sent by Canadian women were like those of American women, with one noteworthy exception. Canada's most popular contribution to the Woman's Pavilion was the operator of its steam engine, Emma Allison, of Grimsby, Ontario. She had grown to like working with engines in her father's grist mills near St Catharines, and found the work less tedious than most women's jobs and less troublesome than child-minding.[68] Visitors praised her neat dress and engine room, highlighting this traditional feminine virtue. Canadian women sent, in addition to the educational exhibits, ninety-five entries, most of them in fancy-work and related fields such as church vestments and basket work.[69] Twenty-six were in fine arts, and included three oil paintings by Cornelia Schreiber. Mrs Jane Taylor showed oats, and Mrs Henrietta Connor of Halifax sent a 'model ladies water proof.' Women won seventeen prizes from the Centennial Commission, but their anonymous contributions were everywhere, in industrial and agricultural departments. One Canadian visitor remarked, 'If you would see a specimen of women's work, come over to Agricultural Hall. I love to go there. Its exhibits are real sterling stuff. No sham, no nonsense, no cheat ... Here is woman's work indeed.'

Canadian women sent little to the international exhibitions held between 1876 and 1893. The Canadian government, which paid transportation costs, refused a Miss Barnum's application to send lace to an exhibition held in Sydney, New South Wales, in 1877 because lace was not 'an article which will promote the manufacturing interest of Canada.'[70] At the Paris Exposition of 1878, five women from Ontario and Prince Edward Island exhibited art, but won only 2 of Canada's 263 prizes.[71] When New Orleans had an exhibition in 1885, only a few private exhibits found their way down from Canada, including fruit from Mrs Kettleworth of London. The young Sara Jeannette Duncan reported on the event for the Toronto *Globe* and London *Advertiser*, and penned frothy, witty descriptions of exotic exhibits and exhibitors. Duncan's popular columns secured her full-time employment in journalism.[72] When London, England, hosted the Colonial and Indian Exhibition in 1886, a Miss McMullin of London proposed that 'a collective Exhibit of Canadian women's work wd be an interesting & attractive feature.' The suggestion came to naught. Sir Charles Tupper discouraged female exhibitors; when a widow tried to send a cigar exhibit that her husband had prepared before his death, Tupper rejected the entry, though in this case the woman was able to exert pressure to reverse his decision.[73] Women did, nevertheless, send various wares, including plants mounted by Catharine Parr Trail and artwork by Mary Dignam. Lucius O'Brien, the president of the Royal Canadian Academy, helped to organize the art show and tried to have amateur art from women's colleges rejected as being a 'mass of bad taste and rubbish.' He had made a similar stipulation when the Ontario Society of Artists was asked to manage the art section of the Toronto Industrial Fair in 1878.[74] Canadian men and women alike were absent from the exposition held in Paris in 1889 to celebrate the centenary of the French Revolution, although a few private exhibitors made their appearance and attended an International Congress of Women.[75] Canadian women did contribute to a small display sent to Jamaica in 1893, exhibiting, among other things, gypsum and a Tomato Chow-Chow.[76] On the whole, however, the cost of transport and the perceived frivolity of women's work prevented women from exhibiting overseas to any great extent.

Not only the exhibits but also the bodies of Canadian women were absent from these gatherings, in contrast to other displays and in defiance of the efforts of some Canadians. Since 1851, exhibitors had hired women as attendants.[77] At New Orleans, Duncan described a 'pretty girl' processing the cotton at the Louisiana exhibit, and added: 'It is a noticeable fact that all the exhibitresses are pretty; there is not an ill-favored one in all this fair centennial multitude.' The same remark was made by the commissioner to the Antwerp Exhibition later that year. Canada's decision to make a small show of resources at Antwerp was, Hector Fabre told Siméon LeSage, a mistake: 'Tu es peut-être sous l'impression comme moi ... que les Expositions dites universelles sont destinées à mettre en relief les produits industriels. Pas le moins de monde ... Les Expositions comme celles d'Amsterdam et d'Anvers sont en réalité des expositions feminines. Les femmes de tous les pays s'y exposent. Elles viennent y cueillir des assauts. Les Folies-Bergères agrandies!' Fabre repeated this remark at the Colonial and Indian Exhibition held in London in 1886: 'Comme à Anvers, la section canadienne manque de femmes ... si le Dominion ne veut rien faire dans cette voie, il est nécessaire que la province de Québec se remue et prenne l'initiative. Qui peut mieux qu'elle nous envoye une petite bataillon de femmes accortes, lestes? Vous garderez leur vertu assez difficilement, bien entendu, dans un dossier poudreux. Elles la repronderont à leur retour.'[78] The proposal by a Winnipeg immigration agent to send a 586-lb native woman 'about forty years of age, good looking for her class and very active,' was probably not what Fabre had in mind.[79]

The next important exhibition for Canadian women was the Chicago World's Fair of 1893. American women decided to use the fair to assert women's role in the life of the nation. American feminist Susan B. Anthony spearheaded the movement, covertly through the agency of more respectable women, and obtained funding from Congress for a Board of Lady Managers (BLM), the first time women had obtained such recognition. At the inauguration, BLM president Bertha Potter Palmer declared that 400 years after Columbus discovered America, the U.S. government had just discovered women.[80] There was some friction among rival groups of women, including even the Lady Managers, though Mrs Palmer, a Chicago society woman, finally squelched the tension. The press took great delight in reporting on outbursts at board meetings. John Dryden, the Ontario minister of agriculture, mentioned them during a legislative debate as an argument against giving women the vote, though he did not fail to add the traditional reason that 'woman always was and always will be dependent upon man.' Descriptions of shrieking and hysterical sobbing at the BLM meetings disappeared from the newspapers after a fistfight broke out in the British House of Commons.[81]

As in 1876, Ottawa took charge of the Canadian exhibit, paying for transportation and attendance on the grounds. Several women asked for jobs at Chicago, among them Mary Holden, a Hamilton woman of social standing, who pointed to extensive American preparations and urged that Canada not be left behind: 'There is a very strong feeling, in fact I may say a universal feeling among the women of the Dominion, which Canadian women claim as a right; and that is, proper and special Representation and Competition with the women of the United States, England, France, Belgium, Austria and Italy.' Women's interests were at stake, she added, for the exhibit would help women obtain equal pay with men. Holden produced references from Hector Langevin, Sir William Dawson, a bishop, a judge, a senator, and dozens more, including Hamilton's leading lady of philanthropy, Adelaide Hoodless, an advocate of practical education for women. Hoodless felt that Canadian women had the right to compete with other women, 'and by competition to prove the value of Women's work when properly understood,' as well as to learn from 'the cleverest Women in the world' at the congress. She added that she was no suffragist, nor did most women seek 'representation providing they receive reasonable recognition when necessary, as in the present instance they feel they have a right to a portion of the Government grant.'[82]

Holden's petitions had some effect. One of her references was Adam Brown, commissioner to Jamaica in 1891. William Saunders, the Canadian commissioner to Chicago asked Brown for advice, declaring himself 'at a loss to know what duties it is proposed a lady Commissioner should discharge.' After consulting with Mrs Potter Palmer, Saunders urged the appointment, so the department hired a woman to visit local fairs and choose goods for Chicago. Holden was edged out for this position by Edith Barrett, the choice of Lady Stanley and, as Holden protested scornfully, '*an expert in Needle-work.*' Women continued to press for a female commissioner at Chicago, but Saunders's successor, J.S. Larke, claimed that the only correspondence came from office seekers (as if this were not true of the male commissioners), and a suggestion to send inventions by women was dismissed as being of a 'trifling nature.'[83]

The work of collecting soon began. Barrett visited exhibitions in central Canada and the Maritimes. The wife of Angus MacKay, head of the Experimental Farm at Indian Head, collected goods from the West and found a dozen pieces at the Regina fair which 'may do for Chicago if not examined too closely at Ottawa.' Manitoba women sent their goods to a separate pavilion erected outside the fair grounds.[84] British Columbia sent a tapestry and preserves. Philanthropic organizations chafed at being vetted by a needlework expert and, instead, the Canadian WCTU exhibited with the World Woman's Christian Temperance Union. Lillian Phelps of St Catharines collected banners, photographs, curios, and leaflets from across Canada, only to learn that the World WCTU had been allocated only 15 square feet. Most of the exhibits had to stay at home, and, moreover, that their leaflets were held up by customs. The Canadian WCTU found the experience disappointing.[85]

Very few Canadian exhibits appeared in the Woman's Pavilion. Saunders learned the display was not competitive and supposed that Canadian women would prefer to exhibit with Canada, so he never applied for space. Mrs Potter Palmer told a Canadian visitor: 'We feel very sore at the way you Canadian women have treated us. We expected so much help from you, our next-door neighbors, in our efforts for the advancement of womanhood.' Canadian women were no less sore at their government, for their exhibit was fitted into the Canadian manufactures display, where it filled two cases largely devoted to needlework. Alice Fenton Freeman reviewed it in the Toronto *Empire*: 'The exhibit is not an extensive one for Canada. We could and should have shown a collection of woman's work more varied and complete; for the women of Canada are not one whit less clever or gifted than those of other nations. But in the department wherein we have competed, that of art needlework, we are inferior to none and excel the majority.'[86] Barrett and her pupils won half Canada's fifty-four prizes in this department. Women took other prizes, for education, design, a patent frying pan, pickles, fruit, butter, and cheese. Sarah Holden won three for art, and a Mrs Morrison won a prize for her Tsimshian collection. As in previous exhibitions, anonymous women contributed to the prizes won by such factories as Penman, a knit-goods company in Paris, Ontario, manned by women.[87] The BLM tried to tally female factory labour without success.

Canada was asked to name judges to the 1893 fair, including lady judges. Ottawa responded that it had 'no names to submit, and would rather not debar the privilege of nominating men' by recommending women, for Canadian judges ensured Canadian prizes, and the government reserved this privilege for economically important exhibits. In the end, the BLM had Eliza Jones named a judge for butter, though she confessed she would rather be judging cows.[88] The BLM could name one woman judge for every five to six men, except in livestock, where the most valuable prizes were distributed.

The 1893 World's Fair was embellished with a series of congresses on politics, economics, and intellectual and religious matters, beginning with a Congress of Women held on 15 May soon after the exposition opened. Several Canadian groups attended, including the WCTU, the Order of the King's Daughters and Sons, missionary societies, and the Women's Enfranchisement Association. The congress was enormously successful, attracting 20,000 women and extensive press coverage, even in Canada. Suffragists described it as 'the greatest single impulse' to their cause: 'The fact that the first congress of the fair was a woman's congress put the woman's rights questions on the front page of every newspaper in the land.'[89] Women journalists flocked to the congress, including Emily Cummings for the Toronto *Globe*, Alice Freeman for the *Empire*, and Marie Lacoste-Guérin and Josephine Dandurand for *Le coin du feu* in Quebec City. Lacoste-Guérin explained the lesson of the exhibition: Women were entering the workforce. It was an important social change, the product of 'un besoin réel,' and the more they earned the better for all.[90]

Canadian women spoke at the congress. Emily Stowe, the president of the Women's Enfranchisement Association, made some remarks, but then fell from a platform and broke her hip.[91] Several women spoke at the symposium, A Century of Progress for Canadian Women, describing suffrage, missionary, and education work.[92] Canadians listened to lectures and discussed them in the streets, the cafeterias, and the women's dormitory. The BLM made special arrangements so that unchaperoned women could visit, and many did. Cummings found that American women 'are possessed of more general information and have a greater desire for knowledge than are their Canadian sisters.' Mrs Alfred Denison described American chauvinism: 'A female shook her umbrella at me and bawled (although I was very near her) "You Canadians are indifferent. You must be aroused. You must canvass. You must *vote.*"'[93]

Canadian women participated in other congresses as well. At a press congress in June, a comment that 'Canadian women have done nothing in art since the squaw maiden of 1400 got up an internecine war through her paints and powders' was refuted by Eve Brodique.[94] The WCTU sent representatives to temperance congresses in June and September. Members had attended a congress at Philadelphia in 1876, where Letitia Youmans expressed pleasure at hearing for the first time a Canadian woman, Mrs Harvey, speak publicly in favour of temperance. In fact, Canadian women had publicly preached temperance since mid-century, and Youmans had begun to do so before mixed audiences in 1875.[95]

The congress and exposition drew attention to the fact and plight of working women. According to Mrs Potter Palmer, the exposition would 'address itself to the formation of a public sentiment which will favor woman's industrial equality, and her receiving just compensation for services rendered. It will try to secure for her work the consideration and respect which it deserves, and establish her importance as an economic factor.' She gave a blistering address at the inauguration of the pavilion on 1 May, declaring the idea of separate spheres appealing but inadequate. There were too many irresponsible husbands, too many spinsters and widows without any male protection, too many women who 'must work or they must starve.' Idealists, usually well-off themselves, must open their eyes to the actual conditions of women. Great applause greeted her remarks.[96]

If women could make their economic activity visible, suffrage had to follow. As economic agents, they had the same right to protect their interests as any man. Goldwin Smith, quintessential Victorian liberal and no friend to female suffrage, admitted in 1872: 'Supposing women to be emancipated legally, conjugally, and industrially ... the objections to their taking part in politics would obviously be diminished. At present, reigning apart in the household the woman does not directly feel those effects of good or bad government which are felt by the man, who goes forth to labour.' He admitted that women were working in greater numbers, but thought the situation 'abnormal and temporary.' Two decades later, the congress speakers corrected him. Mary J. Foster of the Woman's Republican Association remarked that 'in this country a large part of the *wage-earners are women.* The questions of economics which are involved in the present industrial system affect them, and the wives and dependent children of men wage-earners.' Augusta Cooper Bristol stated that 'industrial emancipation broadens by an inevitable law into social and political equality.'[97] Auguste Comte had proclaimed 'a visible tendency towards the removal of women from all industrial occupations,' but women now proved that the opposite was true. Canadian women journalists, even those hostile to suffragism, took this position. At a Manitoba mock parliament, held before the congress and described to it by Manitoba representatives, a Mrs Dolsen invoked the slogan of 'no taxation without representation' to assert: 'Admit in the slightest degree her right to property and education, and she must have the ballot to protect the one and use the other.'[98]

Fine speeches notwithstanding, historians criticize the congress for the bourgeois, white character of the participants. Moreover, the pavilion could not adequately display the work of women, which could not be separated from that of men. The Canadian government lacked the imagination or interest to send more than fancy-work, and its control over the paths to Chicago hindered women from acting themselves. Still, the fair probably did 'enlarge the notion of what was proper activity for ladies.'[99]

The Chicago exposition had important consequences for Canadian women. The Woman's Congress was organized by the International Council of Women, which had been formed five years earlier in Washington. This umbrella group brought women's organizations together to collaborate on goals common to all. Canada was represented at that first congress by Bessie Starr Keefer and Mary McDonell of the WCTU.[100] At that time, the International Council was largely an American entity, but in 1893 it voted to hold the next congress in England. It elected as president Lady Aberdeen, an outspoken feminist, active in middle-class reform movements in Scotland,[101] whose husband had just been named governor general of Canada. Inspired by this connection and by the congress, the Canadian women in Chicago met at Potter Palmer's house to form a provisional National Council of Women for Canada. Though elections were held in Canada that fall, the minute books of the NCW begin on 22 May 1893 in Chicago. Its first act was to petition the government to send a female commissioner to Chicago. Activities of the NCW during the 1890s included efforts to improve conditions for working women.[102] Other organizations benefited from the exposition: Adelaide Hoodless represented the Hamilton YWCA at Chicago and, on her return, wrote to every local YWCA in Canada. As a result, a dominion YWCA was formed in December 1893.[103]

At the first public meeting of the NCW, Lady Aberdeen spoke on the goals of the council, and her remarks were published in the Canadian press. Toronto *Empire* columnist Alice Freeman found this significant: 'Up to the present time it has not been "the correct thing," from a Canadian society standpoint for a woman to speak upon a public platform, no matter how good a word she may have had to say. But since our chiefest lady said her pleasant, earnest motherly words last week there can be no more condemnation ... for the first time within my remembrance a woman's meeting, addressed by women speakers on woman's work was fully reported, even to the extent of two or three columns, in the Toronto daily press. And what greater proof of its importance can we have than that?'[104] The next year, Premier Oliver Mowat supported a bill to enable Clara Brett Martin to become a lawyer, dismissing objections against a woman speaking in a court-room with the remark: 'They went to hear women sing in public, and to speak in public. In fact the highest lady in Canada [Lady Aberdeen] speaks well and ably.'[105] Public speaking by women in Canada thus acquired a heightened respectability.

The Chicago World's Fair demonstrates clearly how exhibitions provided a back door into the public sphere. It helped to make 'woman questions' and suffrage a matter for public debate. Habermas defined the bourgeois public sphere as composed of 'private people making use of their reason' and 'engaged in critical public debate.'[106] Women at the congress were manifestly doing just that. In arguing with their claims, men *ipso facto* invited these women into an enlarged public sphere. Freeman summed up the week of 'womanly intellectuality': 'And thus the philanthropist, the suffragist, the litterateur, the club woman, the exponent of the drama, and the society woman stood side by side in cordial good fellowship; each recognizing the other to be a needful link of the shining silver chain of rounded and complete womanhood.'[107] The suffragist could no more have been included in such a description in 1853 than the prostitute in 1893. The range of public debate had expanded. Expositions first expanded the non-political public sphere: Men had always smiled benevolently on displays of fancy-work, and this reception helped to create a climate sympathetic to the public representation of women's activities. Women cunningly pointed out the economic aspect of these activities and added speeches. Words seemed less substantial than the goods and deeds exhibited, but these speeches were themselves provocative deeds.

The Canadian National Council of Women learned this lesson well and applied it at the next major exhibition, in Paris in 1900. The NCW asked the minster of agriculture, Sydney Fisher, for a special display of women's work. He answered that he could not spare the space: Women's work would enter by the usual channels, and, in the event, twenty-one women sent art, fancy-work, grains, fruit, and minerals. Fisher did offer $2500 to compile a report on 'the organisation of women & women's work in the Dominion of Canada, and any statistics upon economic subjects in which the women of Canada have particularly interested themselves or are particularly engaged,' to be sent to the exposition. This, exclaimed the council, was even better, and it went about assembling the book.[108] By 1900 the NCW boasted seven national and dozens of local member organizations across the dominion, all of which were canvassed for facts about women in Canada. The council called on the leading public women of Canada for essays, including many involved in past exhibition work: Emily Cummings, Lady Aberdeen, Josephine Dandurand, Mary Dignam, Adelaide Hoodless, Clara Brett Martin, and Edith Archibald of the WCTU. The result was a substantial work, published in English and French, entitled *Women of Canada: Their Life and Work*. The English run was soon exhausted and requests had to be refused.

The book addressed economic, religious, artistic, philanthropic, and other activities by women, their legal and political status, their education – virtually every aspect of their lives. There were lists of organizations run by women, occupations open to them and wages they could expect, and even the names of university graduates or women in unusual professions like journalism, the law, and literature. An opening essay rewrote Canadian history along conventional lines, inserting women. Several mentioned the role of exhibitions in bringing women's work before the public. Some of the papers contradicted each other: Hoodless wrote about women's industrial possibilities, but repeated the Immigration Department line that only farm and domestic workers were in demand. The book's editors, Julia Drummond and Lady Aberdeen, claimed that the NCW pursued 'unity of standard, replacing forever that divided ideal which, proclaiming purity to the woman, but to the man truth and honor and strength, does in effect make the man less manly and the woman less womanly, and proves its fallacy by causing half the sins and miseries of life.' Josephine Dandurand, wrote that woman 'transmits from generation to generation, by example and heredity, unaffected goodness, moral and physical health.'[109]

Drummond, Dandurand, and Robertine Barry attended the Paris Exposition for the NCW, but Dandurand also shared in the appointment of her husband, Raoul Dandurand, as dominion commissioner – a reward for his having organized Liberal election campaigns. The daughter of Premier F.-G. Marchand, she was friendly with Wilfrid Laurier, to whom she chatted at a garden party about suitable candidates for Paris, and with Lady Aberdeen, who supported her candidature.[110] Josephine Dandurand was a lively and talented writer. Like Hoodless, she was more interested in reforming women's education than in their political emancipation, and she tried to raise the intellectual content of French-Canadian women's culture, which she found frivolous and barren. Her journal, *Le coin du feu*, pioneered women's journalism in Quebec[111] and, if conservative in outlook, was radical in creating a forum for women in that province.

Dandurand represented Canada at the Congress of Women in Paris in 1900. This congress was not organized by the International Council of Women (which met the year before in London, with forty Canadian delegates) because France did not have a National Council until 1901. Instead, three groups held congresses in 1900: Catholics, Socialists, and Liberals and Protestants.[112] Josephine Dandurand was not a socialist and not particularly religious. She attended the liberal congress, which addressed issues ranging from labour laws to prostitution. Discussion of prostitution took an anti-spectacular tenor. Women who made a notorious spectacle of themselves on Paris streets could be arrested, and speakers denounced the idea that a woman could be an 'immoral exhibition,' the mere sight of whom offended.[113] Prostitution was a private economic transaction, argued some, and women engaging in it were no more on exhibition than anyone else. The delegates invested moral agency not in the spectator, but in the woman herself, decisively refuting the Rousseauian notion that women were a sort of exhibition. Robertine Barry, a columnist for *La Patrie*, briefly noted the furore the debate on prostitution caused.[114] Dandurand did not speak either here or at a session on education, where one lecturer denounced Rousseau's argument that women were destined uniquely to please men. Dandurand lectured on French-Canadian women and on literature, urging that the condition of woman be improved so that she might live up the to lofty mission Rousseau assigned her: 'Qu'elle se renseigne, qu'elle trempe sa raison pour pouvoir guider ses fils et, comme le souhaitait l'auteur de l'*Emile*, "pour savoir conduire les homes qui l'environnent, en les comprennant."'[115] She was not alone in voicing this argument, which had considerable persuasive force. Her argument was not simply Rousseau revived, for the Genevan would have frowned on her writing and speaking, but it did run contrary to the general tenor of the congress. Most speakers denounced the idea of woman as exhibition, a being validated by the male glance and the ennobling qualities she exercised over men. Still, the congress was a public forum for rational debate about these matters.

Many women continued to insist that the role of both women and exhibitions was to ennoble, for this outlook established the fitness of female participation at exhibitions. They hailed exhibitions as factual, quasi-scientific descriptions of the present, with women's presence now properly noted. Exhibitions, they claimed, were educational, though already this view was suspect as fairs took on a more popular and commercial character, complete with lady performers.

Exhibitions and women are no longer expected to ennoble because they no longer represent something greater than themselves. Exhibitions may advance technology, but they do not elevate morals, for cultural representations no longer edify as they once did.[116] Nor do women. Victorian womanhood was an ideological construct, signifying noble ideals of peace and virtue to be fulfilled in a mythological future. As an idea or a 'constitution,' womanhood blinded Victorians to historical reality, so that actual women who worked and sinned were either invisible or dismissed as a transient phenomena that progress would eradicate. 'Woman' was an allegory of ideal history rather than a historical person.[117] Some feminists ruptured this paradigm by insisting that women were not ideological constructs, not transparent vessels pointed towards a future bliss, but beings with mundane material needs, just like men. This approach meant that the idealist view of history as the realization of noble ends had to be replaced by a materialist view, one in which men and women followed no higher purpose than economic necessity.

Early Victorian women replicated the form of the exhibition and could not be both form and content. Representation could not represent itself. Women carved out a place for themselves at exhibitions only gradually and piecemeal. Ironically, only by means of representations could they achieve substantial presence as historical agents. Women could not simply strip away the idealizations to expose a 'real' and unmediated womanhood. They could only construct other kinds of representations. The metonymic images of working women introduced to exhibitions displaced neither domesticity nor the display of female bodies, but they did broaden the discursive field, permitting greater debate about the role of women and with a more varied vocabulary. Exhibitions were, of course, only one forum among many, where women advanced new descriptions of themselves and their work. But because exhibitions were physical representations rather than verbal descriptions, they were particularly effective. Women began the century as mere exhibits, but they ended it as exhibitors.

By the 1890s women had learned to use a technology of the 1850s to gain access to a public sphere that was, according to Habermas, breaking down. He blamed this breakdown on intractable class differences and the creation of the welfare state, which intruded on private life, and made it too public. Long before women obtained the vote, they advocated welfare measures. Every congress ended with demands for reforms, such as state salaries for mothers. Nineteenth-century feminists demanded that women be inserted into the classic liberal public sphere, but they knew this objective would require a radical reformulation of the public sphere, a task that continues to this day.

If some women learned to turn exhibitions to serious purposes, most women in nineteenth-century Canada exhibited for fun and profit. The 'farmer's festival' provided a rare occasion for sociability among rural women, who saw their work honoured. Yet for all women, exhibiting was an act of self-assertion and, thus, a political act. In the days before public expression by women was acceptable, exhibitions gave them a voice to express themselves and it was one they avidly adopted, whether they spoke the language of temperance or tinsel, cows or crochet tidies.

10

Making a Spectacle: Exhibitions of the First Nations

This chapter examines the convergence of two examples of what one might call the 'spectacularesque': exhibitions and Indians. This ungrammatical but useful neologism has been coined from two contradictory theories of action. Bakhtin used the term 'carnivalesque' to refer to discourses and acts that parodied and subverted the dominant order or idiom, while DeBord defined the spectacle as a medium that bolstered the capitalist order by reducing culture to a display at which alienated and isolated consumers passively gazed.[1] This book suggests that the spectacle did not lend itself so easily to hegemonic purposes. Designed to educate popular taste, the spectacle ended by strengthening its impact. One reason for this outcome was the democratic controls that underlay the exhibition system, but another was the medium itself – the content of its form. The exhibition as a genre was supposed to illustrate progress and the superiority of 'civilization' over 'barbarism,' but civilization, unfortunately, was not as spectacular or visually exciting as exotic and gaudy barbarism. A lecture to the Royal Institution in 1855, published in both the *Westminster Review* and the Charlottetown *Islander*, explained that, historically, the ornamental preceded the useful: Savages preferred paint and trinkets to clothing, whereas modern men dressed more for comfort than appearance. Women were not as advanced as men in this respect. The same 'restless craving' to impress others persisted in higher civilization, but in less spectacular forms.[2]

The theory of exhibitions thus contained a contradiction. In their message of triumphant capitalism they pointed to an idealized future, yet their medium of spectacular display seemed to point backwards to barbarism. The term spectacularesque refers to their potential to subvert the sober values they were expected to inculcate. This quality was especially marked in those parts of the exhibition that were most visually exciting, most spectacular: the displays of 'primitive' culture. These exhibits were initially marginal to the larger goal of displaying modern progress, but popular demand and the desire to reinforce this message of progress led to displays of aboriginal artifacts and peoples becoming more prominent. This development destabilized the theory and practice of exhibitions by uncovering their inconsistencies.

By dint of exotic and barbaric traits, 'the Indian' was a kind of spectacle. After America's 'discovery,' explorers exhibited individual natives in European courts. In the nineteenth century, troupes of touring natives attracted huge crowds. The Inuit John Sakeouse performed paddling feats in Leith in 1816; and in 1843 George Catlin toured with some Ojibwa, Ottawa, and Mississauga natives who performed dances and mock scalpings. Peter Jones, a mixed-blood Methodist minister who gave speaking tours in Britain in the 1840s, denounced this tour, yet even he occasionally donned the obligatory 'odious Indian costume.'[3] Four Huron who went to London to present land claims performed dances after a dinner with the lord mayor. Even in Canada, aboriginal people merely going about their business drew crowds when they wore skins, furs, and beads.

There was an affinity between natives and exhibitions: Both of them were spectacles. One represented primitive gaudiness, whereas the other was a scientific representation of the hard-won wealth and knowledge of the world. In fact, from the beginning, whisky stands, gambling stalls, and drunken brawls subverted the exhibition's lesson in sober toil, as did the acrobats and dancers who became midway staples by the end of the century. Earnest agricultural reformers demanded that these carnivals be abolished and that the grant be spent on lectures and schools. They sounded like Indian agents who wanted to strip natives of their exotic, traditional ways and turn them into farmers.

Canadian Exhibitions

Teaching aboriginal peoples to farm was the foundation of assimilation policy in Canada. This sedentary skill would replace hunting as the source of food and lead them towards higher civilization. In 1828 H.C. Darling, the first chief superintendent of Indian affairs, recommended establishing model farms and villages for this purpose.[4] On reserves at Coldwater and near Simcoe, in Ontario, half-hearted instruction was attempted but soon failed, and 'not until the late 1840's were serious attempts made to establish technical and agricultural schools that actually did more than provide an acquaintance with the white man's hand-tools.'[5] Missionaries had only slightly more success with 'manual education,' teaching agricultural and other skills. Near Rice Lake, Rev. William Case taught agriculture and tool making. He exhibited the results as well as the Ojibwa students on tours of the United States. Most of these educational efforts focused on children, for adults were, many believed, beyond assimilation.[6] It was difficult to teach adult whites to be better farmers, and all the more difficult to teach adult Indians, who often did not understand English. Professional teachers were inappropriate because, unless the teacher made a living from farming, he would advertise subsidized rather than subsistence agriculture; moreover, farming was better taught by example than by precept. T.G. Anderson advised the Upper Canada provincial secretary in 1830 that 'our good practicle farmer, with an axe, a hoe or the plough in his hand, who would receive but 2/6 p day, can do more good to the institution than a dozen Gentlemen looking on at 5/- p day. If I say to an Indian *come* help me do this or that, he generally laughs and steps to the spot but if he sees one standing, with his hands in his pockets, who he thinks ought to work and says *go* and do this or that, the Indian stares at him, grins contempt and walks off muttering "do it yourself."' Or, as the *Mail* later remarked, the 'agricultural instructor must not sit on the fence with a rifle across his knee, shouting to the perspiring Indians in the cornfield, "Hoe faster you red fiends, or I'll open out on you."'[7]

Exhibitions overcame these problems in white society by illustrating real farming and providing a form of self-education that did away with the need for an instructor. But the government waited half a century before introducing exhibitions on reserves. One reason may have been concern over the absence of an emulous class, for, unless an agricultural elite already existed, prizes would go to unimproved farmers and might perpetuate the status quo by rewarding them, whether they tried to improve or not. This was the complaint that white agricultural elites and reformers levelled against small township exhibitions in Ontario throughout the century. They complained in vain. The government continued to fund township exhibitions because this was a way of bribing the mass of farmers, but, as natives were not voters, there was little impetus to sponsor a prize system on reserves. The government may also have considered natives unprepared for self-education. Liberal institutions like agricultural societies required members to exercise self-government, but Indians were thought to be incapable of self-discipline because of their undisciplined childhoods and their vagrant and lawless lives.[8]

After Confederation, exhibitions began to appear on Indian reserves.

This development coincided with more strenuous efforts at assimilation once the earlier programs had clearly failed.[9] Some of the exhibitions were mounted by natives and some by missionaries or Indian agents, but many received little external encouragement. Much depended on geography, as was true of the activities of the Department of Indian Affairs (DIA) generally. The DIA was exceedingly parsimonious during these years, especially in eastern and central Canada.[10] Western Canada was another matter. The DIA spent most of its funds on prairie natives, who were suddenly and disastrously affected by the extinction of the buffalo and the spread of immigrants. As Sarah Carter has shown, it tried to convert them to farming, and one means of doing so was sponsoring exhibitions on reserves and giving prizes for Indians at white shows. Individual native farmers on the prairies fared well at these exhibitions, often beating out white farmers, as at Regina in 1886, and some won prizes at international exhibitions from the 1870s on. In 1886 many native farmers sent the barley, oats, and potatoes they had grown to the Colinderies. However, the success of the agricultural program as a whole was undermined by inefficient agricultural instructors, failure to provide supplies, and an insistence that natives should practise 'peasant' agriculture on small plots of land with outmoded techniques.[11] Parsimony and a theory of directed assimilation caused aboriginal peoples to receive a damaging form of special treatment here as elsewhere.

On the Pacific coast, industrious agents like W.H. Lomas at Cowichan laboured to introduce exhibitions in a frank attempt to 'wean the Indians from the celebration of their potlatch feasts' and 'promote a laudable spirit of enterprise and emulation among them.'[12] The potlatch, which missionary William Duncan described as 'giving away property for display,' reversed the principle of accumulation for display that the exhibition tried to instil.[13] An 'Indian Industrial Exhibition' was first established at Cowichan during the late 1860s and revived in 1887, when it drew 300 entries and 1400 aboriginal spectators. The Indian agent crowed in triumph when prizes at a Fraser valley exhibition went to natives. But agriculture made so few inroads in British Columbia that exhibitions could not be widely used.

There was also little scope for agricultural exhibitions in regions of eastern Canada, in the Maritimes, and the northern and eastern parts of Quebec. There, many natives depended for their income on the manufacture and sale of 'Indian curios,' especially baskets and beaded leather. The Abenaki, for example, earned $15,500 from baskets, moccasins, and other items in 1887. Most of their sales were to the United States, where entire native families spent the winter selling to tourists at resorts. The Huron-

Wendat near Quebec City, who numbered only 265 in 1887, earned $46,635 from the sale of curiosities that year. The Mi'kmaq had come to rely on sales of artifacts a century earlier, when white settlement caused game to dwindle, and, in 1852, journalist Joseph Doutre noted the importance of such wares to the Iroquois economy.[14] Government agents had mixed feelings about this industry. It provided the poorest reserves with an income, but it also hindered 'civilized' and manly industry, for it was a woman's trade. A DIA agent among the Mi'kmaq voiced his withering scorn: 'The women, indeed, plaited a few baskets, which were purchased more as toys than as articles designed to serve any useful purpose.' The superintendent of Indian affairs, Sir John A. Macdonald, further complained that 'to dispose of these articles the Indians have to visit numerous places, and thus their old, and to them congenial habit of wandering about the country is fostered, which is attended with evil results to them, morally and materially.'[15] Agriculture, which fostered settlement and manly labour, was much preferred as an economic staple.

Agricultural exhibitions were more successful in Ontario, where, the DIA boasted in 1880, 'a laudable ambition to excel in agricultural pursuits is reported to exist among the majority of bands.' However, the department also boasted that these reserves were 'mainly self-supporting' and that 'these Indians cost the country nothing,' save for a few schools.[16] Thus, while its annual reports were full of references to successful Indian exhibitions, the department provided little encouragement for these activities, and its interventions were often more harmful than useful.

The first native-run show was mounted at the Grand River Reserve in Brant County, where a Six Nations Agricultural Society was formed in 1868 with thirty members. Its first exhibition, held that year, attracted 217 entries.[17] In 1869 it received £20 from an Anglican society; after 1873 it obtained government aid; and, after 1879, it received $270 from the band council. By 1885 membership had risen to 400, entries to 1500, and gate receipts to $1000. Amusements included races on Indian ponies. Aside from sporting events, the competition was restricted to natives, on and off the Six Nations reserve. Visitors attributed a marked increase in the quality and quantity of stock, grains, fruit, and dairy products to the exhibition.

The Chippewa of New Credit, sponsored by Anglicans and Methodists, had their first exhibition in 1870. By 1890 David Boyle remarked with admiration that 'no stranger driving through the settlement could observe anything to indicate that the land was farmed by others than white men.'[18] On Walpole Island, Rev. Jamieson started an agricultural society during the 1870s, but it failed, as did one at Parry Sound, organized by the agent

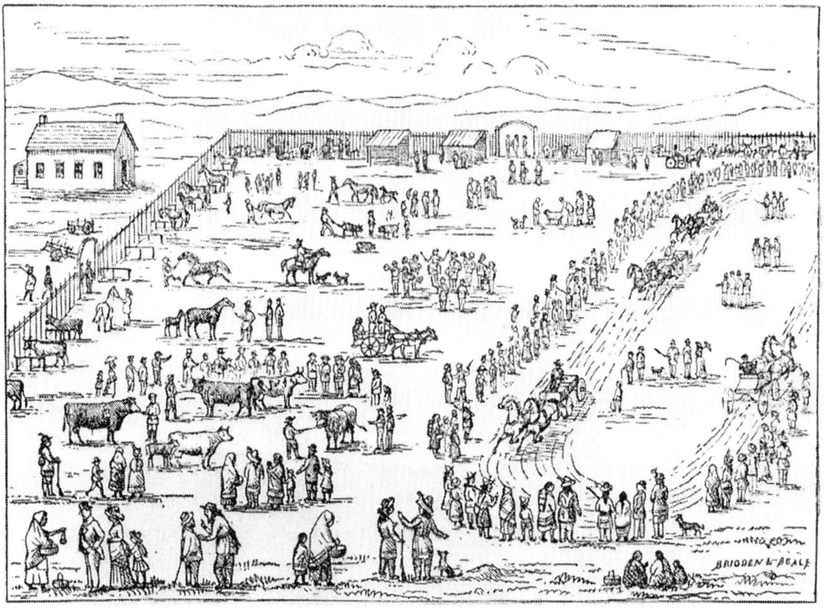

The Oneida exhibition in Delaware Township, Ontario, was a successful annual event that drew whites and aboriginals from all around. William Weld approved of this unpolished, genuine farmers' fair.

in 1883, which attracted only three or four exhibitors.[19] The Oneida of Delaware Township inquired into chartering an agricultural society in 1878, but were told that, because a non-native society already existed in that township, no other could be sanctioned. This refusal did not stop them from organizing a self-supporting show in 1885, with 3000–4000 paying visitors, which the *Farmer's Advocate* praised as a genuine farmers' show with practical livestock.[20]

The Delaware who settled at Moraviantown were traditionally farmers, and their show, which began humbly in 1882, soon prospered. In 1883 thirty-six cattle, a few scrubby sheep, grains, and domestic goods greeted the paying visitors, from whom $80 was extracted at the gate. Chief John Lewis published an account of the exhibition with a few pointed remarks: 'We hope we shall prosper in future and strive to be industrious Indians, and learn to cultivate the soil. Although we find our forty-acre farms too small, we will try and make the best of them by thoroughly cultivating what we have.'[21] Soon the fair attracted competitors from many different reserves, membership doubled, and, by 1885, receipts grew to $175. Bothwell businessmen eyed this profitable enterprise jealously and, after one failed attempt, began to hold annual shows beginning in 1887, when they took in half the gate receipts at Moraviantown. These results turned contemporary cultural theory on its head. Energetic Anglo-Saxons were supposed to inspire nonchalant natives, not the other way around. The Bothwell *Times* ruefully admitted that its 'snoring merchants' were put to shame.

At the Caradoc Reserve, Chippewa, Oneida, and Muncey established fairs. The DIA made small grants of $60–$90 to the Oneida and the Muncey, but turned down requests from the Chippewa on the grounds that the band had enough money of its own. The Chippewa Band Council decided to award $50 to the agricultural society; in 1892 this bounty was divided among thirteen competitors, who won between $2.05 and $6.35 each.[22] In 1917 the department insisted that all three bands collaborate on one united show.

The government was stingy with its own money and with band funds, over which it exercised a paternal supervision. In 1895, when the Sarnia Chippewa applied to spend $500 of its band money to improve the show grounds, Hayter Reed replied that 'the interests of the Indians are better served by competing at the Agricultural Shows organized by white farmers.' Chief Wilson Jacobs answered that this competition was not possible because the Indian farmers were not as advanced as the whites and did not receive justice from white judges. The DIA finally approved the deduction from the band funds, but delayed so much that the work was not completed for either that year's fair or the next. The department scrutinized every penny spent with maddening officiousness: 'I beg to say that the Department approves of an expenditure of $10.00 being incurred in repairing the fence around the Council House. You should report as to the condition of the fence which encloses the Fairgrounds, and state whether you consider it advisable to erect a wire fence in place of the present structure. Such a report should have accompanied the resolution of Council in the matter.'[23] This sort of intrusion would have checked the enterprise of white societies, which paid for improvements through loans that required only members' approval.

The objection in 1895 may have been political. At Sarnia, as elsewhere, the council house doubled as the exhibition hall, and the improvement that had been planned was to move the house from Methodist lands to the fairgrounds. Similarly, when the Alnick band decided that only the Methodist land could serve as the showgrounds, it told the resident missionary to get out, but the department vetoed this decision.[24] In short, exhibitions were soon caught up in local politics. The garrulous manager of the Muncey fair at Caradoc reported that the society's officers in 1894 'are all true conservatives; at our last annual Election – we vote out all the Reformers and place ourselves in power.'[25]

At the Six Nations Reserve, trouble began in 1880, when the band council approved the agricultural society's plans to spend $2500 improving the grounds and building a racetrack.[26] Twenty-three chiefs, most of them Onondaga, complained that this expense would mean a deduction from everyone's annuity. The society offered to borrow the money instead, but the Indian agent objected that this arrangement would put the exhibition under 'the absolute control of an irresponsible body' that might bring in harmful amusements. He wanted the council to own the grounds because its actions, unlike those of the agricultural society, were subject to departmental veto. Eventually he had his way, but the quarrels continued.

The Six Nations Reserve was divided among two groups: traditionalists, who followed old ways, and Christians, who welcomed change. The Six Nations Agricultural Society was run on a party system: every year, two complete slates of candidates presented themselves, all winning or losing together. In 1890, when the traditionalist chiefs lost the election, they formed a short-lived breakaway group called the Tuscarora Township Agricultural Society, which the Indian agent, A. Dingman, actively discouraged.[27] Trouble occurred again in 1895, when the losing party declared itself elected. This dispute was resolved sufficiently to permit regular exhibitions, but a glance ahead to the 1920s shows that deep divisions remained. In 1924 the traditional band council, which 'followed a policy of non-cooperation' with the Canadian government, was locked out of the council house and an elected council replaced it. At the same time, the traditionalist Mohawk 'Workers' took over the agricultural society and used the agricultural grant to fund political activities. In 1927 the 'loyalists' formed their own 'Ohsweken Agricultural Society,' and the band council awarded it the showgrounds. The 'Workers,' objecting to plans to let whites compete, refused to hand over the grounds, until the RCMP was sent in to reclaim them from what the Indian agent described as the 'scum' of the reserve. The 'Workers' relinquished the land and held their own fairs and lacrosse matches on the same day as the Ohsweken exhibition, so that year after year the society lost hundreds of dollars, which the once-niggardly

Department of Indian Affairs freely made up with band money.[28] By this time, natives had begun to object to the goals of agricultural societies. In 1918 a council on Manitoulin Island voted against an exhibition because 'Indians should live as Indians and not follow the example of the white man.'[29] It was no accident that the exhibitions were caught in political disputes, for the principle of self-government and the liberal market values that guided agricultural societies were at the heart of larger political and cultural disputes. The government wanted Indians to act like whites, but would not abandon supervision and paternal guidance of them.

In Quebec the Mohawks at Kahnawake undertook an exhibition in 1883. The chiefs received $100 from the DIA to distribute prizes, 'to encourage as many as possible, even though it be in a small way.'[30] The exhibition was a success, with a profit of $96.41 after prizes of $291.90 and all expenses had been paid. The next year the organizers ambitiously invested $1500 of band funds to build a hall. About 6000 Montreal visitors attended that year, anxious to 'contemplate the red man in his native wigwam,' an activity that was normally forbidden by law. They also watched the parade, pony races, bow and arrow tricks, a lacrosse match, a tug of war between chiefs and warriors (the rope was rotten and broke every time, so they finally divided the prize – cigars – among themselves), a 'Fat Man's Race in Indian costume,' and another for women. Reporters declared the livestock 'rather poor,' except for some fine pigs, but the display of needlework, snowshoes, lacrosse sticks, preserves, vegetables, butter, bread, poultry, and apples was very good. Fairs continued until mid-decade, when railway works diverted the Iroquois from agriculture and the shows lapsed. Meanwhile, the exhibition building, erected on an exposed hill, was destroyed by storms. In 1903, when the band council applied to use $50 of band money to join the Laprairie County Agricultural Society, the DIA refused, and in 1904, when the council asked permission to spend $50 on a ploughing match, the department commented only that it disapproved of the principle involved.[31]

White agricultural societies were funded and largely unsupervised; native societies were supervised and largely unfunded. Missionaries encouraged exhibitions because, as Annie Coombes remarks of English missionary societies, they raised funds by advertising proof of their civilizing efforts.[32] Père Lavigne garnered $405 for his Indian school at St-Boniface with an exhibit, in 1892, that won a prize at the Winnipeg Industrial Exhibition.[33] Some individual Indian agents encouraged shows as well, but the Department of Indian Affairs was more circumspect; officials Hayter Reed and Lawrence Vankoughnet saw them only as a hand-out at odds with their program of encouraging self-sufficiency by refusing 'liberal' aid.[34] They adhered much more stringently to laissez-faire principles than did the Department of Agriculture. There were two standards at work: Indians were pre-political and had to learn private virtues like self-reliance through their own efforts, while white citizens had to learn to direct their private interests to the public good, a lesson best taught through public grants. Furthermore, white societies were given a free hand to run their own affairs, but Indian Affairs insisted on scrutinizing every decision that the native societies made. Such control only hampered native initiative and activity.

Almost the only contribution the Department of Indian Affairs made regularly to aboriginal exhibitions was to supply police officers. There was concern that visitors might form a raucous and rowdy crowd, a fear anticipated whenever natives gathered to hold dances and potlatches. One agent remarked that, even if the potlatch was harmless in itself, 'there are no doubt evils unavoidably connected with the gathering of large bodies of Indians together.'[35] Tony Bennett argues that one of the functions of exhibitions was to transform crowds from 'a compact mass' into an orderly gathering by creating vistas, such as upper-floor galleries in crystal palaces, where the crowd was broken up and could observe itself, thereby learning to govern itself.[36] After the first Kahnawake exhibition, it was reported that residents of nearby parishes 'were surprised to see that the competition led to no act of excess.'[37] In 1883 a local newspaper reported that at the Six Nations show, 'one could not find among the whites a more quiet, orderly, and mind-your-own-business crowd.'[38]

The physical aspect of the fair-going public was only one consideration. Exhibitions were supposed to act through the agency of public opinion by subjecting the individual to its scrutiny and influence, but public opinion was a difficult concept when applied to the First Nations. They had no private property and no newspapers to speak of, no electorate with political rights, and none of the traditional apparatus of the public sphere. White suspicion of the native capacity for citizenship was widely voiced in 1885 when Macdonald introduced a bill to enfranchise Indians. Grit newspapers like the *Manitoba Free Press* saw this initiative as wholesale corruption, introduced 'for the sole purpose of creating enough unintelligent Indian "votes" to render abortive the intelligent vote of the white electorate.'[39] Native public opinion, according to this description, was a form of anarchy, to be trusted only when an Indian agent was on hand to guide it.

For Édouard-A. Barnard in Quebec, exhibitions catered too much to local opinion, and their makeshift form of inspection and instruction fell below ideal standards. On reserves, however, Indian agents exercised a degree of surveillance over the lives of residents which would have been inconceivable among non-natives. At the Peigan Reserve, the agent regularly inspected homes: 'In one house I noticed bedsteads, stoves, tables, chairs, dishes, corn brooms and a lamp. The house was divided into two rooms. The bed had quilts, blankets, sheets, and all perfectly clean. A centre table had a red table-cover, the floors were cleanly scrubbed and the whole place was fit for anyone to occupy.'[40] The exact extent of market activity could also be measured to the last penny, as on the Blood Reserve in the North-West: 'One man, Dead Sarcee, who earns fifty cents a day when butchering and who works for Mr. Pace, the trader, occasionally, has built himself a very nice house. He has also a very nice field neatly fenced. He purchased a stove for $12, pipes and zinc for $8, and he has also a black walnut bedstead and bureau which cost $14. In the house I noticed bracket lamps, looking glasses, pictures, tables, quilts for the bed, the floor had been lately scrubbed and everything as tidy as possible.'[41] To turn from the usual run of sessional papers to the DIA reports is to turn from dry statistics to a vivid description of lives lived. As long as the natives of central Canada were neither starving nor rioting, the government seemed to prefer to act on aboriginal populations through agents rather than through public opinion. The agents may have had little practical effect, but they did create a flow of information that visibly expanded the scope of governmentality.

Exhibitions were supposed to function as instruments of racial hegemony by securing the natives' consent to assimilation and by hastening the process. Certainly, DIA officials flattered themselves that 'keen competition for prizes' among Ontario Indians 'is indicative of their progress towards assimilation of ideas with those of the more advanced level of society.'[42] But it is not always clear what is hegemonic and what is not.[43] Native organizers decided what would be displayed, who would see it, and what visitors would pay for the privilege. The result was no stereotyped cliché but a rich, rounded presentation of native culture. Exhibitions could provide natives with a means of community solidarity and self-expression, and this process could create an autonomous and ungovernable native public opinion.

Another reason for distrusting the exhibitions may lie in the reason for their success. Many reserves held annual exhibitions that were self-supporting or that received a fraction of what township societies were granted because the Indians were so intrinsically spectacular. Native exhibitions displayed the same carnivalesque and spectacularesque qualities that made non-native exhibitions more entertaining than educational, and that tended to undermine the 'civilizing process' rather than to advertise it. Native exhibitions were profitable even without government help because they could draw on fancy dress, dances, lacrosse, horse-riding, and marksmanship to entertain spectators. Evidence of acculturation was mitigated by these traditional and spectacular activities.

Disapproval also surrounded native participation in white exhibitions in Canada and abroad. During the first half-century of white exhibitions, natives rarely competed directly against whites, although exceptions occurred. At the Laprairie County Fair in 1831, for example, Thomas Sauwatsiawané won a prize for corn, and at Toronto in 1852, Peter Jones won small prizes for maple syrup and woodcarving.[44] After 1870, mixed competition became more common on the prairies, but was scarcely seen elsewhere.

More often, natives competed separately in 'Indian manufactures' – basic carpentry and beaded leather or woven artifacts – but prizes and honours in this department were contemptuously small. At the provincial exhibition in Toronto in 1856, £2.15 was distributed to the seventeen entries, out of a total of £1999 in prize money; in 1859, 104 prizes shown by thirty exhibitors were rewarded with $35, accounting for 2 per cent of exhibits and 0.5 per cent of the prize money. Winners' names were not listed because the exhibits 'were so similar in character' and only Christian names were known.[45] This anonymity contradicted the theory of honourable publicity underlying exhibitions. The next year there were only four entries, and, in later years, the 'trinkets' were struck off the prize list entirely, except for isolated references, as in 1867, that '34 small prizes' went to Kahnawake Iroquois for beaded work. Nova Scotia probably went the farthest towards encouraging a good aboriginal display at its first industrial exhibition in 1854, for organizers published in Mi'kmaq the list of prizes for goods like snowshoes ('Ahkumky') and quill boxes ('Ouwe-u-gwolalikunul cow-e-ail').[46] This section of the prize list also disappeared in later years. Nonetheless, Indian curios were always to be found on the fairgrounds, either in the natural history section or for sale. At the Western Fair in 1878, 'nut brown maidens' sold beaded work outside the hall while their 'white sisters' exhibited fancy-work within. At Toronto in 1864, native women 'modestly asked the passers-by to purchase' bead work.[47] Whereas men's work was simultaneously an exhibit and a commodity, women's work was one or the other, but not both.

Although aboriginal artifacts disappeared from prize lists, aboriginal people became part of the spectacle, as lacrosse games and wild west shows became staple fare at the large exhibitions held after mid-century. The first Montreal Provincial, in 1853, was accompanied by a lacrosse game, which turned violent. When the Prince of Wales visited Montreal in 1860, Iroquois and Algonkian played a lacrosse match, followed by another between whites and natives.[48] In 1867 a team of Mohawk visited Europe, and in 1876 Mohawk and whites toured England, performing before Queen Victoria. This tour coincided with the Philadelphia Centennial Exhibition, where Canada tried to impress the world with its modern industrial character. A letter in the Montreal *Gazette* described the tour as exactly the wrong kind of exhibition, advertising Canada as barbarous and uncivilized. The tour's organizer tried to correct this impression with a letter to the English press which described lacrosse as not 'rough and tumble' but 'scientific and graceful.' Whites had already begun to make the game less violent, drawing up rules in 1869. In 1867 the Montreal *Daily News* said: 'Lacrosse, as reduced to rule by the whites, employs the greatest combination of physical and mental activity white men can sustain in amusement, and it is as much superior to the original as civilisation to barbarism, or a pretty Canadian girl to an uncultivated squaw.'[49]

Even as lacrosse was being tamed, the wild west show replaced it as an accompaniment to large exhibitions like the Toronto Industrial. Hundreds of natives, whites, and Mexicans would re-enact old battles like Little Bighorn, raid stagecoaches peopled with tourists, and perform rodeo feats.[50] But even when Indians stood around, their presence was exhibit enough. The Western Fair began to offer dozens of prizes for native productions in 1888, but the real draw was the exhibitors, according to the advertisements: 'It is expected that Indians in charge of their exhibits will be dressed in Native costumes.'[51] At the first private exhibition in Quebec City, in 1894, an encampment of Huron became an attraction. The government began to discourage these activities in the twentieth century, prohibiting 'senseless' drumming, dancing, and Indian costume from exhibitions in 1914, banishing natives from the Calgary Exhibition in 1925, and discouraging speculators who wanted natives to do rodeo acts overseas.[52]

International Exhibitions

At international exhibitions, the tension between the spectacular and the civilized was most acute. As government agents tried to show the world how progressive and civilized Canada was, they thrust Indian artifacts firmly behind the minerals and motors. But most tourists weren't interested in rocks and planks, and they congregated at the more entertaining and visually stimulating displays. At first the battle revolved around the inclusion of artifacts, but, later, it was waged with native bodies.

During the 1850s aboriginal artifacts were a prominent part of the displays sent by Canada to London and Paris. They provided colour and vivacity to an otherwise drab display of natural resources. The centre of the Canadian exhibit was dominated by an enormous canoe, which, with other items, provided a stark contrast with different cultures and enacted, in miniature, the lesson of the Great Exhibition. The London *Art Journal* published this appreciation of the display: 'The most powerful impression I received was on turning down into Canada and wandering among the products of that world we call our colonies; those strange grains, and woods, and animals; those barbarous utensils, arms, and ornaments, mixed up with all the evidences of English civilization; those works of living savage populations – our fellow-subjects! Neither the mass nor the perfection of all that Birmingham, and Sheffield, and Manchester contributed, gave me such an awful sense of the power and the responsibility of England as these contributions of our remotest and earliest settlements to their glorious Mother Country.' The Canadians placed this passage at the front of the pamphlet they distributed at the Canadian court.[53] They wished to show that they had brought British civilization to the wilds of North America.

Native artifacts occupied an ambiguous status in the artistic hierarchy. When the Huron near Quebec City won a prize from their old allies, the French, in 1855 (as they had done in New York in 1853), the judges remarked: 'Ces produits sont l'ouvrage d'une race qui s'éteint et n'a pû se lier à la civilisation moderne. Il y a dans les broderies de ces Indiens un goût particulier et une heureuse disposition de colours.'[54] This view reflected that popularized by Owen Jones, who in 1856 exalted the geometric simplicity and fidelity to nature of 'primitive' art in his *Grammar of Ornament.* The Canadian commissioner to Paris, J.-C. Taché, also commented on the visual appeal of the Huron display: 'The curious and elegant articles of feather work, moose hair, porcupine quills, and bark-work, are attractive to visitors; and it must be confessed that there is in the ornamental articles and those pertaining to the toilet, to be found in this collection, a degree of taste and refinement which excites our wonder, when we consider that all this is the untaught art of the aborigines of the shores of the St. Lawrence.' Taché was more effusive describing the saccharine art Victorian audiences adored, such as a sculpture of a boy rescued by a dog: 'The child in his transports of gratitude holds the dog in his embrace, his beautiful little head intermingles the tresses of his lovely hair with the silky locks of the noble animal; the effect of these groups is really enchanting, innocence and devotion are triumphant. In the contemplation of such objects as these, we recognize the civilizing influence of the arts.'[55]

By this standard of art, as well as by the industrial standards that exhibitions were designed to inculcate, native work was, as a guide to the Crystal Palace remarked, 'substantially the inferior fruits of human industry.'[56] In 1850, speaking at the exhibition in Montreal to collect goods for the Crystal Palace, Justice Day declared that 'the steadfast, resolute application of the mental energies, in a certain direction, and with a fixed object, makes the difference, perhaps all the difference, between a civilized man and a savage man.'[57] Yet even a cursory glance at the wares exhibited belied this claim. An outfit of dressed deerskin, dyed moose hair, and beads made by a Mohawk woman was more elaborate and attractive than other displays, such as the humble nails, planks, ploughs, and grains so proudly displayed, or the view of Quebec done in straw or the portrait of Victoria and Albert in stucco. This discrepancy was a problem at the first Great Exhibition in London, where it became apparent to everyone that mass-produced Western industrial wares were uglier than those from India, Turkey, China, and other 'uncivilized' parts. William Whewell, lecturing to the Society of Arts, asked anxiously: 'Wherein is our superiority? In what do we see the effect, the realization, of all that more advanced stage of art which we conceive ourselves to have attained? ... Surely our imagined superiority is not all imaginary; surely we are really more advanced than they, and this term 'advanced' has a meaning; surely that mighty thought of a PROGRESS in the life of nations is not an empty dream; and surely our progress has carried us beyond them.'[58] His conclusion was that, for eastern and aristocratic nations, 'the arts are mainly exercised to gratify the tastes of the few; with us, to supply the wants of the many.'

This comment reveals just how subversive aboriginal work might be. If it could raise questions about the value of European civilization, how much more easily could it undermine the message of 'civilization' that the Canadian government wished to convey to Europe? In 1851 the most numerous items from Nova Scotia were 'specimens of native manufacture of the usual simple description,' while one visitor from New Brunswick fumed: 'At that ever memorable exhibition, how did New Brunswick figure? *By a lump of asphaltum, the figure of an Indian, and a bark canoe.*'[59] By 1862, when the next and last English international exhibition was held, displays of native artifacts were thought to be more injurious than helpful to advertising Canada, which, with Nova Scotia, refrained from sending any such entities. The Duke of Newcastle, who had toured Canada in 1860, remarked with regret, when he was taken around the Canadian section, on 'the absence of sleighs, furs and Indian curiosities, being of the opinion that the policy which excluded them was not the best.'[60]

Newfoundland, New Brunswick, Prince Edward Island, and British Columbia did send specimens of native work in 1862. In British Columbia, the matter was discussed in a public meeting, where Chief Justice Begbie said that visitors might be lured to the BC court by this stratagem: 'He thought it a very good idea to send to the exhibition Indian curiosities. People would come miles to see Indian curiosities and when they got there, would see, in addition, specimens of our minerals, or our spars, or something else that would be sure to help us along.' Another speaker thought that 'under a judicious system of instruction they might change the grotesque images they at present carved into useful articles,' and still another that these objects would advertise the province's skilled labour pool. Journalist and booster Amor de Cosmos pooh-poohed this suggestion as 'palpably absurd' because, he thundered, Indian indolence, restlessness, and thievishness 'totally disqualifies them from ever becoming either useful or desirable citizens ... It is Caucasian – Anglo-Saxon bone, muscle and intellect we want.' Further, to create space for them, he added, natives should be shunted onto reserves and punished severely for trespassing on white land.[61]

In 1865 Canadian natives sent work to the Dublin Exhibition, but when D'Arcy McGee saw an elaborate chief's dress on display, he banished it, exclaiming, 'That was very well as showing what the inhabitants of Canada were 200 years ago, but not now.'[62] The Nova Scotia commissioner to Dublin included some Mi'kmaq work as a curiosity, and he was surprised when it obtained a prize for workmanship, but a Nova Scotian visitor complained that these 'Indian knickknacks' were too prominently displayed.[63] For the Paris Exposition of 1867, Nova Scotia and Canada both took a hard line against Indian artifacts. Thomas Annand commented: 'We trust that the "irrepressible" Indian will be kept in the background. His rule has passed away for ever, and his useless, though pretty toys, should disappear also.'[64] Canadian collectors also determined 'that valuable space would not be wasted on Indian curiosities and articles calculated to make us appear semi-barbarous here.'[65] However, a historian of Kahnawake records that, in 1866, government agents came to the reserve for wares to send to Paris; he also notes that two Mohawk visited the exposition and returned greatly pleased by the display, but 'scandalized at the way Paris observed Sunday and the laws of fasting.'[66] Equally scandalized was a Canadian visitor to the exhibit who found that the commissioners had gone away and left the Iroquois in charge of the display. Visitors could also admire

Inuit artifacts on display at the Newfoundland court, including some striking art work, although a Halifax newspaper objected that 'it was sadly out of place to send Esquimaux hats, coats, and boots, which have a tendency to do anything but tempt Frenchmen to emigrate to Newfoundland.'[67] In fact, Newfoundland was not advertising for immigrants, but trying to sell goods and attract capital.

During the 1870s native artifacts began to reappear, sent by private speculators or anthropologists. In the early 1870s an attempt to exhibit the 'declining races of the empire' provoked little interest in Canada, which sent only photographs. The Colonial Office predicted a lack of enthusiasm among colonial governments, which would have little to gain from such a display.[68] Doubtless it recalled its own attempt in 1869 to obtain ethnographic photographs, when it received a regretful reply from the governor of British Columbia that 'no Indians here will consent to be photographed in a state of nudity, although reward has been offered.'[69] For the Philadelphia Centennial Exhibition of 1876, artifacts from the Pacific coast figured prominently in the Smithsonian display, and individuals also sent collections: Robert Bell of the Geological Survey and A. St-Amant of Lorette sent collections, but the bulk of the Canadian display was meant to emphasize its distance from the primitive exhibit Canada would have shown a century earlier.[70] Two Quebec exhibitors sent Indian work to the Paris Exposition of 1878, and interested visitors could read in the handbook that the interests of the natives are 'scrupulously guarded, and they are treated with uniform kindness and good faith.'[71]

Ethnographic displays began to appear in earnest during the 1880s, beginning with the Colonial and Indian Exhibition, when many aboriginal people made and sold crafts on the grounds. Canadian promoters suggested sending 'civilized' Indians to advertise assimilation policies, but John Carling and Sir Charles Tupper rejected this suggestion.[72] The exhibition was graced by a visit from Delaware chief Waubuno, who was touring England to raise money for a church in Muncey, Ontario.[73] Two medical men suggested that Poundmaker, Big Bear, and Crowfoot be sent to England for display, where the spectacle of Victoria's majesty and Britain's vast military forces would inspire such awe that, once home again, they would never again consider rebellion. Sir John A. Macdonald dismissed the idea with a jibe: 'Well, gentlemen, the plan from a money-making point, is a good one. Do you propose to have these chiefs stuffed or take them over alive?' Poundmaker did not make it over, but collector Newton Chittenden sent the chief's elaborate garb, complete with feathers, ermine, racoon tails, and flowered beadwork, after first trying it on himself.[74]

By 1886 the Canadian government no longer purchased exhibits but merely ferried them for private exhibitors. It might be thought that natives had the same rights as other exhibitors, especially as their work could be sold to advantage overseas: Britain had been taxing this lucrative industry since 1854.[75] However, Ottawa reassured an anxious J.B. Hurlburt that although it had let some Indian articles go over, the display was not very large.[76] The press emphasized the contrast between Canada's practical display and the glittering India court: 'The India show may be the more attractive to the eye of the professional sightseers of London and of the world, but mainly Canada wishes for recognition at the hands of people who desire to do business with us.'[77] Nonetheless, the DIA solicited exhibits from its agents across Canada, ranging from agricultural produce to carvings, beadwork, and totem poles. The Thompson Indian Bazaar Company of Saint John, New Brunswick, did a roaring business in sweet-smelling hay baskets.[78] Drawings by Ariwakenha Sowatis from Kahnawake were exhibited in the art department.

Canada did not attend the French Exposition of 1889, where visitors were entranced by a staged replication of Cairo street life. The Chicago fair of 1893 accommodated many such recreations of exotic lifestyles, including encampments of Inuit from Labrador and Kwagiulth from Fort Rupert on Vancouver Island. The Inuit soon fell out with the speculators who had brought them there. John Sugarloaf (who, with his four sons, trounced 'a dozen Arabs' in a fight) was thrown out of the encampment for refusing to wear his hot, sealskin suit all day, but he quickly found work with Buffalo Bill at $1.50 per day.[79] The Inuit had been coaxed to Chicago with an offer of $40 per family for two years' work, but Sugarloaf convinced them they could do better. They left the midway to set up their own show nearby, where they remained until autumn. They toured the Canadian fairs on their way home.

The Kwagiulth group included ten men, four women, and two small children. One man sported mutton-chop whiskers, and the women were described as 'intelligent' and 'inclined to wear gaudy colours and heavy bracelets of quaint design.'[80] Douglas Cole reports that they did not make much of a stir amid all the other exotic peoples until August, when they performed the Tamanawas dance, which was banned in Canada. The London *Times* reported on the gory scene. Two natives had their backs slit, and heavy twine was passed through the hole in their skin. They raced around the platform, with others holding the twine: 'Around and around they ran, leaping, twisting, and diving till it seemed to the horror-stricken spectators that each instant would see the flesh torn from their bodies.' They snarled

The Inuit encampment at the Chicago World's Fair in 1893. The clothing was hot and uncomfortable in the Chicago sun.

and snapped their teeth, one taking a bite from the arm of the Kwagiulth agent who had brought them down.[81] Indignant Canadians complained to the DIA, which asked fair officials to prevent any recurrence. However, knowledgeable British Columbians insisted, and American authorities confirmed, that the gore had been feigned.[82]

The Canadian government had its own exhibit of native culture which, like that of the American government, highlighted its educational efforts. Hayter Reed initially advised against exhibiting because the Americans could do a better job. The midway encampment with natives from across the United States would represent most Canadian Indian 'types,' he argued, and their model school, where thirty boys and girls would go through lessons before the eyes of visitors, would eclipse the newer Canadian residential schools.[83] But Canada had to be represented, so $2000 was allocated for Indian artifacts. When the BC office received its cheque for $500 to collect West Coast artifacts, it advised abandoning the plan, for American agents had spent $4000 buying up the best goods; an agent in Nova Scotia made the same complaint.[84] By going over budget, Indian Affairs was able to gather a collection of artifacts from across the country: $200 in fancy-work from Brantford; baskets and beaded wares from the Abenaki and Iroquois in Quebec; fancy-work and a canoe and paddles worth $17 from the Mi'kmaq; grass-work from the prairies; and harpoons, canoes, embroidered skins, dancing masks, cedar bark blankets, and other artifacts, old and new, from British Columbia. Many bands sent in vegetables, grains, and preserved fruits, some of them collected at local shows. These exhibits represented substantial economic activity, though profits had been declining since the mid-1880s owing to white competition, American tariffs, and economic recession.

Missionaries had a hand in these exhibits, but they also determined to show their educational efforts. In the end, with Archbishop A.-A. Taché insisting, the DIA sponsored a display by the residential schools. A small contingent of boys and girls went to Chicago to show their skill at making baskets, sewing, carpentry, even printing: boys from St Joseph's Industrial School near Winnipeg composed and printed in the Canadian court a small journal called *Canadian Indian*, which described some of the exhibits and told visitors about the school system.[85] Another pamphlet, printed at the school presses and distributed at Chicago, explained Canada's Indian policy. Faced with a choice between letting Indians die, maintaining them in idleness, or teaching them industry, it explained, the government had taken the noble course of educating them in spite of themselves: 'It may be true that, when viewed in prospect by the Indians, they may not covet the

The Kwagiulth encampment at the Chicago World's Fair in 1892. Members are shown in a quiet moment, but it was their reputation for wild dances that brought in the crowds. The public was as fascinated with West Coast native art as it was with the artists, and the hut shown in the background was one of a great many items that found their way to American exhibitions and museums.

advantages of civilization for themselves, nor for their children, but no civilized nation can justify itself in leaving ignorant savages, whose country it has taken possession of, to determine their own course and follow the blind promptings of their natural impulses.'[86] Presumably, the boys who set the type for these stirring thoughts were, in the course of their work, brought around to this enlarged view of the matter.

Visitors cooed over the display: 'Here is a coffee-colored cousin of some mute, inglorious "Sitting Bull" working away in an improved and modern style with a printing press. There a dark-eyed Penthesila with raven locks is weaving a homely mat with the swiftness and skill of an expert.'[87] Judges were no less impressed, awarding a collective medal to the Department of Indian Affairs for the 'excellent manner in which the Indian is educated and trained and the evidence of the good results thereby attained.' Prizes also went to eight of the nine schools exhibiting: one from Kuper Island, British Columbia, three from the Territories, three from Manitoba, and Shingwauk School in Sault Ste Marie, Ontario. Only Qu'Appelle school, an important Oblate establishment near Regina, failed to receive a prize, to the distress of the DIA.[88] The judges could know little of the conditions at these schools: at Battleford (rewarded for fancy-work and first-class blacksmithing by the children), the inspector had instituted a 'reign of terror and despotism' after some boys were caught escaping, beating them and locking them in special cells, to the detriment of one boy's health. An inspector complained that the boys were poorly taught, that industrial education was defective, and that agricultural education was 'a total failure.' Later that year still more misdeeds resulted in a change of management.[89] At St Paul's School, which did not exhibit, children were ill-clad against the cold and covered in vermin, leading the native council to threaten to remove every child and close the school. The inspector concurred with the council: 'The management of the Saint Paul School is perfectly disgraceful, and makes our boasted civilization in this training school stench in the nostrils of the Indian whom we endeavor to raise to our standard of civilization and enlightenment.'[90]

The exhibit did not ward off funding cuts introduced in July 1893, but doubtless helped to legitimize residential schooling. The pleasing spectacle of civilizing Indians prompted visitors to remark that 'no thoughtful person can examine the workmanship and ingenuity displayed in the manufacture of many of the things that belong to the red men, and still say of those who made them, as has been said too often alas, "the only good Indian is a dead Indian."'[91] Nonetheless, the American display was considered a failure by the Smithsonian because the native encampments made a more lasting impression on the public.[92]

Natives were exhibited in a way that white people were not. Ontario and Quebec battled for educational prizes, but neither sent the schoolchildren themselves, for they were children first and objects of knowledge second, unlike native children. Yet the Canadian government did not encourage the exhibition of native adults, as the United States did. Parsimony was one reason; another was the fear that such an exhibition would reveal the failure of assimilationist policies. Canada was probably more anxious than the United States in this respect because the forces of Nature, under which white Canadians subsumed native groups, seemed always ready to overpower a small European population thinly spread over the land.

These same Canadians also believed that representations could remake the world in their own image. Exhibitions measured, and, by the very act of measurement, increased the productive and refining forces at work in society. But the paying public demanded amusements, and, reformers complained, this trend lowered rather than elevated public opinion, reinforcing its most vulgar aspects. Speeches at Canadian agricultural exhibitions at mid-century played on a 'lonely wigwam to bustling city' theme, but, by the end of the century, wigwams once again occupied fairgrounds. If exhibitions were supposed to be microcosms and markers of world history, this presence did not bode well for the future. Indians were supposed to be a dying race, but their growing presence on exhibition grounds suggested that 'civilization' was not going to win such a clean victory over them after all. Indeed, in 1877, after the Centennial Exhibition, the American government discovered that native populations were in fact increasing.[93] The Canadian government refused to extend political representation – the vote – to unassimilated natives, and actively discouraged their material representation. Even agricultural exhibitions, though they fostered self-improvement, threatened to perpetuate and even celebrate native culture, and to reinforce the reserve as a distinct geographical and political entity run by natives.

The deeper problem was the shifting boundary of the spectacle. Indians were a spectacle to whites even when they were not behaving spectacularly. The category of the spectacle seemed to leak out of its containment, until it became indistinguishable from culture itself. If all of native culture was a spectacle to whites, then wasn't all of white culture a spectacle to natives? Indeed, in 1865, the *British Columbian*, published in New Westminster, speculated that the annual Victoria Day ceremonies, when hundreds of natives came to accept their gifts from the Canadian authorities, functioned as an exhibition:

> What else was the World's Fair of 1851 but a social gathering, affording the various members of the human family an opportunity to get rid of some of those prejudices and misapprehensions which an imperfect acquaintance had given rise to. The gathering of last week may not unfitly be considered as a world's fair to the aboriginal inhabitants of this colony, leading to a better acquaintance with the whites, and to intertribal intercourse never before existing amongst themselves. The representatives of the semi-civilized nations who visited the great fair alluded to in the old world were perhaps less astonished at what they saw than were the representatives of our interior tribes last week. The mountaineer who had never seen anything more formidable than a canoe upon the inland waters must indeed have been struck with the sight of a sea war ship, with yards manned and guns belching forth fire and smoke. The great difficulty in dealing with the Indians has been the absence of any means by which they could be made acquainted with the feelings and intentions of the governing power towards them, and by which they might be impressed with the superiority of the whites and the advantages of the civilization which we offer to them.[94]

This reflection followed several articles in which the writer observed the natives observing white society. They stared at the buildings with 'curious wonderment' and lingered 'at every shop window to feast their uncivilized eyes upon the rich and showy wares exhibited therein.' He advocated a show of force to seal the superiority issue, arguing that a display of shelling by warships would make 'a most salutary impression' and arouse 'a superstitious respect for us, which would quickly be communicated to other tribes, and would operate as a wholesome preventative of future insurrection.'[95]

White culture was, it seemed, a permanent spectacle too. But this admission raised problems. Are all cultures also spectacles? Or are spectacles a subcategory of culture, considered under a particular description? The exhibition was a representation of culture, formalized to make it more effective: machines, minerals, and manufactures were all to be found outside the crystal palace walls as well as inside, but with the spectator now more aware of their use and value. Exhibitions were, supposedly, culture made more graphic, to concentrate and speed up historical processes. But what difference was there between the spectacle of white culture and the spectacle of native culture? Could one theory of the spectacle, a theory of aesthetic reception, suffice to cover both cases? Canadians dodged this issue by denying that Indians had culture. The school pamphlet of 1893 contrasted Canadians, whose Indian policy was dictated by the burden of being a 'civilized nation,' with Indians, who were guided only by 'the blind promptings of their natural impulses.' White society represented triumphant civilization, but Indians were unaccommodated man, 'the thing itself.' Likewise, in describing the annual gift giving as a world fair, the *British Columbian* characterized Western civilization (which New Westminster, by some stretch of the imagination, epitomized) as the spectacle and Indians as the spectators. But the real point of an exhibition was to put every culture on display. Indeed, the Victoria *British Colonist* objected to the annual gift-giving ceremony precisely because it exposed whites to a seamy, aboriginal culture: 'Vagrancy, filth, disease, drunkenness, larceny, maiming, murder, prostitution, in a multiplied form, are the invariable results of an annual visit from the Northern Tribes.'[96]

This account raised a troubling question: If white society made such an impressive spectacle, why weren't the Indians noticeably improved by it? Some argued that Indians should be settled among white farmers, to learn civilized manners and a settled life.[97] The *British Colonist* disagreed: 'They appear to imbibe our vices – which are many – while being proof against our virtues, – which are fewer.'[98] Its editors favoured settling natives on reserves where they would be exposed only to good whites – missionaries and Indian agents, carefully chosen for their morality and industry. In fact, Indian agents were bureaucrats, chosen for party loyalty, and their corruption or incompetence was, rightly or wrongly, blamed for the failure of reserves. Nor were the missionaries necessarily superior to the culture that produced them. Culture, it seemed, could not be distilled into its best qualities.

Could whites be certain that their culture would 'win' the confrontation of races? They believed that, historically, their culture was the best, for Anglo-Saxon values of individualism, industry, and thriftiness had made that group the most powerful and wealthy in the world. Exhibitions were supposed to be an inventory of power and wealth, to demonstrate and reinforce the hierarchy of the world and the processes that determined history. It was war by other means, intended to produce its effects without all the messy bloodshed. But when culture took concrete shape in the form of an exhibition, new rules were brought into play. Pleasure became important as well as power. Imagination counted for more than drab industry, and Canada's much vaunted practical exhibits paled before contributions from more spectacularesque exhibitors. When the battle for cultural supremacy was waged on the exhibition grounds, Anglo-Saxon communities found themselves at a disadvantage.

The problem was systemic. The country hosting a world's fair was not exotic and spectacular to most visitors. Even the bewildering array of commodities, which so struck visitors to the Crystal Palace, became standard fare in shop windows and department stores in industrialized countries after mid-century. One correspondent remarked at Philadelphia in 1876: 'England and America have the largest displays, and the most valuable goods, but they are, for the most part, the best of such articles as we see daily, as can be seen in any large city.'[99] Exotic displays from foreign countries were more popular, especially when they played up their foreignness, or their indigenous traditions. Moreover, mere visual inspection revealed, as William Whewell admitted, that much of this 'primitive' indigenous work was far more finely made than the shoddy, mass-produced works of industrial societies. Exhibitions were designed to convey the lessons of history to the broad public, to people who did not read history, but this lesson was, ultimately, ambiguous.

Exhibitions had a dual role: They encouraged the display of cultural diversity, often to the point of caricature, and they also seriously impaired this diversity. Exhibitions were supposed to increase universal assimilation by bringing all cultures into contact in the world's market. These cross-cultural transactions could never be wholly honest. Displaying Indian culture meant displaying their bodies and also their sacred artifacts, some of them centuries old. Douglas Cole describes the underhanded methods used, in some cases, to 'persuade' native peoples to relinquish cultural treasures.[100] Many of these items were not made to be sold, and their transformation into commodities signalled a cultural change. Exhibitions speeded up this process because the goods and people displayed were, simultaneously, objects of knowledge and art as well as commodities. In fact, the exhibition was, from the beginning, as Walter Benjamin remarked, a temple to the commodity fetish.[101] The accompanying intellectual lessons had to be explained by intermediaries in special sites like educational departments, just as agents were needed to explain culture properly to Indians on reserves.

Culture cannot be reduced to an exhibition. Europeans thought they put the most important parts on display and accomplished something by having Indians exhibit agricultural and industrial wares. But native culture was far more than what whites saw. Though missionaries and agents tried to make it visible and to root it out, they could not. Culture was also resistance, just as strong when it was silent and hidden.

Conclusion

The concept of the 'exhibition' is a reasonably coherent and bounded one, but exhibitions took many different historical forms, even in the confines of the nineteenth century. Some were small, rural affairs, with half a dozen exhibitors and a sparse crowd of farmers, while others were gigantic extravaganzas with millions of participants from all walks of life. For the historian, exhibitions offer a magnificent prospect for studying the material culture, symbols, values, and activities of nineteenth-century people, but their significance is difficult to quantify. They were idealized representations of the world which were expected to reshape the world in their own image. Unfortunately, they fulfilled only the baser hopes of their founders by encouraging economic exchange and commodification without heralding a new era of peace.

As descriptions of the world, exhibitions were neither true nor false. They were designed as an encyclopedia of the world's products, worked into a classification system that was intellectually rigorous and economically useful.[1] But even more vividly than they illustrated the world, these exhibitions illustrated Victorian epistemology: the urge to categorize all things and peoples accordingly to starkly utilitarian criteria. They also revealed the widespread belief that if the forces represented by knowledge and money could only be loosened on the world, free to work progress untrammelled, they would inaugurate a new society that would owe something to the model of the learned society and something to that of a private corporation.

Exhibitions were supposed to serve as exhaustive representations of the world. They did not include everything in their classifications, but they did try to encompass everything relating to modern industry and, increasingly, the arts and sciences. Inevitably, they were as incomplete and unsatisfactory to the Victorians as they are to modern scholars, albeit for different reasons. Organizers always complained that full representation had not occurred because farmers or manufacturers had failed to participate. The true wealth of the land was never adequately represented on the fairgrounds. Modern scholars are more likely to remark the absence or distortion of the subaltern – especially labourers, minority groups, or women. Scanty evidence records the presence of some of these groups, such as an offhand reference in 1885 to 'an oil portrait, painted by a Chinaman,' or the carved walnut exhibited by a fugitive slave in the Canadian court in 1851.[2]

The project of physically recreating the world in an exhibition was novel enough, but even more novel was its intended purpose: to secure consent and precipitate action. Exhibitions were one of the first great modern efforts to arouse and manage public opinion. This project of secular mass communication had begun in the eighteenth century, but only in the nineteenth were mechanisms developed capable of realizing it. Rousseau lauded as the ideal a *polis* small enough that citizens could meet face to face, but soon the only possible standard became what Benedict Anderson calls the 'imagined community': one that imagined itself into existence based on books and newspapers read by the members of the community.[3] Enlightenment writers imagined themselves to form part of a republic of letters, supported by a network of learned societies. The democratizing tendencies of the nineteenth century led their heirs to wonder if far-flung geographical communities could be shaped to resemble learned societies, sharing knowledge and values. Print alone could not create this community, for the nation's well-being rested upon its values being disseminated even to the illiterate members of society who formed the backbone of the economic body.

Designed to generate and disseminate common values, exhibitions were ambiguous institutions that both empowered local communities and imposed centralized standards. However hegemonic in intent, exhibitions became, in fact, more liberal institutions that subaltern groups could, at least part of the time, turn to their own advantage. To some extent, the exhibitions teach the impossibility of hegemony, at least as it is commonly treated by historians, as a set of values and intentions embodied in texts. In theory, the wealthy, the powerful, and the educated collaborated to produce a hegemonic discourse that would bolster their values, social standing, and general prosperity. In fact, this collaboration produced an incoherent cacophony of conflicting values that missed its target. No historical group can create a monologue to impose on other groups its values and its knowledge. If these values and this knowledge are indeed alien to the audience, then the lesson will be lost. To make it familiar, the message must be translated into more popular terms and, in the process, its meaning will be altered, as, indeed, will be the original group itself. Moreover, in a liberal society, the audience, rather than the author, exercises greater control over the means of communication, for it can always choose not to listen or to remain aloof from educational or entertaining forums. Audiences transformed the form and the content of exhibitions. Exhibitions were intended to shape history from above by imposing elite direction more thoroughly on daily life; instead, inevitably, they worked to the opposite effect by helping to unleash popular culture. Nonetheless, throughout this struggle for control of agency and meaning, certain principles were not open to debate. Exhibitions accommodated considerable diversity within an explicitly capitalistic framework.

So what effect did the exhibitions have on history? The agricultural exhibitions held in Upper and Lower Canada, later in Ontario and Quebec, can be said to have aided the development of agricultural industries by encouraging specialization and facilitating the dissemination of products across the country. But they could not create an industry where it did not exist already. Those industries they did encourage, such as livestock, fruit, and agricultural implements, were hampered by an exaggerated importance attached to showmanship. Sometimes exhibitions served as markets, where farmers sold their surplus stock, and buyers came from a distance. On other occasions, exhibitions failed to attract enough farmers or buyers because of political disputes or the poverty of the region. Such an exhibition was a blot on a community rather than a source of pride and confidence. But most exhibitions, even in Quebec, were happy occasions where farming families came together, pocketed a few dollars, and made merry alongside the larger public. Exhibitions helped to create an economic public sphere, a place for reflecting on the role of production, and later consumption, in the life of the nation.

In Quebec, exhibitions conspicuously failed to fulfil the goal assigned to them: to transform agriculture. Severe economic pressure and soil exhaustion, not exhibitions, caused Quebec farmers to move out of wheat into other grains, and finally into dairying. Quebec farmers largely abstained from competing in exhibitions, a trend already in evidence before mid-century, but one reinforced by the hostile attack on exhibitions launched by the agricultural reformers William Evans, Joseph Perrault, and Édouard-A. Barnard. For these men, exhibitions moved too slowly and, because they relied on public opinion, they reinforced popular prejudices that prevented the adoption of modern farming. Although many of their objections to exhibitions – such as the concentration of prizes among the rich farmers – were well founded, the solutions they proposed often exacerbated the problem and further alienated farmers from organized agricultural improvement.

In Ontario, exhibitions enjoyed greater success, especially to the west, where climate and soil were most suited to wheat and cattle raising. Farmers made use of exhibitions, to win prizes, do business, and enjoy a festival. Here, they were able to take control of exhibitions and mould them to their purposes. Large breeders and farmers dominated the bigger shows, while small-scale farmers ran local exhibitions. Some agricultural reformers grumbled that the local shows set their standards too low, but, unlike their Quebec counterparts, they were unable to force legislative change.

Some of the same reformers objected to the industrial exhibitions that emerged in Ontario's larger towns after Confederation, where agricultural exhibits paled before paid amusements like balloonists and lady riders. But this was a logical evolution for exhibitions; they had been designed to attract the public and give its members a say in setting standards. In due course, public taste would be elevated and directed to the pursuit of excellence. Instead, exhibitions amplified existing public opinion and its importance in the cultural life of the nation. Life outside the exhibition began to look more like life inside the exhibition in all its commercial and carnivalesque aspects. The great Enlightenment project of introducing rationality into daily life foundered.

This evolution was apparent in the great international exhibitions. They originated as concrete encyclopedias and temples of reason, and ended as phantasmagorias – temples to frivolity and the hyperreality of advertising. The pursuit of paid admissions, by offering popular amusement, caused this change. Another cause was a decline in referentiality, which occurred on two fronts. The early exhibitions were meant to be transparently educational, through a direct appeal to the senses, the building blocks of knowledge. By the end of the century, commentators conceded that most of the exhibition was not educational but consisted of items that one could find on a smaller scale in any department store. Education was concentrated in special departments, where professional educators and 'object lessons' mediated overtly between the learning subject and the object studied.

The erosion of the truth claim at the heart of public display was another example of the decline of referentiality. Advertising techniques became increasingly sophisticated and detached from the object advertised. Andrew Wernick argues that communication is both referential and performative or rhetorical: It describes and it persuades.[4] The exhibition was designed to root the rhetorical strongly in the referential: Reality itself would persuade. But even by 1862, buxom young ladies proffering cocoa appeared in the Official Catalogue. Exhibitors created special exhibit objects far removed from what they actually sold. This trend was pervasive, from agricultural implements done up in silk, precious woods, and metals to displays by government departments: the deceptive Ontario educational exhibit, the Indian schools display, the many false descriptions of Canada's climate and resources, and the attempts to display Canada as a united nation. The result was suspicion: suspicion of advertising, of authoritative statements by government, and even of the concept of authoritative statements.[5]

Conceived as a metonymic index of the world, the exhibition became something else, a world apart. Even as the status of the exhibit as a referent declined, its status as an object in its own right seems to have increased. The public demand for exhibitions grew, and with it the determination by nation states to participate in these extravaganzas. Without exception, the costly and even lavish displays sent from Canada to England, France, and the United States had disappointing results, failing to produce the economic benefits that promoters predicted would follow. But the Canadian government persisted in sending ever grander collections, deploying ever greater resources in the attempt. The enthusiasm of Sir Charles Tupper more accurately reflected public sentiment than the suspicion of Sir John A. Macdonald. This outcome was not simply a descent into empty discourse but a retreat into representation. Once the project of materially representing the world, and Canada within it, at a world's fair had been undertaken, nothing short of full presence could satisfy, and so the project expanded almost endlessly. This expansion was in several directions. Science, in the form of geology, for example, enlarged the realm of the visible and representable. Evolutionary theory added a historical dimension to the spatial arrangement, with different peoples organized according to their level of development. The addition of categories like art and education provided new ways of organizing the exhibits to illustrate a greater intellectual or practical mastery over material life and human history. Through such acts, the project of representation acquired greater urgency and substance. Nineteenth-century governments, no less than avant-garde artists and writers, tried to push the boundary of representativity. Although they did so in the name of realism, they became only more entangled in layers of representation. Representing was a political and a discursive project, although it is unclear whether democratic theory was a cause or an effect of this will to represent. The agricultural societies merit particular attention because they served both the material and the political representation of farmers. Sometimes one form was privileged at the expense of the other, as when complaints arose that the shows were a distraction and a bribe, but farmers were also able to use one form of representation to bolster the other. The descent into representation could be an ascent towards power.

The Canada advertised at international exhibitions was in some respects as irrational and phantasmagoric as the exhibitions themselves. Perhaps that explains why exhibitions never quite convinced the millions of visitors who passed through them to emigrate *en masse*. But just as the exhibitions became an entrenched presence on the cultural landscape of nineteenth-century Canada, so the concept of Canada acquired substance at a popular level, as well as in political and financial circles. Exhibitions helped to construct a sense of Canada – its resources and its people. These displays inspired Canadians with pride and confidence in themselves and their country, even when they failed to bring financial benefits. At the very least, they were a mirror held up to society, and few people can gaze at a mirror without some slight satisfaction.

Appendix: Tables

TABLE 1 Ontario Exhibitions by County

	1869 and 1870		1889 and 1890		
	Exhibitions	Prizes ($)	Exhibitions	Prizes ($)	Receipts ($)
Algoma	2	482	18	2,867	510
Brant	11	5,378	11	9,168	9,117
Bothwell	9	2,552	–	–	–
Bruce	21	4,181	16	5,497	1,458
Cardwell	10	2,904	6	3,071	1,104
Carleton	7	2,060	10	3,833	1,047
Dufferin	–	–	4	2,713	1,664
Dundas	8	2,022	8	4,087	3,260
Durham	17	562	11	6,591	2,144
Elgin	16	3,923	14	6,506	5,333
Essex	7	853	14	5,803	3,317
Frontenac	7	1,756	8	9,408	9,485
Glengarry	7	1,607	2	1,772	1,042
Grey	27	4,070	36	10,540	3,025
Haldimand	10	2,785	8	3,614	1,700
Halton	10	4,185	9	7,066	2,843
Hastings	13	3,705	27	6,887	6,289
Huron	23	5,647	27	13,212	8,281
Kent	10	2,249	22	10,484	12,586
Lambton	13	2,704	18	8,711	4,004
Lanark	11	2,108	16	6,963	5,257
Leeds & Grenville	20	4,947	22	9,033	6,531
Lennox & Addington	18	2,402	13	5,003	1,713
Lincoln	8	2,746	5	4,360	1,714
Middlesex	30	5,922	30	10,630	5,017
Monck	11	2,352	13	4,790	2,664
Northumberland	20	4,526	15	4,952	3,327
Norfolk	14	3,473	16	8,724	5,817
Ontario	18	7,831	15	8,927	5,382
Oxford	21	7,374	21	12,422	6,652
Peel	7	4,101	5	4,441	2,879
Perth	16	5,014	17	6,214	4,272
Peterborough	15	3,293	17	7,331	4,900
Prescott*	10	2,618	9	3,095	1,685
Prince Edward	11	2,788	10	4,941	3,624
Renfrew	13	2,421	16	4,010	2,326
Russell*	9	2,465	10	3,430	1,227
Simcoe	28	6,001	22	16,722	11,860
Stormont	7	1,776	8	2,365	704
Victoria*	20	3,062	20	6,490	4,343
Waterloo	9	5,209	14	8,196	4,657
Welland	14	2,843	12	3,345	1,428
Wellington	27	8,033	30	12,786	5,320
Wentworth†	14	5,657	18	12,302	2,376
York	16	10,520	18	16,635	8,580

*These counties awarded substantial prizes for crops in the field.
†Collaborated on Hamilton Central Exhibition (only their contribution is noted).
Source: Ontario, *Sessional Papers*, Agricultural Society Reports.

TABLE 2 Prize Money Distribution in Ontario

	County		Township	
1868	Amount ($)	Percentage	Amount ($)	Percentage
livestock	21,681	50.8	20,750	54.7
crops	5,263	12.3	6,829	16.6
dairy	1,807	4.2	1,564	4.1
horticulture	3,109	7.3	1,678	4.4
implements	3,299	7.7	2,048	5.4
arts	5,849	13.6	4,195	11.1
plough match	1,508	3.5	1,362	3.6
miscellaneous	375	0.9	–	–
Totals	42,891		38,421	
1878	County and Township		Provincial Exhibition	
horses	22,908	20.2	1,914	12.8
cattle	16,931	15.0	2,941	19.6
sheep	11,821	10.4	1,310	8.7
swine	4,557	4.0	854	6.0
poultry	3,455	3.0	329	2.2
grain	8,496	7.5	626	4.2
roots	4,390	3.9	201	1.3
plough match	1,013	0.9	–	–
dairy	5,010	4.4	446	3.0
fruit	4,181	3.7	848	5.7
vegetables	2,870	2.5	187.50	1.3
flowers	2,496	2.2	308.50	2.0
implements	6,171	5.4	474	3.2
manufactures	7,198	6.3	2,371	15.9
arts	1,936	1.7	841	5.6
ladies	8,459	7.4	343	2.3
miscellaneous	851	0.7	1,000	6.7
Totals	113,722		14,994	

TABLE 2 (*concluded*)

1879 (original categories)	County		Townships	
	Amount ($)	Percentage	Amount ($)	Percentage
livestock	41,830	53.1	31,706	54.6
agricultural produce	18,143	23.1	15,510	26.7
arts & manufactures	18,654	23.7	11,016	18.9
Totals	136,498		58,232	

1890 (original categories)	County and Townships	
	Amount ($)	Percentage
livestock & dairy	107,336	63
agricultural products	28,027	17
manufactures	8,423	5
ladies & art	26,179	15
Totals	169,965	

TABLE 3 Quebec Agricultural Societies, 1872 and 1880

	% of Farmers Who Joined in 1872†	% of Farmers Who Joined in 1880†	Fr. Cdn Membership, 1872 (%)	Fr. Cdns Living in County, 1871 (%)
Argenteuil	4 (4)	(8)	7	30
Arthabaska	0 (0)	0	n/a	n/a
Bagot	20 (20)	(23)	100	98
Beauce	7 (7)	(12.5)	90	96
Beauharnois	14 (14.5)	(26)	73	90
Bellechasse	2 (2)	(2)	100	98
Berthier	5 (5)	(5)	98	98
Bonaventure*	5 (5.5)	(5)	12	60
Brome	n/a	(8)	n/a	25
Chambly	9.5 (7.5)	(10)	93	93
Champlain	9 (8.5)	(11.5)	99	98
Charlevoix*	4 (3.5)	(12.5)	99	98
Chateauguay	7 (7)	(6)	40	70
Chicoutimi & Saguenay*	n/a	(11.5)	n/a	95
Compton*	8 (8)	(6.5)	2	28
Dorchester	2 (2)	(1.5)	99	85
Drummond*	3 (3)	(3)	77	74
Gaspé*	6 (6)	(5)	21	68
Hochelaga	11 (11)	(25.5)	31	79
Huntingdon*	16 (10)	(34)	2	30
Iberville	2 (2)	(4)	97	90
Jacques-Cartier	10 (10)	(16)	77	88
Joliette	5 (5)	(13.5)	95	96
Kamouraska	2 (1.5)	(3)	99	99
L'Assomption	15 (15)	(22)	94	97
L'Islet	5 (5)	(8)	100	99
Laprairie	11 (10)	(10)	100	85
Laval	6.5 (5.5)	(11.5)	99	99
Lévis	3 (2.5)	(4)	97	91
Lotbinière*	7 (6.5)	(13.5)	75	84
Maskinongé	8 (8)	(9)	99	98
Megantic*	2 (2)	(4.5)	1	64
Missisquoi	5 (5)	(20)	6	42
Montcalm	10 (10)	(13.5)	93	85
Montmagny	3 (2.5)	(4.5)	100	99
Montmorency*	3 (2.5)	(24)	99	91
Montreal	55		7	53
Napierville	5 (5.5)	(8)	88	93
Nicolet	11 (11)	(13)	97	97
Ottawa*	1 (1)	(2.5)	1	56
Pontiac	3 (3)	(4.5)	0	22

TABLE 3 (*concluded*)

	% of Farmers Who Joined in 1872†	% of Farmers Who Joined in 1880†	Fr. Cdn Membership, 1872 (%)	Fr. Cdns Living in County, 1871 (%)
Portneuf	7 (7)	(11.5)	98	90
Quebec City	n/a	(67)		
Quebec County	n/a	(6.5)	n/a	72
Richelieu	10 (10.5)	(19)	98	97
Richmond & Wolfe	6 & 10 (6 & 9)	(13.5)*	38	59
Rimouski	2 (2.5)	(3)	95	95
Rouville*	12 (11.5)	(25)	93	96
Shefford	2 (2)	(3.5)	12	96
Sherbrooke	12 (12.5)	(23)	14	42
Soulanges	7 (6.5)	(11)	85	90
Stanstead	10 (10)	(8.5)	0	24
St-Hyacinthe	19 (19)	(31)	100	99
St-Jean	7 (7)	(11)	71	78
St-Maurice	8.5 (7.5)	(8.5)	99	97
Témiscouata	6 (6)	(11)	98	97
Terrebonne*	1 (1)	(3)	74	77
Three Rivers	29	(45.5)	93	92
Two Mountains	5 (5)	(12)	77	90
Vaudreuil	3 (2.5)	(18)	87	85
Verchères	18 (18)	(28.5)	100	99
Yamaska	8 (8)	(14)	99	97

*More than one Agricultural Society in the county. Figures represent members in all societies.

†Numbers in parentheses indicate percentages arrived at using total membership rather than simply farmers. Figures on farmers alone were not available for 1880, however, as can be seen from the figures for 1872, total membership statistics closely reflect those for farmers alone. Approximately 11.5 per cent of rural agricultural society members were not farmers in 1872.

Sources: Quebec, *Sessional Papers*, and Census of Canada

TABLE 4 Montmagny Farm Inspection Prizes, 1871–1893*

1871	J.-O. Beaubien	L.-H. Blais	A.-O. Talbot	H. Talbot	P. Becer
1873	L.-H. Blais	A. Talbot	J. Collin	E. Talbot	H. Talbot
1875	A. Talbot	X. Letourneau	J. Collin	H. Talbot	E. Bernatchez
1883	A. Talbot	X. Letourneau	J. Collin	M. Blais	G. Cloutier
1885	X. Letourneau†	J. Collin	E. Blais	M. Blais	F. Blais
1887	$40	$30	$20	$14	$13
	A. Talbot	E. Bernatchez	E. Blais	F. Boufard	J. Collin
1889	$40	$30	$20	$14	$13
	F.-X. Letourneau	E. Bernatchez	E. Blais	J. Collin	E. Blais
1891	$100	$60	$40	$30	$20
	F.-X. Letourneau	E. Bernatchez	J. Collin	S. Letourneau	E. Blais
1893	$100	$60	$40	$30	$20
	F.-X. Letourneau	E. Bernatchez	J. Collin	E. Blais	L. Gaze

*Names not listed in the minutes in 1877, 1879, 1881.
†E. Bernatchez won, but was ruled *hors concours*.
Prize amounts were dictated by the Council of Agriculture as $50, $40, $30, $20, and $10. Exceptions are noted.
Source: Minutes of the Montmagny Agricultural Society, Landry Papers, NA, MG 30 E17.

TABLE 5 Ontario Provincial Exhibition Livestock Winners

	Number of Competitors with Total Winnings as Follows			
	1851–2	1860–1	1870–1	1888–9
$1–10	59	112	38	70
$11–20	55	105	48	62
$21–50	28	67	50	61
$51–100	7	31	21	28
$101–200	6	17	14	12
$201–300	1	2	8	12
$300+	–	6	7	7
Totals	156	340	185	242
	Sample Winnings of the Top Competitors			
$51–100	$444	–	–	–
$101–200	$806	$2,192	$2,098	$1,760
$201–300	$211	$487	$1,644	$3,137
$300+	–	$2,591	$4,041	$3,782
Total prizes paid	$3,801	$12,190	$12,812	$13,796

Sources: Ontario, *Sessional Papers*, and *Journal and Transactions of the Board of Agriculture of Upper Canada*.

Notes

Abbreviations

ANQ	Archives Nationales du Québec
ANQM	Archives Nationales du Québec à Montréal
AO	Archives of Ontario
CIHM	Canadian Institute for Historical Microreproductions
DCB	*Dictionary of Canadian Biography*
DIA	Department of Indian Affairs
DLAUC	*Debates of the Legislative Assembly of United Canada*
JBAMUC	*Journal of the Board of Arts and Manufactures of Upper Canada*
JLALC	*Journals of the Legislative Assembly of Lower Canada*
JLAPC	*Journals of the Legislative Assembly of the Province of Canada*
MG	Manuscript Group
MTRL, BR	Metropolitan Toronto Reference Library, Baldwin Room
NA	National Archives of Canada
RG	Record Group
TMI	Toronto Mechanics' Institute

Introduction

1 John Ralston Saul, *Voltaire's Bastards: The Dictatorship of Reason in the West* (London: Sinclair Stevenson, 1992).

2 For example, Olivia Smith, *The Politics of Language* (Oxford: Clarendon Press, 1984).

3 Jürgen Habermas, *The Structural Transformation of the Public Sphere: An Inquiry into a Category of Bourgeois Society* (Cambridge, Mass.: MIT Press, 1989); see the essays in Craig Calhoun, ed., *Habermas and the Public Sphere* (Cambridge, Mass.: MIT Press, 1993).

4 M.M. Bakhtin, *Rabelais and His World,* trans. Helen Iwolsky (Cambridge, Mass.: MIT Press, 1968; see also Emmanuel Le Roy Ladurie, *Carnival at Romans,* trans. Mary Feeney (New York: G. Brazillier, 1979); Peter Stalleybrass and Allon White, *The Politics and Poetics of Transgression* (Ithaca: Cornell University Press, 1986).
5 E.J. Hobsbawm, *The Age of Capital, 1848–1875* (London: Cardinal, 1988), 47.
6 New Westminster *British Columbian,* 3 Oct. 1861.

Chapter 1: The Theory of the Exhibition

1 This distinction is taken from L. de Ligt, *Fairs and Markets in the Roman Empire: Economic and Social Aspects of Periodic Trade in a Pre-Industrial Society* (Amsterdam: J.C. Gieben, 1993), 14. See also Ellen Wedemeyer Moore, *The Fairs of Medieval England: An Introductory Survey* (Toronto: Pontifical Institute, 1985); William Addison, *English Fairs and Markets* (London: B.T. Batsford, 1953); Fernand Braudel, *Civilization and Capitalism, 15th–18th Century,* vol. 2: *The Wheels of Commerce,* trans. Siân Reynolds (New York: Harper & Row, 1982): 81–114; Richard Perren, 'Markets and Marketing,' in G.E. Mingay, ed., *The Agrarian History of England and Wales,* vol. 6: 1750–1850 (Cambridge: Cambridge University Press, 1989) 223.
2 Thomas Frost, *The Old Showmen and the Old London Fairs* (1881; Gryphon: Ann Arbor, 1971), 55–6; Richard D. Altick, *The Shows of London* (Cambridge, Mass.: Belknap Press, 1978).
3 James M. Mayor, *The American Grocery Store: The Business Evolution of an Architectural Space* (Westport: Greenwood Press, 1993), 40.
4 Louise Dechêne, *Habitants and Merchants in Seventeenth-Century Montreal,* trans. Liana Vardi (Montreal: McGill-Queen's University Press), 192; Louise Dechêne, *Le partage des subsistances au Canada sous le régime français* (Montreal: Boréal, 1994); John Dickenson notes that 'il y eut toujours un magistrat sur les marchés' in New France: 'La Police en Nouvelle-France,' *McGill Law Journal* 32 (1987): 513.
5 E.A. Cruickshank, ed., *The Correspondence of Lieut. Governor John Graves Simcoe* (Toronto: Ontario Historical Society, 1923–31), 4: 551–2; Brian Osborne, 'Trading on a Frontier: The Function of Pedlars, Markets, and Fairs in Nineteenth-Century Ontario,' *Canadian Papers in Rural History* 2 (1980): 59–82.
6 Osborne, 'Trading on a Frontier,' 73–5; Ontario Association of Agricultural Societies, *Story of Ontario Agricultural Fairs and Exhibitions 1792–1967* (Picton: Gazette, 1967), 13.
7 National Archives of Canada (NA), RG 5 A1, Provincial Secretary, Upper Canada, 98963–4.

8 NA, RG 4 C1, Provincial Secretary, Lower Canada, file 611 for 1847.

9 'Annual Report of the Minister of Agriculture for 1856,' Province of Canada, *Sessional Papers* 2, 1857; Kenneth Kelly, 'The Development of Farm Produce Marketing Agencies and Competition between Market Centres in Eastern Simcoe County, 1850–1875,' *Canadian Papers in Rural History* 1 (1978): 67–88.

10 The first fair with prizes in what is now Canada was in Windsor, Nova Scotia, in 1765, and prizes were awarded at Niagara fairs from the turn of the century. Bill Russell, 'First Agricultural Fair in Canada,' in *Miscellaneous Historical Studies* (Parks Canada Manuscript Report 165), 113–24; Janet Carnochan, *History of Niagara* (Toronto: W. Briggs, 1914), 268; *Niagara Gleaner*, 18 Oct. 1827.

11 Robert Sellar, *The History of the County of Huntingdon and of the Seignories of Chateauguay and Beauharnois* (Huntingdon: Canadian Gleaner, 1888), 338–9.

12 Kenneth Hudson, *Patriotism with Profit: British Agricultural Societies in the Eighteenth and Nineteenth Centuries* (London: Hugh Evelyn, 1972), 53; James Meenan and Desmond Clarke, *The Royal Dublin Society, 1731–1981* (Dublin: Gill & McMillan, 1981); Kenneth Hudson, *The Bath and West: A Bicentenary History* (Wiltshire: Moonraker Press, 1976).

13 Rosalind Mitchison, 'The Old Board of Agriculture (1793–1822),' *English Historical Review* 74 (1959): 41–69.

14 Frederick B. Artz, *Development of Technical Education in France* (Cambridge, Mass.: MIT Press, 1966), 79; Margaret W. Rossiter, 'The Organization of Agricultural Improvement in the United States,' in Alexandra Oleson and Sanborn C. Brown, eds., *The Pursuit of Knowledge in the Early American Republic: American Scientific and Learned Societies from Colonial Times to the Civil War* (Baltimore: Johns Hopkins University Press, 1976), 279–98.

15 Derek Hudson and Kenneth Luckhurst, *The Royal Society of Arts, 1754–1954* (London: Murray, 1954); John Styles, 'Manufacturing, Consumption and Design in Eighteenth-Century England,' in John Brewer and Roy Porter, eds., *Consumption and the World of Goods* (London: Routledge, 1993), 527–54; D.C.G. Allen and John L. Abbott, eds., *The Virtuoso Tribe of Arts and Sciences: Studies in the Eighteenth-Century Work and Membership of the London Society of Arts* (Athens, Ga: University of Georgia Press, 1992).

16 Patricia Mainardi, *Art and Politics of the Second Empire: The Universal Expositions of 1855 and 1867* (New Haven: Yale University Press, 1987); Artz, *Development of Technical Education*; Charles Babbage, *The Exposition of 1851* (1851; London: Frank Cass, 1968), 18–20; Jack Morrell and Arnold Thackray, *Gentlemen of Science: Early Years of the British Association for the Advancement of Science* (Oxford: Oxford University Press, 1981). Much has been written on world exhibitions: see Brigitte Schroeder-Gudehus and Anne Rasmussen, *Les fastes*

du progrès: Le guide des expositions universelles, 1851–1992 (Paris: Flammarion, 1992); also John E. Findling, *Historical Dictionary of World's Fairs and Expositions, 1851–1988* (New York: Greenwood Press, 1990).

17 Jean-Jacques Rousseau, *Politics and the Arts: Letter to M. D'Alembert on the Theatre*, trans. Allan Bloom (Ithaca: Cornell University Press, 1991), 126–30; Mona Ozouf, *Festivals and the French Revolution*, trans. Alan Sheridan (Cambridge, Mass.: Harvard University Press, 1988).

18 London (Ontario) *Advertiser*, 29 Sept. 1871.

19 J.E. Hodgetts, *Pioneer Public Service: An Administrative History of the United Canadas, 1841–1867* (Toronto: University of Toronto Press, 1955), 229.

20 Morris Berman, *Social Change and Scientific Organization: The Royal Institution, 1799–1844* (London: Heinemann, 1978).

21 Ontario, *Sessional Papers* 1, 12 (1868–9), appendix C, 50–1.

22 In Arthur M. Melzer, *The Natural Goodness of Man: On the System of Rousseau's Thought* (Chicago: University of Chicago Press, 1990), 38.

23 Thomas Hobbes, *Leviathan* (Oxford: Basil Blackwell, 1955), 83.

24 Quoted in Albert O. Hirschman, *The Passions and the Interests: Political Arguments for Capitalism before Its Triumph* (Princeton: Princeton University Press, 1977), 58. On the debate surrounding commercial and traditional values, see J.G.A. Pocock, *The Machiavellian Moment: Florentine Political Thought and the Atlantic Republican Tradition* (Princeton: Princeton University Press, 1975); Milton L. Myers, *The Soul of Modern Economic Man: Ideas of Self-Interest, Thomas Hobbes to Adam Smith* (Chicago: University of Chicago Press, 1983); Thomas Pangle, *The Spirit of Modern Republicanism: The Moral Vision of the American Founders and the Philosophy of Locke* (Chicago: University of Chicago Press, 1988); Joyce Appleby, 'Consumption in Modern Social Thought,' in Brewer and Porter, *Consumption*, 162–73, and her *Capitalism and a New Social Order: The Republican Vision of the 1790s* (New York: New York University Press, 1984).

25 Jules Simon, 'Introduction,' in Ministère de l'Agriculture et du Commerce, *Rapports du jury international* (Paris: Imprimerie nationale, 1880), 279 and passim.

26 *Canadian Agriculturist* 2 (1850): 121–2.

27 *Canada Farmer* 1 (1847): 11.

28 Adam Smith, *The Theory of Moral Sentiments*, ed. D.D. Raphael and A.C. Macfie (Oxford: Clarendon Press, 1976), 50, 65; see James Edward Alvey, 'A New Adam Smith Problem: The Teleological Basis of the Commercial Society' (PhD dissertation, University of Toronto, 1996); also E.G. Hundert, *The Enlightenment's Fable: Bernard Mandeville and the Discovery of Society* (Cambridge: Cambridge University Press, 1994), 53, 172.

29 *Colonial Advocate,* 18 March 1830; emphasis in the original unless otherwise stated.

30 *Canadian Agriculturist* 4 (1852): 193–201. On Hutton, see Bruce Curtis, *True Government by Choice Men? Inspection, Education, and State Formation in Canada West* (Toronto: University of Toronto Press, 1992).

31 *Canadian Stock Raiser's Journal* 1 (1884): 62.

32 Clifford Orwin, *The Humanity of Thucydides* (Princeton: Princeton University Press, 1994), 124.

33 Hobbes, *Leviathan,* Part I, ch. 10.

34 Jean-Jacques Rousseau, *Emile, or On Education,* trans. Allan Bloom (New York: Basic Books, 1979), 213–14.

35 Alexander Ramsay, *History of the Highland and Agricultural Society of Scotland* (Edinburgh: William Blackwood and Sons, 1879), 28–9.

36 Smith, *Theory of Moral Sentiments,* 110.

37 Ibid., 160.

38 Ibid., 61. His critique of the aristocracy is in *The Wealth of Nations.*

39 Donald Winch, *Adam Smith's Politics: An Essay in Historiographical Revision* (Cambridge: Cambridge University Press, 1978), 168–70.

40 Bianca Fontana, 'Whigs and Liberals: The *Edinburgh Review* and the "Liberal Movement" in Nineteenth-Century Britain,' in Richard Bellamy, ed., *Victorian Liberalism: Nineteenth-Century Political Thought and Practice* (London: Routledge, 1990), 52–3. Other recent discussions of early public opinion are Jonathan Parry, *The Rise and Fall of Liberal Government in Victorian Britain* (Princeton: Yale University Press, 1993), 27–49, and Dror Wahrman, *Imagining the Middle Class: The Political Representation of Class in Britain, c. 1780–1840* (Cambridge: Cambridge University Press, 1995).

41 Toronto *Leader,* 9 Oct. 1866, from the *Journal of the Board of Arts and Manufactures of Upper Canada* (*JBAMUC*) 6 (1866): 288.

42 In William Kilbourn, *The Firebrand: William Lyon Mackenzie and the Rebellion in Upper Canada* (Toronto: Clarke, Irwin, 1956), 118.

43 G.P. Gooch, *History and Historians in the Nineteenth Century* (1913; New York: Peter Smith, 1949), 572, on Voltaire. See David Hume, *The History of England* (London: A. Millar, 1754–62).

44 John Stuart Mill, *A System of Logic, Ratiocinative and Inductive,* ed. J.M. Robson (1843; Toronto: University of Toronto Press, 1974), 829.

45 Founders of the Toronto Mechanics' Institute included John Ewart, Tiger Dunlop, Sheriff Jarvis, Dr John Rolph, and Robert Baldwin and his father, William. Metropolitan Toronto Reference Library, Baldwin Room (MTRL, BR), Toronto Mechanics' Institute Papaers, 'The Toronto Mechanics Institute.' The first Montreal Mechanics' Institute was summoned by Rev.

Esson in 1828; when it was re-established in 1840, John Redpath presided, and other founders included Benjamin Holmes and John Lovell. Atwater Library, Montreal, Minute Books of the Montreal Mechanics' Institute.

46 In 1848 A.A. Riddel asked permission to lecture to the Toronto Mechanics' Institute on 'The Rights of Labour.' He was refused on the grounds that parts of his speech were 'objectionable.' MTRL, BR, Toronto Mechanics' Institute Papers, vol. B2, Board Minutes 1848–54, 14 Nov. and 5 Dec. 1848. Most of the lectures given were scientific, but in 1842–3 Rev. D. Rintoul lectured to the Toronto Group on 'Political Economy,' 'Productive and Unproductive Labour,' and 'The Theory of Rent of Land.'

47 *British American Cultivator* 3 (1844): 12.

48 *JBAMUC* 1 (1861): 211.

49 Montreal *Gazette*, 5 Dec. 1833.

50 Antonio Gramsci, *Selections from the Prison Notebooks*, ed. and trans. Quintin Hoare and Geoffrey Nowell Smith (New York: International Publishers, 1971), 12.

51 Association of Arts and Agriculture of Ontario, in Province of Ontario, *Sessional Papers* 20, 10 (1888): 51.

52 *Canadian Agriculturist* 10 (1858): 11.

53 John Young, *Letters of Agricola on the Principles of Vegetation and Tillage* (1822; Halifax: Blackader Brothers, 1922), 225; Graeme Wynn, 'Exciting a Spirit of Emulation among the "Plodholes": Agricultural Reform in Pre-Confederation Nova Scotia,' *Acadiensis* 20 (1991): 5–50.

54 *British American Cultivator* 1 (April 1842). Evans quoted this passage again in 1844, as 'men who are not well educated, are generally prone ...' *Canadian Agricultural Journal* 1 (1844): 72.

55 *Canadian Agriculturist* 1 (1849): 293.

56 *Agricultural Journal* 2 (1849): 109; in French: *Journal d'agriculture* 2 (1849): 101. 'Too wise' was Young's description of the French in Nova Scotia, in *Letters of Agricola*, 342. Another example of hostility to improvers in Quebec during the 1830s is cited in H.A. Innis and A.R.M. Lower, eds., *Select Documents in Canadian Economic History, 1783–1885* (Toronto: University of Toronto Press, 1933), 85.

57 *L'Agriculteur* 13 (1860–1): 97–100.

58 Quebec *Gazette*, 6 Dec. 1820.

59 Samuel Johnson, *A Dictionary of the English Language* (1755; New York: Arno Press, 1979).

60 Smith, *Theory of Moral Sentiments*, 114. See also the discussion of Hobbes's definition of emulation as 'griefe arising from seeing ones self exceeded or excelled by his Concurrent, together with hope to equall or exceed him in

tyme to come.' In Quentin Skinner, *Reason and Rhetoric in the Philosophy of Thomas Hobbes* (Cambridge: Cambridge University Press, 1996), 39.

61 He listed prize-giving as one means of securing emulation: Gramsci, *Selections from the Prison Notebooks*, 246–7, 260; John Patrick Diggins also describes hegemony as an invention of the Scottish Enlightenment: 'Gramsci and the Intellectuals,' *Raritan* 9, 2 (1989): 129–52.

62 S.F. Wise, 'Sermon Literature and Canadian Intellectual History,' in A.B. McKillop and Paul Romney, eds., *God's Peculiar Peoples: Essays on Political Culture in Nineteenth-Century Canada* (Ottawa: Carleton University Press, 1993), 16–17.

63 Karl Marx, 'The Eighteenth Brumaire of Louis Bonaparte,' in Karl Marx and Friedrich Engels, *Selected Works* (Moscow: Foreign Languages Publishing House, 1958), 305.

64 John Locke, *Essay Concerning Human Understanding* (Oxford: Clarendon Press, 1975), 305–22, 642–3. See also David Hume *Treatise on Human Nature* (Oxford: Clarendon Press, 1978), 73. Sergio Moravia argues that '*Comparaison* is one of the principal cognitive acts of the modern human sciences. It is quite significant that it was worked out during the Enlightenment.' In 'The Enlightenment and the Sciences of Man,' *History of Science* 18 (1980): 250.

65 Mill, *System of Logic*, 379.

66 William Whewell, 'The General Bearing of the Great Exhibition,' in Society of Arts, *Lectures on the Results of the Great Exhibition of 1851* (London: David Gogue, 1852), 3–34; on Whewell, see Richard Yeo, *Defining Science: William Whewell, Natural Knowledge, and Public Debate in Early Victorian Britain* (Cambridge: Cambridge University Press, 1993).

67 James A. Schmiechen, 'The Victorians, the Historians, and the Idea of Modernism,' *American Historical Review* 93 (1988): 287–316; Brian Simon, *The Two Nations and the Educational Structure, 1780–1870* (London: Lawrence & Wishart, 1974).

68 Jean Le Rond D'Alembert, *Preliminary Discourse to the Encyclopedia of Diderot*, trans. Richard N. Schwab (New York: Bobbs-Merrill, 1963), 31.

69 *British American Cultivator* 2 (1843): 22. This article is not his own, but it is one of several that appeared in his journal that made this point. Elsewhere, Evans himself stated: 'It was for the general good of society that the Creator of all things endowed a few with superior intellect, that they might employ it for the good of their fellow men as well as for their own; and they make but a poor return to the Giver of these blessings when they only employ them for their own purposes and benefit.' *Canadian Agricultural Journal* 2 (1845): 33; and he quoted an economist estimating that one-twentieth of the population can derive pleasure from science and the rest 'will be accessible only to

physical enjoyments or excitation of the fancy. This is not peculiar to the lower orders, it pervades alike every walk of life.' *Canadian Agricultural Journal* 1 (1844): 184–5. His views were less conservative before the rebellion: see Evans, *A Treatise on the Theory and Practice of Agriculture* (Montreal: Fabre, Perreault, 1835), 293; and the discussion in Tom Nesmith, 'The Philosophy of Agriculture: The Promise of the Intellect in Ontario Farming, 1835–1914' (PhD dissertation, Carleton University, 1988), 50.

70 *Canadian Agriculturist* 1 (1849): 330.

71 On earlier attempts to mechanize thinking, see William J. Ashworth, 'Memory, Efficiency, and Symbolic Analysis: Charles Babbage, John Herschel, and the Industrial Mind,' *Isis* 87 (1996): 629–53.

72 J. Tallis, *History and Description of the Crystal Palace and the Exhibition of the World's Industry in 1851* (London: J. Tallis, [1851]), 101–2; *JBAMUC* 1 (1861): 59.

73 Carl Woodring, *Nature into Art: Cultural Transformations in Nineteenth-Century Britain* (Cambridge: Harvard University Press, 1989), 64.

74 Kate Silber, *Pestalozzi: The Man and His Work* (London: Routledge & Kegan Paul, 1960); on Pestalozzi in North America, see Lucille Wrubel Grinhammer, *Art and the Public: The Democratization of the Fine Arts in the United States, 1830–1860* (Stuttgart: J.B. Metzlesche, 1975), ch. 1; also Barbara Maria Stafford *Artful Science: Enlightenment Entertainment and the Eclipse of Visual Education* (Cambridge, Mass.: MIT Press, 1994).

75 Amos Funkenstein, *Theology and the Scientific Imagination from the Middle Ages to the Seventeenth Century* (Princeton: Princeton University Press, 1986), 297–8. See the general discussion in Paolo Rossi, *Philosophy, Technology, and the Arts in the Early Modern Era*, trans. Salvator Attanasio (1962; New York: Harper & Row, 1970).

76 J. George Hodgins, *Documentary History of Education in Upper Canada* (Toronto: Department of Education, 1894), 288–308, has Duncombe's report. See Michael S. Cross, 'Charles Duncombe,' *Dictionary of Canadian Biography* (*DCB*), vol. 9 (Toronto: University of Toronto Press, 1976), 230.

77 *Journal of Education for Upper Canada* 1 (1848): 301.

78 John Langton, 'The Importance of Scientific Studies to Practical Men,' *Canadian Journal* 3 (1854): 201–4. On his education, see W.A. Langton, ed., *Early Days in Upper Canada: Letters of John Langton* (Toronto: Macmillan, 1926), xiv.

79 Gerald Killam, *David Boyle: From Artisan to Archaeologist* (Toronto: University of Toronto Press, 1983).

80 *Canadian Journal*, ns 1 (1856): 105–26; Toronto *Globe*, 10 April 1847.

81 NA, MG 24 D16, Chamberlin Papers, vol. 4, 22 Nov. 1846. See A.B. McKillop, *A Disciplined Intelligence: Critical Inquiry and Canadian Thought in the Victorian Era* (Montreal: McGill-Queen's University Press, 1979).

82 *Canadian Journal of Industry, Science, and Art* 3 (1854): 9 and 65.
83 Peter Dear, 'Totius in verba: Rhetoric and Authority in the Early Royal Society,' *Isis* 76 (1985): 145–61, and Richard F. Jones, 'Science and English Prose Style in the Third Quarter of the Seventeenth Century,' *Publications of the Modern Languages Association* 45 (1930): 976–1009.
84 D'Alembert, *Preliminary Discourse,* 47. See Daniel Brewer, *The Discourse of Enlightenment in Eighteenth-Century France: Diderot and the Art of Philosophizing* (Cambridge: Cambridge University Press, 1993).
85 In Nancy Partner, 'History without Empiricism / Truth without Facts,' in Christie MacDonald and Gary Wihl, eds., *Transformations: Languages of Culture and Personhood after Theory* (Pennsylvania: Pennsylvania State University Press, 1994), 9.
86 M.M. Bakhtin, *The Dialogic Imagination: Four Essays,* trans. Caryl Emerson and Michael Holquist (Austin: Texas University Press, 1981).
87 *Household Words,* 3 (1851): 519–23; and *Official Catalogue of the Great Exhibition,* 4th ed. (London: Spicer, September 1851).
88 *Household Words* 3 (1851): 121–4 and 356–60.
89 Martin Jay, *Downcast Eyes: The Denigration of Vision in Twentieth-Century French Thought* (Berkeley: University of California Press, 1993).
90 The problem recurs through all Foucault's *oeuvre,* but is most explicitly stated in *Discipline and Punish: The Birth of the Prison,* trans. Alan Sheridan (New York: Vintage, 1979). Bentham's theory was that, 'to be incessantly under the eyes of the inspector is to lose in effect the power to do evil and almost the thought of wanting to do it.' In Eli Halévy, *The Growth of Philosophical Radicalism,* trans. Mary Morris (London: Faber & Faber, 1934),183.
91 See Tony Bennett, *The Birth of the Museum: History, Theory, Politics* (London: Routledge, 1995).
92 Guy Debord, *Society of the Spectacle* (Detroit: Red & Black, 1983).
93 See Raymond Williams on hegemony in *Marxism and Literature* (Oxford: Oxford University Press, 1977), 110.
94 *JBAMUC* 4 (1864): 141–7; R.G.W. Anderson, 'What Is Technology? Education through Museums in the Mid-Nineteenth Century,' *British Journal for the History of Science* 25 (1992): 169–84. George Wilson was the brother of Daniel Wilson of the University of Toronto.

Chapter 2: The Early History of Exhibitions in Canada, 1789–1837

1 David S. Landes, *The Unbound Prometheus: Technological Change and Industrial Development in Western Europe from 1750 to the Present* (Cambridge: Cambridge University Press, 1989), 69.

2 B.A. Holderness, 'Prices, Productivity, and Output,' in Mingay, *Agrarian History of England*, vol. 4, 140.

3 R. Marvin McInnis, *Perspectives in Ontario Agriculture* (Gananoque: Langdale Press, 1992); John McCallum, *Unequal Beginnings: Agriculture and Economic Development in Quebec and Ontario until 1870* (Toronto: University of Toronto Press, 1980).

4 See Robert Lochiel Fraser, 'Like Eden in Her Summer Dress: Gentry, Economy and Society in Upper Canada, 1812–40' (PhD dissertation, University of Toronto, 1979).

5 Edward Lytton Bulwer, *England and the English*, ed. Standish Meacham (Chicago: University of Chicago Press, 1970), 30.

6 Paul Langford, *Public Life and the Propertied Englishman, 1689–1798* (Oxford: Clarendon Press, 1991), 560–1; see also G.E. Mingay, *The Gentry: The Rise and Fall of a Ruling Class* (London: Longman, 1976).

7 Peter Mandler, *Aristocratic Government in the Age of Reform: Whigs and Liberals, 1830–1852* (Oxford: Clarendon Press, 1990), 88–91.

8 Habermas, *Structural Transformation of the Public Sphere*; Foucault, *Discipline and Punish.*

9 *Canadian Farmer and Mechanic* 1 (1841): 1.

10 Bennett, *Birth of the Museum*, 65.

11 Rosalind Mitchison, *Agricultural Sir John: The Life of Sir John Sinclair of Ulbster, 1754–1835* (London: Geoffroy Bles, 1962), 185.

12 G.E. Mingay, *Arthur Young and His Times* (London: Macmillan, 1975).

13 Archives nationales du Québec (ANQ), P450–5: Literary and Historical Society of Quebec; Records of the Agricultural Society of Quebec, 1789–99, Minutes, 6 Jan. 1789; C.C. Jones, 'The First Agricultural Societies,' *Queen's Quarterly* 10 (1903): 218–25.

14 Agricultural Society of Quebec, *Papers and Letters on Agriculture* (Quebec: Neilson, 1790).

15 ANQ, P 450–5, Letterbook, 55; Minutes, 207–9.

16 In F. Murray Greenwood, *Legacies of Fear: Law and Politics in Quebec in the Era of the French Revolution* (Toronto: University of Toronto Press, 1993), 52; and see Mason Wade, *The French Canadians, 1760–1945* (London: Macmillan, 1955), 93.

17 Cruickshank, *Correspondence of Simcoe*, I, 17–19 and 318; Robert Leslie Jones, *History of Agriculture in Ontario, 1613–1880* (Toronto: University of Toronto Press, 1946), ch. 10. On Banks, see John Gascoigne, *Joseph Banks and the English Enlightenment: Useful Knowledge and Polite Culture* (Cambridge: Cambridge University Press, 1994).

18 Talman, 'Agricultural Societies of Upper Canada,' *Ontario History* 27 (1931):

545; Robert W. Carbert, 'Agricultural and Horticultural Societies and Fairs in the Niagara Peninsula,' in John Burtniak and Wesley B. Turner, eds., *Agriculture and Farm Life in the Niagara Peninsula* (St Catharines: Brock University, 1983), 47–62. Carnochan, *History of Niagara,* 47.

19 Jones, *History of Agriculture,* 158.

20 An overview is provided in Vernon C. Fowke, *Canadian Agricultural Policy: The Historical Pattern* (1946; Toronto: University of Toronto Press, 1978); Wynn, 'Exciting a Spirit of Emulation.'

21 Fernand Ouellet, *Economic and Social History of Quebec, 1760–1850: Structures and Conjonctures,* trans. (1966; Toronto: Macmillan, 1980); *Lower Canada, 1791–1840: Social Change and Nationalism,* trans. Patricia Claxton (Toronto: McClelland & Stewart, 1980); and *Economy, Class, and Nation in Quebec: Interpretive Essays,* trans. Jacques A. Barbier (Toronto: Copp Clark Pitman, 1991), are the major works.

22 Gilles Pacquet and Jean-Pierre Wallot, 'Crise agricole et tensions socio-ethniques dans le Bas-Canada, 1802–1812: Eléments pour une réinterprétation,' *Revue d'histoire de l'Amérique française* 26 (1972): 185–237.

23 Serge Courville, 'La crise agricole du Bas-Canada: éléments d'une réflexion géographique,' *Cahiers de géographie du Québec* 24 (1980): 193–224, 385–428.

24 R.M. McInnis, 'A Reconsideration of the State of Agriculture in Lower Canada in the First Half of the Nineteenth Century,' *Canadian Papers in Rural History* 3 (1982): 9–49. See also Frank D. Lewis and Marvin McInnis, 'Agricultural Output and Efficiency in Lower Canada, 1851,' *Research in Economic History* 9 (1984): 45–88; Frank Lewis and Marvin McInnis, 'The Efficiency of the French-Canadian Farmer in the Nineteenth Century,' *Journal of Economic History* 40 (1980): 497–515. Criticisms are in Louise Dechêne, 'Observations sur l'agriculture du Bas-Canada au début du XIXe siècle,' in Jean-Pierre Wallot and Joseph Goy, eds., *Evolution et éclatement du monde rural* (Montreal: Les Presses de l'Université de Montréal, 1986), 189–202; J.I. Little, 'Agricultural Progress in Canada East/Quebec: Problems in Measuring Relative Productivity during the Grain-Dairy Transition,' *Histoire sociale/Social History* 18 (1985): 425–32; and Robert Armstrong, 'The Efficiency of Quebec Farmers in 1851,' *Histoire sociale/Social History* 17 (1984): 149–65.

25 Allan Greer, *Peasant, Lord, and Merchant: Rural Society in Three Quebec Parishes, 1740–1840* (Toronto: University of Toronto Press, 1985); Christian Dessureault, 'Crise ou modernisation. La société maskoutaine durant le premier tiers du XIXe siècle,' *Revue d'histoire de l'Amérique française* 42 (1989): 359–88.

26 F.J. Fisher, cited in Braudel, *Wheels of Commerce,* 183.

27 Dechêne, 'Observations sur l'agriculture,' 192.

28 *Journals of the Legislative Assembly of Lower Canada* (*JLALC*), 1816, appendix E.
29 Montreal *Gazette*, 24 Feb. and 3 March 1817.
30 Ibid., 20 Jan. 1817.
31 Charles Frederick Grece, *Essays on Practical Husbandry, Addressed to the Canadian Farmers* (Montreal: William Gray, 1817), iv–v; J.P. Wallot, 'Charles Frederick Grece,' *DCB*, vol. 7, 358.
32 Montreal *Gazette*, 24 Jan. and 4 Feb. 1818. See also a circular issued by the Agricultural Society of Quebec (CIHM 54858), also its *Règles de la Société* (1818; Canadian Institute for Historical Microreproductions [CIHM] 57306) and proceedings of June 1819 (CIHM 54838). There already existed, at least since 1812, a Montreal Florists' Society, which offered prizes for essays; in 1818 it transformed itself into a Horticultural Society and began to hold exhibitions for fruits, roots, vegetables, and other productions. Montreal *Gazette*, 4 Feb. 1818.
33 58 George III, c. 6.
34 National Archives of Canada (NA), MG 24 B1, Neilson Papers, vols. 3–4 passim.
35 *JLALC*, 1819, appendix D, and 1820, appendix F (unpaginated).
36 The entry book is NA, MG 24 I37.
37 *JLALC*, 1832, appendix L.
38 *JLALC*, 1824, appendix F.
39 Halévy, *Growth of Philosophical Radicalism*, 119.
40 Evans, *A Treatise on the Theory and Practice of Agriculture.*
41 See the essays in Graham Burchell, Colin Gordon, and Peter Miller, eds., *The Foucault Effect: Studies in Governmentality* (Hemel Hempstead: Harvester Wheatsheaf, 1991), and further commentaries in Andrew Barry, Thomas Osborne, and Nikolas Rose, eds., *Foucault and Political Reason: Liberalism, Neo-liberalism and Rationalities of Government* (Chicago: University of Chicago Press, 1996).
42 Michel Foucault, *The Order of Things: An Archaeology of the Human Sciences* (London: Tavistock, 1970); see also *The Birth of the Clinic: An Archaeology of Medical Perception*, trans. A.M. Sheridan Smith (New York: Pantheon Books,1973).
43 Axel Honneth, *The Critique of Power: Reflective Stages in a Critical Social Theory*, trans. Kenneth Baynes (Cambridge: MIT Press, 1991); also Lois McNay, *Foucault: A Critical Introduction* (Cambridge: Polity Press, 1994), 112.
44 T.J. Jackson Lears, 'The Concept of Cultural Hegemony: Problems and Possibilities,' *American Historical Review* 90 (1985): 567–93.
45 See the discussion by Ronald Beiner in the volume he edited: Hannah Arendt, *Lectures on Kant's Political Philosophy* (Chicago: University of Chicago Press, 1982).
46 *Journal d'agriculture* 6 (1853): 92.

47 *JLALC*, 1821–2, appendix F.
48 Ouellet, *Lower Canada*.
49 NA, MG 11, Colonial Office, Q Series, vol. 151A, 314–15, Bathurst to Richmond, 31 Aug. 1819.
50 Montreal *Gazette*, 19 Jan. 1820.
51 *Canadian Courant*, 23 March 1833.
52 NA, RG 4 A3, Lower Canada Provincial Secretary Correspondence, vol. 248, Address of the Quebec Society, 21 Jan. 1827; Dalhousie's response is on the document.
53 NA, RG 1 E15A, Receiver General, Lower Canada, Correspondence, vol. 44, 14 Oct. 1823; vol. 51, 6 Oct. 1825; NA, RG 4 A3, vol. 295, 1 Aug. 1829; Jules Bélanger et al., *Histoire de la Gaspésie* (Montreal: Boréal Express/IQRC, 1981).
54 Allan Greer, *The Patriots and the People: The Rebellion of 1837 in Rural Lower Canada* (Toronto: University of Toronto Press, 1993), 47–8.
55 NA, Neilson Papers, vol. 8, 347–9; RG 1 E15A, vol. 87.
56 Montreal *Minerve*, 11 March 1833; *Gazette*, 5 April 1832, 15 March 1831.
57 *Minerve*, 14 Feb. 1833, from the Montreal *Vindicator*.
58 Sir C.P. Lucas, ed., *Lord Durham's Report on the Affairs of British North America* (Oxford: Clarendon Press, 1912), 2: 43.
59 *Minerve*, 10 Nov. 1831, 15 March and 8 Oct. 1832.
60 Evans wrote to the *Vindicator* in February 1833, and to the Canadian *Courant* in March and April; he also carried on a controversy with Charles Grece as to whether the Canadian cow was more profitable than imported breeds. Duvernay is in the *Minerve*, 20 and 27 June 1833; 19 March and 25 Oct. 1832.
61 *Missiskoi Standard*, 8 Nov. 1837. Membership lists are in NA, RG 1 E15A, filed by year, alphabetically, under 'Agricultural Societies.' Prize lists and elections are in the *Missiskoi Standard* and the Montreal press.
62 *Journals of the Legislative Assembly of the Province of Canada* (*JLAPC*), 1835, appendix M.
63 *Ordinances Made and Passed by the Administrator of the Government and Special Council for the Affairs of the Province of Lower Canada*, 1838–41. In 1838 the total was £750; in 1839, £394; in 1840–1, £560.
64 NA, RG 1 E15A, vols. 87 and 88.
65 Brome County Historical Society, Knowlton, Quebec, Minute Book.
66 *JLAPC*, 1846, appendix J.
67 Montreal *Vindicator*, 5 Jan. 1830.
68 *Missiskoi Standard*, 26 Sept. 1837.
69 Archives of Ontario (AO), MS520, Solomon Jones Papers, 17 Dec. 1818.
70 Lois Darroch Milani, *Robert Gourlay, Gadfly: The Biography of Robert (Fleming)*

Gourlay, 1778–1863, Forerunner of the Rebellion in Upper Canada (Thornhill: Ampersand Press, 1974); his remark is in his *Statistical Account of Upper Canada* (1822; Yorkshire: S.R. Publishers, 1966), 2: viii.

71 James Strachan (pseud. of John Strachan), *A Visit to the Province of Canada in 1819* (New York: Johnson Reprint Company, 1968), 187.

72 Kingston *Chronicle*, 26 May 1820, 22 Oct. 1819, 5 May 1820; 11 Oct. 1822. On the branch societies, see Quebec *Gazette*, 1 Nov. 1821, and NA, MG 9 D818, Lennox and Addington Historical Society, William Bell Papers, Markland to Bell, 1 June 1821.

73 Kingston *Chronicle*, 26 Oct. 1821.

74 In Talman, 'Agricultural Societies,' 551.

75 NA, RG 5 A1, 1828, 483771–81; Talman, 'Agricultural Societies,' 552–3; AO, MU2883, John Steele Papers, 30 July 1829, has proceedings for that year.

76 NA, RG 5 A1, 1839, 87150–3 and 125674–86. Fred Coyne Hamil, *The Valley of the Lower Thames, 1640–1850* (Toronto: University of Toronto Press, 1978), 135–6.

77 The debate is in the *Colonial Advocate*, 21 Jan. 1830.

78 Paul Romney, 'A Man Out of Place: The Life of Charles Fothergill, Naturalist, Businessman, Journalist' (PhD dissertation, University of Toronto, 1982), 516–17.

79 On patronage in Upper Canada, see S.J.R. Noel, *Patrons, Clients, Brokers: Ontario Society and Politics, 1791–1896* (Toronto: University of Toronto Press, 1990).

80 *Colonial Advocate*, 7 July, 1 Oct., and 5 Nov. 1829.

81 NA, RG 5 A1, 1830, passim.

82 Ibid., 56560–7 and 65790–3.

83 C.C.J. Bond, 'Alexander James Christie, Bytown Pioneer, His Life and Times 1787–1843,' *Ontario History* 56 (1964): 17–36; NA, MG 24 I9, Hill Papers, 589–90.

84 Michael S. Cross, 'The Shiners' War: Social Violence in the Ottawa Valley in the 1830s,' *Canadian Historical Review* 54 (1973): 1–47.

85 NA, MG 24 D8, Wright Papers, vol. 113; vol. 124, 66332–7.

86 *Colonial Advocate*, 1 April and 8 July 1830; Audrey Saunders Miller, 'Yonge Street Politics, 1828 to 1832,' *Ontario History* 62 (1970): 101–18.

87 David Gagan, *The Denison Family of Toronto, 1792–1925* (Toronto: University of Toronto Press, 1973), 18–19.

88 Romney, 'A Man Out of Place.'

89 *Canada Freeman*, 14 Oct., 19 May 1831; *Colonial Advocate*, 27 Oct. 1831; *Canada Freeman*, 24 May 1832.

90 Toronto *Patriot*, 26 and 30 Sept. 1834; see F.H. Armstrong, 'Toronto in Transition: The Emergence of a City, 1828–1838' (PhD dissertation, University of Toronto, 1965), and Pleasance Crawford, 'The Roots of the Toronto Horticultural Society,' *Ontario History* 89 (1997): 125–40.

91 NA, RG 5 A1, 87150–3 and 125674–86.

92 Ibid., 86237–49, 61305; *Hallowell Free Press*, 31 May 1831, 28 May and 28 Oct. 1833.

93 Ibid., 85938–42, 111597–600, and 124254–7 for Ottawa; 96889–94 and 115035–7 for Bathurst; and 129171–6 for the Eastern District.

94 Ibid., 125674–86, and RG 5 C1, Provincial Secretary, Canada West, Correspondence, 4394–97, 1708–13, and 20519–30.

95 James Young, *Reminiscences of the Early History of Galt and the Settlement of Dumfries* (Toronto: Hunter & Rose, 1880), 182–3.

96 Macaulay wrote regularly to the provincial secretary to demand the moneys for Prince Edward and Hastings, or to insist that Lennox and Addington were two counties entitled to £50 each. NA, RG 5 A1, passim. See S.F. Wise, 'John Macaulay: Tory for all Seasons,' in McKillop and Romney, *God's Peculiar Peoples*, 73–90.

97 Information is taken from the *DCB*.

98 Niagara *Mail*, 15 Oct. 1851. Early membership lists do not survive.

99 NA, RG 5 A1, 113748–52; *JLALC*, 1831, appendix D; and passim for the annual reports of the Beauharnois Society. Greer, *The Patriots and the People*, 269–70, and Graham D. Taylor and Peter A. Baskerville, *A Concise History of Canadian Business* (Toronto: University of Toronto Press, 1994), 209. Ellice resided in England and his agents acted for him.

100 NA, Neilson Papers, vol. 12, 496–510. The memo was presented to the British authorities as an argument for timber preferences.

101 Janet Azjenstat, *The Political Thought of Lord Durham* (Kingston: McGill-Queen's University Press, 1988). Talk of 'political apprenticeship' long predated Durham's visit.

102 Habermas, *Structural Transformation of the Public Sphere*; *Missiskoi Standard*, 26 Jan. 1836. The Montreal *Gazette* frequently stated that the ignorance of the habitants made them prey to demagogues. *Hallowell Free Press*, 5 Aug. 1834, discussing Papineau's Montreal Agricultural Society, decries the 'grossly illitterate' [sic] habitants.

103 *JLAPC*, 1844–5, appendix Z.

104 Charlottetown *Islander*, 2 Nov. 1853, 14 Feb. 1861.

105 *Manitoba Free Press*, 28 Sept. 1886.

Chapter 3: Exhibitions as Politics in Central Canada, 1841–1891

1 An overview is in J.M.S. Careless, *The Union of the Canadas: The Growth of Canadian Institutions, 1841–1857* (Toronto: McClelland & Stewart, 1967).
2 Figures are taken from the provincial *Sessional Papers.*
3 Toronto *Globe,* 8 and 28 Sept., 9 Oct. 1847, 1 Sept. 1855.
4 Newcastle *Farmer* 2 (1847–8): 67–8.
5 Careless, *The Union of the Canadas*; Gordon T. Stewart, *The Origins of Canadian Politics: A Comparative Approach* (Vancouver: UBC Press, 1986).
6 *JBAMUC* 2 (1862): 338–9.
7 McInnis, *Perspectives on Ontario Agriculture*; Douglas McCalla, *Planting the Province: The Economic History of Upper Canada, 1784–1870* (Toronto: University of Toronto Press, 1993); McCallum, *Unequal Beginnings.*
8 For example, when John Neilson thanked farmers for voting for him in 1818, the year the first agricultural society appeared in Quebec, *Le Canadien* was concerned at this distinction among voters: 'c'est une politique nouvelle d'un nouvel homme public.' Quoted in Montreal *Gazette,* 15 April 1818.
9 Harvey C. Mansfield Jr, 'Hobbes and the Science of Indirect Government,' *American Political Science Review* 65 (1971): 97–110.
10 Fowke, *Canadian Agricultural Policy.*
11 *British American Cultivator* 2 (1843): 87.
12 George Metcalf, 'William Henry Draper,' in J.M.S. Careless, *The Pre-Confederation Premiers: Ontario Government Leaders, 1841–1867* (Toronto: University of Toronto Press, 1980), 62.
13 13 & 14 Victoria, 73.
14 Elizabeth Nish and D. Blais, eds., *Debates of the Legislative Assembly of United Canada, 1841–1867* (Montreal: Presses de l'École des hautes études commerciales, 1970–95); *DLAUC* 11 (1852–3): 1193 (Macdonald speaking).
15 Sir Francis Hincks, *Reminiscences of His Public Life* (Montreal: William Drysdale, 1884), 255–6.
16 *DLAUC* 11 (1852–3): 850.
17 NA, RG 17, vol. 1, file 1. It was also specified that Colonel Rhodes should be added, but he had garnered some votes, while De Blois appeared on no one's list of recommendations.
18 Niagara *Mail,* 6 Oct. 1858.
19 Ian Drummond, *Progress without Planning: The Economic History of Ontario from Confederation to the Second World War* (Toronto: University of Toronto Press, 1987); see also J.E. Hodgetts, *From Arm's-Length to Hands-On: The Formative Years of Ontario's Public Service, 1867–1940* (Toronto: University of Toronto Press, 1995); Paul-André Linteau, René Durocher, and Jean-Claude

Robert, *Quebec: A History*, trans. Robert Chodos (Toronto: James Lorimer, 1983).

20 Jones, *History of Agriculture*; G. Elmore Reaman, *A History of Agriculture in Ontario* (Toronto: Saunders, 1970); McCalla, *Planting the Province*; Drummond, *Progress without Planning*.

21 Lewis and McInnis, 'The Efficiency of the French-Canadian Farmer'; Armstrong, 'The Efficiency of Quebec Farmers'; Jean Hamelin and Yves Roby, *Histoire économique du Québec, 1851–1896* (Montreal: Fides, 1971); J.I. Little, *Crofters and Habitants: Settler Society, Economy and Culture in a Quebec Township, 1848–1881* (Montreal: McGill-Queen's University Press, 1991); Firmin Letourneau, *Histoire de l'agriculture (Canada français)* (Montreal: Imprimerie populaire, 1950); Normand Séguin, *Agriculture et colonisation au Québec: aspects historiques* (Montreal: Boréal, 1980).

22 *JLAPC* 1850, appendix TT. Eveline Bossé, *Joseph-Charles Taché (1820–1894): un grand répresentant de l'élite canadienne-française* (Quebec: Garneau, 1974).

23 Quebec, Legislative Assembly, *Journals* 1 (1868), appendix 12, 'Second Report of the Select Standing Committee of Agriculture, Immigration, and Colonisation.'

24 Total reported membership was 36,298 that year. I have divided the amounts subscribed by membership, arriving at an average subscription of $1.17 for township societies and $1.25 for the counties. I then divided total subscription amounts by these figures. However, because the grant was obtained by subscription rather than number of members, some societies not reporting memberships may have wished to hide low memberships behind high subscriptions by local elites. The figures are taken from the Ontario *Sessional Papers*.

25 *Canada Farmer* 2 (1865): 142–3. There were nearly 227,000 farmers in Ontario in 1871, but as many as one-quarter were farmers' sons. (See Gordon Darroch and Lee Soltow, *Property and Inequality in Victorian Ontario: Structural Patterns and Cultural Communities in the 1871 Census* (Toronto: University of Toronto Press, 1994), ch. 1. Therefore, there were 170,250 farming households. One can only guess at the number of non-farmers who subscribed to agricultural societies. If the number was as high as 25 per cent of membership, then 18 per cent of farmer households subscribed to agricultural societies.

26 Figures are in *Revue agricole* 6 (1866–7): 101, and in the Quebec *Sessional Papers* for the 1890s.

27 Ontario, *Sessional Papers* 3, 5 (1870): 66–7; *Canadian Agriculturist* 12 (1860): 658–9.

28 *Canada Farmer* 4 (1867): 361–2.

29 *Farmer's Advocate* 2 (1867): 95–7, 100; *Canada Farmer* 5 (1868): 5, 40.

30 *Farmer's Advocate* 25 (1890): 370–1.
31 *Canadian Agriculturist* 6 (1854): 289–94.
32 *Canada Farmer*, ns 3 (1871): 25; *Farmer's Advocate* 14 (1879): 200–1; *Canadian Agriculturist* 12 (1860): 561–2.
33 *Farmer's Advocate* 16 (1881): 130; 13 (1878): 2; *Canadian Agriculturist* 5 (1853): 225; Ontario, *Sessional Papers* 5, 1 (1872–3), appendix A, 163.
34 *Canada Farmer* 2 (1865): 142–3.
35 Ibid., 14–15.
36 *Canadian Agriculturist* 5 (1853): 36–40; *Farmer's Advocate* 19 (1884): 37, and 18 (1883): 368.
37 *Canada Farmer* 2 (1865): 142–3.
38 Ontario, *Sessional Papers* 6, 1 (1874), appendix A, 10; *Owen Sound Comet*, 19 Oct. 1860 and 10 Oct. 1863.
39 Ontario, *Sessional Papers* 3, 5 (1871).
40 NA, RG 4 C1, Provincial Secretary for 1850, passim. I have deducted wives from two lists and used Missisquoi's report for 1849. RG 5 C1 (file 970).
41 NA, MG 28 I277, Missisquoi Agricultural Society Papers. This collection contains complete prize lists, but it is only for the 1880s that the treasurers' lists of accounts paid have survived, obviating any confusion over names and households.
42 NA, MG 30 E17, Philippe Landry Papers, records of the Montmagny Agricultural Society.
43 *Gore District Agricultural Advertiser* (Hamilton, 1847), in AO Pamphlets 1847, no. 22; AO, MU2086, Addington County Agricultural Society; MU2092, Darlington Agricultural Society.
44 J.K. Galbraith, *The Scotch* (Cambridge: Riverside, 1964): 59–62.
45 Some statistics are provided in chapter 4.
46 *Gazette des campagnes* 11 (1872–3): 111, and 19 (1881–2): 130–1.
47 Ibid., 24 (1886–7): 23; 27 (1889–90): 245; 30 (1894–5): 287.
48 Ibid., 27 (1889): 119 and passim through 1892–3.
49 Quebec, *Sessional Papers* 8, 4 (1874): xlii.
50 Jean-Jacques Jolois, *J.F. Perrault, 1753–1834, et les origines de l'enseignement laique au Bas-Canada* (Montreal: Les Presses de l'Université de Montréal, 1969).
51 Numbers are recorded in the annual reports in the Quebec *Sessional Papers*, Reports of the Commissioner of Agriculture. See Honoré Mercier's assessments in Reports 24, 2 (1890), and 25, 2 (1891).
52 *Gazette des campagnes* 22 (1884–5): 209–14; Quebec, *Sessional Papers* 18, 2 (1884–5): 103.
53 *Gazette des campagnes* 25 (1887–8): 245.

54 Robert Rumilly, *Histoire de la Province de Québec* (Montreal: Bernard Valiquette, nd), I, 286.

55 *Gazette des campagnes* 23 (1885–6): 226.

56 *Revue agricole* 3 (1863–4): 328, and 4 (1864–5): 203–5; *Gazette des campagnes* 4 (1864–5): 57–8; *L'Agriculteur* 12 (1859–60): 49–50 and 99–101.

57 *Journal d'agriculture* 2 (April 1849): 162.

58 Quebec, *Sessional Papers* 8, 4 (1874), Report by Barnard, cxlviii; reports by county societies of Champlain, Montmorency, and Bagot.

59 NA, MG 28 I277, Missisquoi County Agricultural Society, vol. 3, Minutes, 8 Jan. 1871, 29 Dec. 1869. See Ashley Sheltus, 'The Early History of Bedford Fair,' *Missisquoi County Historical Society Report* 20 (1990): 125–9.

60 Quebec, *Sessional Papers* 6, 4 (1872), and 7, 4 (1873).

61 ANQ, E25, vol. 52, file 36645. Reports for 1875, vol. 53, file 130. J.M. Browning of the Council to Édouard Barnard, 17 Jan. 1877, remarked that 'it would not be advisable to make public opinions held by some of the Societies.' The manuscript reports are in vol. 52, file 36645. Usually the reports were published as an appendix to the Report of the Commissioner of Agriculture in the *Sessional Papers.*

62 Ibid., vol. 65, no. 358 of 1879.

63 Quebec, *Sessional Papers* 24, 2 (1890), Report by Barnard, 20.

64 See Jean-Baptiste Roy, *Histoire du mérite agricole* (Quebec: Government of Quebec, 1985).

65 Quebec, *Sessional Papers* 24, 2 (1890), Council of Agriculture, Feb. 1890; ANQ, E25, vol. 83, 3575 and 3509.

66 Institut canadian de Québec, *Concours d'éloquence sur l'agriculture (Éloge de l'agriculture, ce qu'est l'art agricole au Canada, des moyens de l'y faire progresser)* (Quebec: A. Coté, 1879).

67 Quebec, *Sessional Papers* 28, 2 (1894). In addition, many had priests as honorary presidents.

68 Ibid., 18, 2 (1884–5), 191.

69 ANQ, E9, vol. 2, 2444.

70 *Gazette des campagnes* 19 (1882–3): 98.

71 Quebec, *Sessional Papers* 28, 2 (1894): 123–4.

72 Ibid., 27 (1893) and 28 (1894).

73 *Journal d'agriculture illustré* 16 (1893): 13.

74 Their reports are in ANQ, E9, Letters Received for 1893.

75 *Revue agricole* 5 (1865–6): 168–70 and 233–6; memberships during the 1890s are in the *Sessional Papers.*

76 ANQ, E25, vol. 88, 2928, has complete subscription lists.

77 Alain Ménard, *La Société d'agriculture du Comté de Rouville: son histoire* (Quebec: SAR, 1993): 81.
78 ANQ, E25, vol. 71, 583.
79 ANQ, E9, vol. 299, Saguenay Agricultural Society no. 2, 17 July 1893.
80 *Revue agricole* 7 (1867–8): 275.
81 Ibid. Membership is for 1866.
82 ANQ, E25, vol. 87, file 2258, Barnard to E. Gagnon, 3 Sept. 1882.
83 ANQ, E9, vol. 299, Saguenay Society, 10 and 17 Feb. 1895.
84 ANQ, E25, vol. 52, 36645.
85 ANQ, E25, vol. 58, 2972, Barnard memo, 12 Sept. 1877; vol. 61, 872, memo, 9 April 1878. For sympathetic accounts of Barnard, see Marc-A. Perron, *Un grand éducateur agricole: Édouard-A. Barnard, 1858–1898. Etude historique sur l'agriculture de 1760 à 1960* (Montreal: n.p., 1955); Bruno Jean, 'Les idéologies éducatives agricoles (1860–1890) et l'origine de l'agronomie québécoise' (MA thesis, Laval University, 1977): and his article in the *DCB*, vol. 12, 57–60.
86 ANQ, E25, vol. 81, 2825, and vol. 78, 1097. Usually the president's remarks were published in the annual departmental report.
87 ANQ, E25, vol. 101, 2032.
88 Foucault, *Discipline and Punish*, 205.
89 Ibid., 217; also Burchell et al., *The Foucault Effect*. Bentham is quoted in 'The Eye of Power,' in Colin Gordon, ed., *Power/Knowledge: Selected Interviews and Other Writings, 1972–1977* (New York: Pantheon Books, 1980): 152.
90 James Mill, 'Government,' *Encyclopædia Britannica*, 5th ed., Supplement (1820): 496. See the discussion by Sir Leslie Stephen, 'Mill on Government,' in *The English Utilitarians* (London: Duckworth, 1900), 2: 74–97.
91 *Canadian Agriculturist* 3 (1851): 11–14; see the discussion of McDougall's politics in Suzanne Zeller, *DCB*, vol. 13, 632–6.
92 Douglas Archie Lawr, 'The Development of Agricultural Education in Ontario, 1870–1910' (PhD dissertation, University of Toronto, 1972).
93 *Farmer's Advocate* 14 (1878): 278.
94 See Marjorie Griffin Cohen, 'The Decline of Women in Canadian Dairying,' *Histoire sociale/Social History* 17 (1984): 307–34.
95 Ontario, *Sessional Papers* 9, 33 (1877): 142–8.
96 Debord, *Society of the Spectacle*.
97 Ontario *Sessional Papers* 6, 1 (1874): 79.
98 ANQ, E25, vol. 95, 3608. This case went to the courts, and copious correspondence continued into 1884.
99 *Revue agricole* 7 (1862–3): 218–19; 4 (1864–5): 239–41; 7 (1867–8): 172; *Lower Canada Agriculturist* 1 (1861–2): 41.

100 Gramsci, *Prison Notebooks,* 116–17.
101 Quebec *Verité,* 7 Jan. 1893; quoting the Sherbrooke *Pionnier.*

Chapter 4: The Provincial Exhibitions and Economic Development, 1846–1893

1 Vance Hall, *A History of the Yorkshire Agricultural Society, 1837–1987* (London: B.T. Batsford, 1987); J.A. Scott Watson, *The History of the Royal Agricultural Society of England, 1839–1939* (London: Royal Agricultural Society, 1939).
2 See Donald B. Marti, *To Improve the Mind and the Soil: Agricultural Societies, Journals, and Schools in the Northeastern States, 1791–1865* (Ann Arbor: University Microfilms International, 1979); *Transactions of the New York State Agricultural Society.*
3 *Bytown Packet,* 30 Sept. 1848.
4 *British American Cultivator* 2 (1843): 16.
5 Ibid., ns 2 (1846): 10, 242, 262–3, and 321–47; Toronto *British Colonist,* 23 and 30 Oct. 1846. See also the manuscript at the AO, V.M. Roberts, 'The Canadian National Exhibition, Its Origin and History,' and the account in the *Journal and Transactions of the Board of Agriculture of Upper Canada* (*JTBAUC*) 1 (1855–6). Finances are in the *JTBAUC.*
6 Patrick Brode, *Sir John Beverly Robinson: Bone and Sinew of the Family Compact* (Toronto: University of Toronto Press, 1984).
7 Toronto *Globe,* 20 Oct. 1847.
8 *Canadian Agriculturist* 1 (1849): 253.
9 AO, MU1761, McDonald Papers, D. McDonald to Dr Mitchell, 28 Sept. 1852.
10 *Canadian Journal* 1 (1852): 51–67.
11 ANQ, E9, Minister of Agriculture, vol. 232, Minutes of the Board of Agriculture, 28 March 1853.
12 Montreal *Gazette,* 27 Sept. 1853; *Minerve,* 29 Sept. 1853; Hamilton *Spectator,* 6 Oct. 1853; Montreal *Gazette* Supplement, 11 Oct. 1853.
13 MTRL, BR, TMI, Minutes, May 1857. At the Mechanics' Festival in 1853, President Henry Bulmer declared the 'aim and object' of the Montreal Mechanics' Institute to be '*the Elevation of the Moral and Intellectual Condition of the Working Classes*': *Gazette,* 9 Feb. 1853; also NA, MG 24 D16, Chamberlin Papers, vol. 11, 1–33, on the Granby Mechanics' Institute.
14 Vernon, 'Development of Adult Education,' 309.
15 J.W. Hudson, *The History of Adult Education* (1851; New York: Augustus M. Kelley, 1969) 218–19; and James Hole, *An Essay on the History and Management of Literary, Scientific & Mechanics' Institutions* (1853; London: Frank Cass, 1970) 81.
16 AO, MU2020, TMI, Notice of Fourth Annual Exhibition.
17 *JBAMUC* 6 (1866): 97.

18 MTRL, BR, TMI, 10 Feb. and 28 Aug. 1876.
19 Montreal *Gazette*, 9 Feb. 1853.
20 *Stanstead Journal*, 14 Feb. 1850.
21 Ontario, *Sessional Papers* 8, 1 (1875–6): 217 and 222. Simcoe had an exhibition of flowers and minerals, but provided no financial statement.
22 Discussed in Bryan D. Palmer, *A Culture in Conflict: Skilled Workers and Industrial Capitalism in Hamilton, Ontario, 1860–1914* (Montreal: McGill-Queen's University Press, 1979): 49–57.
23 Quoted in Home, *An Essay*, 107.
24 Its papers are at ANQ, Montreal Branch. On the council and subsequent reform of adult education, see Ruby Heap, 'Un chapitre dans l'histoire de l'éducation des adultes au Québec: les écoles du soir, 1889–1892,' *Revue d'histoire de l'Amérique française* 34 (1981): 597–626. The Upper Canada Board's Papers are at AO, MU279 and 280.
25 *Nova Scotian*, 23 Oct. 1854.
26 Ibid., 9 Oct. 1854; classification is 3 April 1854.
27 Victoria *British Colonist*, 6 Feb. 1861; *Mainland Guardian*, 21 Nov. 1861.
28 *JTBAUC* 2 (1858): 297–300; Niagara *Mail*, 7 Oct. 1857; NA, MG 24 D2, Buchanan Papers, 73053; see Marjorie Freeman Campbell, *A Mountain and a City: The Story of Hamilton* (Toronto: McClelland & Stewart, 1966).
29 Toronto *Leader*, 30 Sept. 1865.
30 London *Advertiser*, 21 Sept. 1870. See the *History of the County of Middlesex* (Belleville: Mika Studio, 1972), ch. 12.
31 London *Advertiser*, 23 Sept. 1878; Toronto *Mail*, 17 Sept. 1879; Ontario, *Sessional Papers*, Minister of Agriculture reports, provide some details on all the shows discussed here.
32 Peterborough *Times*, 11 Oct. 1880; Elwood H. Jones, *Winners: 150 Years of the Peterborough Exhibition* (Peterborough: Peterborough Agricultural Society, 1995).
33 NA, MG 28 I27, Ottawa Agricultural Society, Minute Book.
34 *History of the Central Canada Exhibition from 1888* (Ottawa: Central Canada Exhibition Association, 1976).
35 Keith Walden, *Becoming Modern in Toronto: The Industrial Exhibition and the Shaping of a Late Victorian Culture* (Toronto: University of Toronto Press, 1997).
36 *Once upon a Century: 100 Year History of the 'Ex'* (Toronto: John Robinson, 1878): 14.
37 Toronto *Leader*, 16, 25 Aug. and 14 Oct. 1859.
38 Toronto *Mail*, 26 June 1879.
39 *Farmer's Advocate* 19 (1884): 292–3.

40 *Canadian Live Stock Journal* 3 (1886): 110.
41 *Farmer's Advocate* 25 (1890): 15–16.
42 Ibid., 14 (1879): 221.
43 Ibid., 16 (1881): 55.
44 Ontario, *Sessional Papers* 21, 8 (1889): 29.
45 Toronto *Leader*, 19 Sept. 1853.
46 Ibid.; *Canadian Agriculturist* 6 (1854): 313; *Journal de Québec*, 16 Sept. 1853; *Journal du Cultivateur* 2 (1854–5): 185–6.
47 Three Rivers *Le Bas-Canada*, 23 and 30 Sept. 1856. Normand Séguin remarks that the region was not strong in livestock: 'L'agriculture de la Maurice et du Québec, 1850–1950,' *Revue d'histoire de l'Amérique française* 35 (1982): 548; *La conquête du sol au 19e siècle* (Montreal: Boréal Express, 1977).
48 Brantford *Expositor*, 26 Sept. 1856.
49 Quebec City Archives, vol. 53, dossier 7: Expositions, Conseil et comités, 1833–65; *Journal de Québec*, 12 June 1860.
50 Sources are Montreal newspapers for April 1860; Montreal *Gazette*, 6 July 1860, for the minutes of the Board of Arts for April; and the board's papers in the ANQM. Also Raymond Montpetit, 'Fêtes et société au Québec: la visite du Prince de Galles et la construction du Crystal Palace à Montreal, en 1860,' in Groupe de Recherche en Art Populaire, *Travaux et conférences, 1975–1979* (Université de Québec à Montréal: Département Histoire de l'Art, 1979): 258–80.
51 Montreal *Gazette*, 16 and 31 May, 5 and 13 June 1860; ANQM, Board of Arts Letterbook, 13 June 1860, Chamberlin to Alleyn.
52 Quebec *Morning Chronicle*, 11 June 1860; Montreal *Gazette*, 27 April 1860 (a correspondent).
53 Montreal *Gazette*, 14 June 1860.
54 British Canadian, *The Tour of His Royal Highness the Prince of Wales through British America and the United States* (Montreal: John Lovell, 1860): 95; the *Times* is quoted in the Montreal *Gazette*, 29 Sept. 1860; published slightly altered in Nicholas Augustus Woods, *The Prince of Wales in Canada and the United States* (London: Bradbury & Evans, 1861): 117–18.
55 *Journal de Québec*, 2 Oct., and Quebec *Morning Chronicle*, 1, 2, 3 Oct. 1860.
56 ANQM, Board of Arts Letterbook, 24 July 1862.
57 Charles D. Day, *Objections of the Royal Institution to Any Legislation Affecting the Property Known as the Crystal Place Property, Montreal* (Montreal, 1873) (CIHM no. 58891).
58 ANQ, E25, vols. 22–3, 19533, 19751, 19955, 20623½; see also the extensive coverage and correspondence in the daily press.

59 ANQ, E25, vols. 43 to 49, files 32260, 32864, 32953, 33151, 34063, 34439, 34821, 35273, 36868.

60 *Revue agricole* 2 (1862–3): 11–14.

61 *Lower Canada Agriculturist* 1 (1861–2): 298–300.

62 Quebec City Archives, Expositions, vol. 53, dossier 7, Report of Finance Committee, 9 May 1867.

63 Quebec, *Sessional Papers* 1, 4 (1869): 74. The profit and loss was apparently after the government grant had been paid. The Sherbrooke exhibition cost $7372, of which the government paid $4434. For Montreal in 1863 and 1865, the respective figures were $13,073 and $3389; and $12,996 and $2540. NA, MG 27 II E4, Joseph Landry Papers, vol. 21, Statistics and Plans (loose papers).

64 *Revue agricole* 5 (1865–6): 4–5 and 12–19.

65 *Le monde illustré*, 20 Sept. 1884.

66 Michael Bliss, *Plague: A Story of Smallpox in Montreal* (Toronto: HarperCollins, 1991).

67 *Stanstead Journal*, 13 Nov. 1885; P.B. Waite, 'John Henry Pope,' *DCB*, vol. 11, 705–8.

68 *Gazette des campagnes* 24 (1886–7): 187–8; *L'électeur*, 25 Aug. 1887.

69 *L'Électeur*, 28 Aug. 1887; *Progrès du Saguenay*, 18 Aug. 1887.

70 NA, Joseph Landry Papers, vols. 20–5, have early correspondence and financial records; the minutes of the company are in the Quebec City Archives, Commission de l'Exposition Provinciale, reel 3569; correspondence with the city is in the same archives, Sous-Serie Conseil et Comités, 1835–1965, vol. 53, dossier 8: Cie. de l'Exposition.

71 *Soleil*, 11 June 1898.

72 See Christopher Armstrong and H.V. Nelles, *The Revenge of the Methodist Bicycle Company* (Toronto: Peter Martin Associates, 1977).

73 Joseph Landry Papers, vol. 21, 'Place de gérant,' Stevenson to Landry, 7 March 1894; Montreal *Minerve*, 2 Dec. 1893.

74 *L'Étendard*, 7 Dec. 1889.

75 Montreal *Gazette*, 13 Dec. 1892.

76 Ibid., 20 Sept. 1880.

77 Conrad Archambeault, *La Province*, 8 May 1937.

78 NA, RG 17, vol 606, 68572; vol. 386, 41596; vol. 568, 64041; vol. 869, 106395; RG72, vol. 119, 91908.

79 Toronto *Leader*, 28 Sept. 1864; Barrie *Northern Advance*, 5 Oct. 1864.

80 *Canada Farmer* 4 (1867): 323; *Farmer's Advocate* 2 (1867): 90, and 9 (1874): 9.

81 Grant MacEwan, *Highlights of Shorthorn History* (Winnipeg: Comprint, 1982):

75–9; *Canada Farmer* 12 (1877): 231; *Canadian Livestock Journal* 3 (1886): 279–81.

82 *Journal d'agriculture illustré* 11 (1888): 124–5.

83 Quebec, *Sessional Papers* 1, 4 (1869); London *Advertiser* 22 Sept. 1868; *Farmer's Advocate* 6 (1868): 146–7; 17 (1882): 315–16; 19 (1884): 170; *Gazette des campagnes* 9 (1870–1): 199.

84 J.L. Gourlay, *History of the Ottawa Valley* (Np, 1896): 153.

85 *Canada Farmer* 15 (1880): 148.

86 Ontario Agricultural Commission, *Report of the Commissioners* (Toronto: C. Blackett Robinson, 1881), 1: 231–2. Reports by township, used for this paragraph, are alphabetical by county in vol. 2.

87 Ibid., 2: 616.

88 Figures are from M.C. Urquhart and K.A.H. Buckley, eds., *Historical Statistics of Canada* (Toronto: Macmillan, 1965): 369–70.

89 *Canadian Agriculturist* 13 (1861): 722.

90 George Elmore Reaman, *History of the Holstein-Friesian Breed in Canada* (Toronto: Collins, 1946): 41.

91 *Canadian Livestock and Farm Journal* 3 (1886): 345.

92 *Canadian Agriculturist* 13 (1861): 290–2; Christabel S. Orwin and Edith H. Wetham, *A History of English Agriculture, 1846–1914* (London: Areham, 1964): ch. 5.

93 *Canada Farmer* 2 (1865): 296, and 5 (1868): 294.

94 *Farmer's Advocate* 28 (1893): 476.

95 Ibid., 22 (1887): 109; 23 (1888): 49.

96 *Journal of the Royal Agricultural Society of England*, ns 17 (1881): 593.

97 See the report by the Bureau of Industries in Ontario, *Sessional Papers* 22, 50 (1890): 75.

98 *Canada Farmer* 4 (1867): 290.

99 *Farmer's Advocate* 11 (1876): 126.

100 Ibid., 20 (1885): 8 and 298–9.

101 OAC *Report*, section 1; Ontario, *Sessional Papers* 22, 22 (1890): 84–5.

102 *Canadian Agriculturist* 11 (1859): 17–18.

103 Toronto *Mail*, 9 Sept. 1879; OAC *Report* 1: 228; Toronto *Globe*, 20 Sept. 1883.

104 See Gordon M. Winder, 'Following America into Corporate Capitalism: Technology and Organization of the Ontario Agricultural Implements Industry to 1930' (PhD dissertation, University of Toronto, 1991); also Harold S. Turner and Ross W. Irwin, *Ontario's Threshing Machine Industry: A Short History of These Pioneer Companies and Their Contribution to Ontario Agriculture* (Guelph: Ontario Agricultural College, 1974), 126–30.

105 *Canadian Agriculturist* 12 (1860): 565.

106 Ibid., 5 (1853): 271–2.
107 Ibid., 10 (1858): 122–3; Barrie *Northern Advocate*, 5 Oct. 1859; Toronto *Leader*, 29 Sept. 1859.
108 *Canada Farmer*, ns 4 (1872): 252–3.
109 NA, RG 72, vol. 26, 21858; vol. 192, unpaginated account book for the Paris Exposition; and London *Advertiser*, 2 Oct. 1878.
110 *Canada Farmer* 4 (1867): 8; Merrill Denison, *Harvest Triumphant: The Story of Massey-Harris* (Toronto: McClelland & Stewart, 1948): 96.
111 Ben Forster, *A Conjunction of Interests: Business, Politics, and Tariffs, 1825–1879* (Toronto: University of Toronto Press, 1986).
112 John Rae, *Statement of Some New Principles on the Subject of Political Economy* (1834): republished as R. Warren James, *John Rae, Political Economist* (Toronto: University of Toronto Press, 1965), 2: 70 and passim.
113 The letters are in the *Gazette*, January and March 1833, and are republished in James, *John Rae*, 1.
114 Robin Neill, *A History of Canadian Economic Thought* (London: Routledge, 1991), ch. 4.

Chapter 5: Exhibition Culture

1 Michel de Certeau, *The Practice of Everyday Life*, trans. Steven F. Rendall (Berkeley: University of California Press, 1984), xiii.
2 John Fiske expands on this point to distinguish between objective and subjective culture, arguing that cultural consumers are empowered when they feel themselves to be resisting an ideology, even if their objective circumstances do not change. Critics have been quick to distance themselves from this 'subjective idealism.' John Fiske, *Reading the Popular* and *Understanding Popular Culture* (Boston: Unwin Hyman, 1989); see John Storey, *An Introductory Guide to Cultural Theory and Popular Culture* (Hampstead: Harvester Wheatsheaf, 1993), 185.
3 Ontario, *Sessional Papers* 16, 6 (1884): 20.
4 Ibid., 34. A general survey of developing commercial and popular culture in nineteenth-century England is Peter Bailey, *Leisure and Class in Victorian England: Rational Recreation and the Contest for Control, 1830–1885* (London: Routledge & Kegan Paul, 1978). Although exhibitions were no longer hegemonic in the strict Gramscian sense, Richard Butsch argues for a hegemony based on commercialization, but admits it operates only when coupled with 'centralization and standardization of leisure practices' that operates through 'the impersonal, even apparently "natural" structures of corporate industries and mass markets.' 'Introduction: Leisure and He-

gemony in America,' in Butsch ed., *For Fun and Profit: The Transformation of Leisure into Consumption* (Philadelphia: Temple University Press, 1990), 19. These conditions do not apply at the exhibition.

5 City Hall Archives, Toronto, has a file with attendance figures.

6 Richard A. Lanham, *The Motives of Eloquence: Literary Rhetoric in the Renaissance* (New Haven: Yale University Press, 1976), has a chapter on Rabelais entitled 'The War between Play and Purpose.'

7 Peterborough *Review*, 12 Sept. 1862.

8 Karal Ann Marling, *Blue Ribbon: A Social and Pictorial History of the Minnesota State Fair* (St Paul: Minnesota Historical Society Press, 1990), 275.

9 Hans Magnus Enzensberger, 'Constituents of a Theory of the Media,' in Enzensberger, *The Consciousness Industry: On Literature, Politics, and the Media*, ed. Michael Reloff (New York: Seabury, 1974), 111. See also Alan Trachtenberg, *The Incorporation of America: Culture and Society in the Gilded Age* (New York: Hill and Wang, 1982).

10 Friedrich Nietzsche, *On the Advantage and Disadvantage of History for Life*, trans. Peter Preuss (Indianapolis: Hackett, 1980), 28; also his 'The Case of Wagner,' in *The Birth of Tragedy and the Case of Wagner*, trans. Walter Kaufmann (New York: Vintage, 1967); Theodore W. Adorno and Max Horkheimer, 'The Culture Industry: Enlightenment as Mass Deception,' in their *Dialectic of Enlightenment*, trans. John Cumming (1944; London: Verso, 1979), 120–67.

11 James B. Twitchell, *Carnival Culture: The Trashing of Taste in America* (New York: Columbia University Press, 1992); Patrick Brantlinger, *Bread and Circuses: Theories of Mass Culture as Social Decay* (Ithaca: Cornell University Press, 1983); Lawrence W. Levine, *Highbrow/Lowbrow: The Emergence of Cultural Hierarchy in America* (Cambridge, Mass.: Harvard University Press, 1988).

12 Ontario, *Sessional Papers* 16, 6 (1884), Report on Prize Farms.

13 Raymond Williams, *Culture and Society, 1750–1950* (London: Chatto and Windus, 1958).

14 *Canada in 1880: Reports of Tenant Farmers' Delegates on the Dominion of Canada*, in Canada, *Sessional Papers* 14, 12 (1880–1): 122.

15 *Canadian Agriculturist* 12 (1860): 657.

16 Agricultural Association meeting, Toronto *Leader*, 26 Sept. 1863.

17 Grant McCracken, *Culture and Communication: New Approaches to the Symbolic Character of Consumer Goods and Activities* (Bloomington: Indiana University Press, 1990); Rosalind H. Williams, *Dream Worlds: Mass Consumption in Late Nineteenth-Century France* (Berkeley: University of California Press, 1982); Neil Harris, 'Museums, Merchandising, and Popular Taste: The Struggle for Influence,' in his *Cultural Excursions: Marketing Appetites and Cultural Tastes in*

Modern America (Chicago: University of Chicago Press, 1990), 56–82, and other essays in the same collection.

18 NA, MG 29 C75, Scrapbook of Thomas C. Watkins. A Methodist minister lectured on Wanamaker's store in Brantford. *Expositor,* 17 Nov. 1876. On Wanamaker, see William Leach, *Land of Desire: Merchants, Power, and the Rise of a New American Culture* (New York: Vintage, 1993), and S.J. Bronner, 'Reading Consumer Culture,' in S.J. Bronner, ed., *Consuming Visions: Accumulation and Display of Goods in America, 1880–1920* (New York: W.W. Norton, 1989), 26–7.

19 NA, MG 28 I230, Canadian Manufacturers' Association, vol. 1, Minutes, 12 Jan. 1887.

20 David Morgan, *The Morgans of Montreal* (Np, 1992), 62; Huguette Filteau, 'Henry Morgan,' *DCB,* vol. 12, 760–1.

21 Keith Walden, 'Speaking Modern: Language, Culture, and Hegemony in Grocery Window Displays, 1887–1920,' *Canadian Historical Review* 70 (1989): 285–310.

22 Toronto *World,* 14 Sept. 1888.

23 London *Advertiser,* 3 Oct. 1881.

24 Toronto *Mail,* 26 Sept. 1880.

25 On liminality, see Victor Turner, 'Variations on a Theme of Liminality,' in Sally F. Moore and Barbara G. Meyerhoff, eds., *Secular Ritual* (Amsterdam: VanGoram, 1977), 37–9.

26 H. Trueman Wood, 'Exhibitions,' *The Nineteenth Century* 20 (1886): 365.

27 Toronto *Mail,* 6 Sept. 1881 and 6 Sept. 1882.

28 *Saturday Night,* 27 Aug. 1898, 9.

29 See Neil Harris, *Humbug: The Art of P.T. Barnum* (Chicago: University of Chicago Press, 1981). The Broadsides Collection, BR, MTRL, has a note from a Toronto divine. Jackson Lears argues in his survey of American advertising; *Fables of Abundance: A Cultural History of Advertising in America* (New York: Basic Books, 1996), 265–7, that, after mid-century, Barnum began to seek respect as a cultural impressario. When it was decided to prolong the New York International Exhibition of 1853 another year, Barnum took over as its manager. On exhibitions as heralds of commodity culture and advertising, see Thomas Richards, *The Commodity Culture of Victorian England: Advertising and Spectacle, 1851–1904* (Stanford: Stanford University Press, 1990).

30 London *Advertiser,* 29 Sept. 1880.

31 Montreal *Gazette,* 17 Sept. 1882.

32 Toronto *Mail* and Montreal *Gazette,* 15 Sept. 1880.

33 Montreal *Gazette,* 18 Sept. 1891.

34 NA, RG 72, vol. 46, 2070; Toronto *World,* 10 Sept. 1891.

35 Montreal *Gazette,* 18 Sept. 1891; Toronto *World,* 27 Sept. 1891.

36 *Canadian Livestock and Farmer's Journal* 3 (1886): 345.
37 *Canadian Manufacturer* 24 (1893): 223–5, 267–8.
38 The paraphrase of Barthes is taken from Andrew Wernick, *Promotional Culture: Advertising, Ideology, and Symbolic Expression* (London: Sage, 1991).
39 Leach, *Land of Desire*, 205. Lears surveys anxieties over advertising and commodification in his *Fables of Abundance*.
40 In D.J. Cochrane, 'Agricultural Societies, Cattle Fairs, Agricultural Shows and Exhibitions of Upper Canada prior to 1867' (Milton Agricultural Museum, unpublished paper, 1976), 9.
41 Hamilton *Spectator*, 23 Sept. 1887.
42 Bothwell *Times*, 17 Oct. 1889.
43 *Minerve*, 20 Sept. 1882.
44 London *Advertiser*, 28 Sept. 1869.
45 Peterborough *Daily Evening Review*, 24 Sept. 1891.
46 Ontario Agricultural Commission, *Report*, 1: 435.
47 *Canadian Agriculturist* 13 (1862): 606. See Paul Bernier, *Le cheval canadien* (Quebec: Sentrentrion, 1992).
48 *Canadian Agriculturist* 5 (1852): 295–6.
49 For example, see Bouchard, 'L'agriculture saguenayenne.'
50 Toronto *World*, 20 Sept. 1888.
51 *Canadian Agriculturist* 13 (1862): 643–4; *Mail*, 15 Sept. 1880.
52 Harriet Ritvo, *The Animal Estate: The English and Other Creatures in the Victorian Age* (Cambridge: Harvard University Press, 1987), 56; *Canada Farmer* 1 (1864), 308.
53 *Canadian Agriculturist* 9 (1857), 256.
54 On British taste, see Jeremy Rifkin, *Beyond Beef: The Rise and Fall of the Cattle Culture* (New York: Dutton/Penguin, 1992), ch. 10.
55 London *Advertiser*, 28 Sept. 1877.
56 Newcastle *Farmer* 2 (1847): 22–3; quoted in Alan Skeoch, 'Changes in Lincoln County Agriculture, 1793–1881,' in John Burtniak and Wesley B. Turner, eds., *Agriculture and Farm Life in the Niagara Peninsula* (St Catharines: Brock University Press, 1983), 41.
57 Marling, *Blue Ribbon*, 81.
58 See Michael Bliss, *A Canadian Millionaire: The Life and Business Times of Sir Joseph Flavelle, Bart., 1858–1919* (Toronto: Macmillan, 1978), ch. 2.
59 London *Advertiser*, 22 Sept. 1881.
60 Ibid., 24 Sept. 1877.
61 Carl Berger, *Science, God, and Nature in Victorian Canada* (Toronto: University of Toronto Press, 1983); R. Glen, 'A History of Entomology in Canada up to 1956,' *Canadian Entomologist* 88 (1956): 290–371.

62 Ted Ferguson, ed., *Kit Coleman, Queen of Hearts* (Toronto: Doubleday, 1970), 46.
63 Brampton *Times*, 22 Oct. 1880.
64 Toronto *Leader*, 21 Sept. 1860.
65 Ontario, *Sessional Papers* 21, 8 (1889): 163.
66 Montreal *Gazette*, 22 Sept. 1891.
67 Toronto *Mail*, 16 Sept. 1881.
68 *Canada Farmer* 6 (1869): 429.
69 Bothwell *Times*, 13 Oct. 1887; Brantford *Expositor*, 19 Oct. 1860; Napanee *Standard*, 24 Oct. 1861.
70 *Canada Farmer* 7 (1870): 135.
71 London *Advertiser*, 3 and 4 Oct. 1882 and passim.
72 *Canada Farmer* 2 (1865): 230–1, and 4 (1867): 277–8.
73 Ibid., 3 (1866): 294; NA, MG 29 A19, James Crawford Papers.
74 William Arthur Deacon, *The Four Jameses* (Ottawa: Graphic Publishers, 1927). See also Edward S. Moore, *When Cheese Was King: A History of the Cheese Factories in Oxford County* (Aylmer: Aylmer Press, 1987), and Heather S. Menzies, *By the Labour of Their Hands: The Story of Ontario Cheddar Cheese* (Kingston: Quarry Press, 1994).
75 *Story of Ontario Agricultural Fairs*, 30.
76 Barrie *Northern Advocate*, 17 Aug. 1855; Ontario, Unpublished Sessional Papers, 1891 and 1892; Peterborough *Review*, 14 Oct. 1864.
77 Toronto *Leader*, 8 Oct. 1861; London *Advertiser*, 22 Sept. 1869; *L'Évenément*, 20 Sept. 1880.
78 *Canada Farmer* 9 (1872): 374.
79 Dennis Reid, *A Concise History of Canadian Painting*, 2nd ed. (Toronto: University of Toronto Press, 1988), 75.
80 London *Advertiser*, 27 Sept. 1876.
81 Ibid., 11 Oct. 1881.
82 Toronto *Mail*, 1 Oct. 1878.
83 *JBAMUC* 1 (1861): 309–12; 4 (1864): 310–12; 5 (1865): 36.
84 Toronto *Mail*, 9 Sept. 1881.
85 *Canadian Manufacturer* 24 (1893): 179–80, 223–5.
86 David G. Burley, 'Charles Horatio Waterous,' *DCB*, vol. 12, 1090–2; Brantford *Expositor*, 1 Oct. 1868 and 7 Oct. 1869.
87 *Canada Farmer* 9 (1872): 370–1.
88 Montreal *Minerve*, 16 Sept. 1882.
89 *Canadian Agriculturist* 5 (1852): 291; Toronto *Mail*, 28 Sept. 1882; Montreal *Gazette*, 16 Sept. 1882.
90 See Bettina Bradbury, *Working Families: Age, Gender, and Daily Survival in Industrializing Montreal* (Toronto: McClelland & Stewart, 1993).
91 Toronto *Mail*, 22 Oct. 1881.

92 Ibid., 9 Sept. 1879; *Canada Farmer*, ns 3 (1871): 397, and ns 4 (1872): 370.
93 NA, RG 4 C1, files 1135 of 1848 and 447 of 1849.
94 Peterborough *Daily Evening Review*, 29 Sept. 1891.
95 *Canada Farmer* 4 (1867): 329–30.
96 Ontario, *Sessional Papers* 22, 22 (1890).
97 Brampton *Conservator*, 26 Oct. 1877. At Canadian fairs, first prize was a red ribbon, second prize, blue.
98 *New Brunswick Reporter and Fredericton Advertiser*, 24 Oct. 1851.
99 *Canada Farmer* 10 (1873): 182.
100 London *Advertiser*, 18 Oct. 1879.
101 Ibid., 28 Sept. 1877.
102 *Farmers' Advocate* 16 (1881): 94.
103 *Canada Farmer* 4 (1867): 292.
104 Toronto *Leader*, 21 Sept. 1863; D'Arcy McGee, *A Collection of Speeches and Addresses*, ed., C. Murphy (Toronto: Macmillan, 1937): 168–90.
105 Ontario, *Sessional Papers* 11, 3 (1878), appendix B, 191–6.
106 Toronto *World*, 12 Sept. 1888.
107 John P. Diggins, 'The Misuses of Gramsci,' *Journal of American History* 75 (1988): 141.
108 Warren J. Gates, 'Modernization as a Function of an Agricultural Fair: The Great Grangers' Picnic Exhibition at Williams Grove, Pennsylvania, 1873–1916,' *Agricultural History* 58 (1984): 271–2.
109 Brantford *Expositor*, 30 Sept. 1856.
110 C.O. Jones, 'Missisquoi Agricultural Society,' *Missisquoi County Historical Society Report* 7 (1961): 25–31. This was not strictly true, for two years later another oration was given: L.S. Huntingdon, *The Independence of Canada: The Annual Address Delivered before the Agricultural Society of the County of Missisquoi at Bedford, September 8, 1869* (Montreal: Herald, 1869).
111 Toronto *Globe*, 11 Sept. 1889.
112 Barrie *Northern Advance*, 26 Sept. 1860; Toronto *Leader*, 21 Sept. 1860.
113 London *Advertiser*, 25 Sept. 1869.
114 *Patriot and Farmer's Monitor*, 9 Aug. 1833; quoting in part Edward Lytton Bulwer, *England and the English*, 36. On another Tory's dislike of 'all the shoemakers and tailors in town' talking about politics, see Bode, *Sir John Beverly Robinson*, 192.
115 *Patriot and Farmer's Monitor*, 1 May 1832.
116 Lynne Marks, '"The Hallelujah Lasses": Working-Class Women in the Salvation Army in English Canada, 1882–92,' in Franca Iacovetta and Mariana Valverde, eds., *Gender Conflicts: New Essays in Women's History* (Toronto: University of Toronto Press, 1992): 67–117.
117 *Canadian Live Stock Journal* 3 (1886): 175.

118 *Owen Sound Comet*, 13 Nov. 1863.
119 Quoted in London *Advertiser*, 20 Sept. 1881.
120 Toronto *World*, 15 Sept. 1888.
121 Montreal *Gazette*, 21 Sept. 1880 and 18 Oct. 1881.
122 Toronto *Globe*, 27 May 1893.
123 Toronto *World*, 13 Sept. 1889.
124 Toronto *Globe*, 19 Sept. 1883.
125 London *Advertiser*, 18 Oct. 1879; Robert Bogdon, *Freak Show: Presenting Human Oddities for Amusement and Profit* (Chicago: University of Chicago Press, 1988).
126 Toronto *World*, 17 Sept. 1890.
127 London *Advertiser*, 27 Sept. 1888.
128 Ibid., 28 Sept. 1888.
129 Toronto *World*, 15 Sept. 1890.
130 Toronto *Leader*, 28 Sept. 1863.
131 Hamilton *Spectator*, 30 Sept. 1887.
132 London *Advertiser*, 18 Sept. 1875.
133 Montreal *Gazette*, 17 Sept. 1892.
134 Hamilton *Spectator*, 5 Oct. 1887.
135 Brantford *Expositor*, 24 Sept. 1868.
136 Toronto *Mail*, 24 Sept. 1880.
137 Brantford *Expositor*, 20 Oct. 1871.
138 Constance Backhouse, *Petticoats and Prejudice: Women and the Law in Nineteenth-Century Canada* (Toronto: Women's Press, 1991).
139 NA, MG 28 I277, vol. 3.
140 24 Sept. 1880.
141 Montreal *Gazette*, 17 Sept. 1892.
142 Toronto *Leader*, 24 Sept. 1862.
143 Toronto *Globe*, 10 Sept. 1884.
144 Peterborough *Daily Evening Review*, 18 Oct. 1882.
145 London *Advertiser*, 14 Oct. 1878.
146 Brantford *Expositor*, 26 Oct. 1883.
147 Jean S. McGill, *A Pioneer History of the County of Lanark* (Toronto: T.H. Best, 1968): 127–8.
148 NA, MG 29 D106, Woodburn Papers, 64; Nick and Helma Mika, *Bytown: The Early Days of Ottawa* (Belleville: Mika, 1982): 103.
149 Milton *Canadian Champion*, 20 Oct. 1887.
150 London *Advertiser*, 3 and 18 Oct. 1881.
151 Montreal *Gazette*, 10 and 30 March, 7 and 30 April 1860. On earlier quarrels in that society, see *Missiskoi Standard*, 4 and 18 Aug. 1835. Fisticuffs at

agricultural society meetings occurred from time to time throughout the century.

152 London *Advertiser*, 3 Oct. 1888.

153 Young, *Reminiscences of the Early History of Galt*, 69; M.A. Garland and J.J. Talman, 'Pioneer Drinking Habits and the Rise of Temperance Agitation in Upper Canada prior to 1840,' in F.H. Armstrong, H.A. Stevenson, and J.D. Wilson, eds., *Aspects of Nineteenth-Century Ontario: Essays Presented to James J. Talman* (Toronto: University of Toronto Press, 1974), 171–93, and see Jan Noel, *Canada Dry: Temperance Crusades before Confederation* (Toronto: University of Toronto Press, 1995).

154 Bahktin, *Rabelais and His World.*

155 Walrond, *Letters and Journals of James Eighth Earl of Elgin*, 48–9.

156 London *Advertiser*, 27 Sept. 1888.

157 NA, RG 17, vol. 113, 10040; vol. 117, 11401.

158 Toronto *World*, 21 Sept. 1888.

159 AO, MU2823, David Smith Papers, file 9, Accounts, 1895.

160 Montreal *Gazette*, 17 Sept. 1892.

161 Peterborough *Daily Evening Review*, 28 Sept. 1892.

162 London *Advertiser*, 21 Sept. 1872.

163 Chris Allen Rasmussen, 'State Fair: Culture and Agriculture in Iowa, 1854–1941' (PhD dissertation, Rutgers University, 1992): 294.

164 *Canadian Agricultural Reader* (Niagara: John Simpson, 1845): 225–7.

165 London *Advertiser*, 5 Oct. 1870.

166 Peterborough *Times*, 30 Oct. 1880.

167 Milton *Canadian Champion*, 1 Nov. 1866.

168 Brantford *Expositor*, 22 Oct. 1886.

Chapter 6: International Exhibitions and Canadian Nationality, 1851–1867

1 *Canadian Agriculturist* 3 (1850): 151–4.

2 D.S.L. Cardwell, *The Organization of Science in England* (London: Heinemann, 1957); Michael Argles, *South Kensington to Robbins: An Account of English Technology and Science since 1851* (London: Longman's, 1968); C.R. Fay, *Palace of Industry, 1851: A Study of the Great Exhibition and Its Fruits* (Cambridge: Cambridge University Press, 1951); Paul Greenhalgh, *Ephemeral Vistas: The Expositions Universelles, Great Exhibitions, and World's Fairs 1851–1939* (Manchester: Manchester University Press, 1988).

3 Montreal *Gazette*, 14 June 1851; see also Audrey Short, 'Canada Exhibited, 1851–1867,' *Canadian Historical Review* 48 (1967): 353–64; Merle Curti, 'America at the World Fairs, 1851–1893,' *American Historical Review* 55

(1949–50): 833–56; Nathan Rosenberg, *The American System of Manufactures* (Edinburgh: University of Edinburgh Press, 1969).

4 Montreal *Minerve*, 28 March 1850. See Jacques Monet, *The Last Cannon Shot: A Study of French-Canadian Nationalism, 1837–1850* (Toronto: University of Toronto Press, 1969), 352; NA, MG 24 B2, 3457–60, Papineau Papers, Papineau to his wife, 23 June 1839.

5 Montreal *Gazette*, 27 March 1850.

6 MTRL, BR, Broadsides; 'Report of the Committee,' *JLAPC*, appendix L, 1850.

7 Quebec *Morning Chronicle*, 21 Oct. 1850.

8 Stanstead *Journal*, 20 June 1850.

9 Charlottetown *Islander*, 10 Jan. and 7 Feb. 1851.

10 *New Brunswick Reporter and Fredericton Advertiser*, 17 Jan. and 31 Oct. 1851.

11 *Nova Scotian*, 13 Jan. and 18 Aug. 1851.

12 Doughty, ed., *Elgin-Grey Correspondence* 2: 726–33.

13 Montreal *Witness*, 4 Nov. 1850.

14 *La Minerve*'s articles on the exhibition are held at Montreal City Hall Archives.

15 Charles Dewey Day, *Address Delivered at the Provincial Industrial Exhibition, Montreal, on Saturday, October 19, 1850* (Montreal: R. Campbell, 1850).

16 M.H. Abrams, *The Mirror and the Lamp: Romantic Theory and the Critical Tradition* (New York: Oxford University Press, 1953); William H. Galpern, *The Return of the Visible in British Romanticism* (Baltimore: Johns Hopkins University Press, 1993); see also Jan Mukarovsky, *Structure, Sign, and Function: Selected Essays*, trans. John Burbank and Peter Steiner (West Haven: Yale University Press, 1978): 89; Northrop Frye, *The Educated Imagination* (Montreal: CBC, 1963), 30.

17 Speech given 21 March 1850, in *The Principal Speeches and Addresses of His Royal Highness the Prince Consort* (London: John Murray, 1862), 109–14.

18 In Morrell and Thackray, *Gentlemen of Science*, 3.

19 And again, 'The Crystal Palace contained the best collection ever made of the products of art and the works of nature. Therefore, it was the best school ever formed for studying the connection of art and nature with revealed religion.' George Troup, *Art and Faith, in Fragments from the Great Exhibition of Arts and Manufactures in 1851* (London: Partridge and Oakley, 1852), 317.

20 United Church of Canada Archives, Samuel S. Nelles Papers, box 8, file 168, 'Thoughts on the Great Exhibition,' read before the Hamilton Merchants' Association, 14 April 1851; Mrs Col. Savage, *Watch, the Prophecy of the Scripture and Truth: Which Came to Pass in the Year 1851* (Toronto: Privately printed, 1857) (CIHM 40661).

21 *Illustrated London News*, supplement to 21 June 1851, 597–8.
22 NA, MG 24 D16, Buchanan Papers, vol. 32, 1881–8; vol. 34, 28049–57.
23 Montreal *Gazette*, 26 May 1851.
24 An unrepentant reminiscence is in NA, MG 29 D21, Perry Papers.
25 NA, MG 24 A16, Elgin Papers, Frederick Cumberland to Elgin, 7 Nov. 1851.
26 Theodore Walrond, *Letters and Journal of James, Eighth Earl of Elgin* (London: John Ming, 1873), 167.
27 Montreal *Gazette*, 28 May 1851.
28 Montreal *True Witness*, 30 May 1851.
29 *Illustrated London News*, supplement to 3 May 1851, and *Tallis's History and Description of the Crystal Palace*, 1: 51–3.
30 Bernard J. Harrington, *Life of Sir William E. Logan* (Montreal: Dawson Brothers, 1883), 272.
31 Her Majesty's Commissioners, *Official Catalogue of the Great Exhibition* (London: Spicer, 1851), 963; *A Few Words upon Canada, and her Productions in the Great Exhibition*, published by Authority (London: Piper, 1851), 4.
32 NA, MG 24 B14, LaFontaine Papers, 2522–6, 4 Nov. 1851.
33 *Exhibition of the Works of Industry of All Nations, 1851 Reports by the Juries* (London: William Clowes & Sons, 1852), 15–16; *Nova Scotian*, 8 Sept. 1851; *British Columbian*, 3 Oct. 1861.
34 In Robert C. Post, 'Reflections of American Science and Technology at the New York Crystal Palace Exhibition of 1853,' *Journal of American Studies* 17 (1983): 338.
35 NA, RG 17, Department of Agriculture, vol. 2314, 37; Donald C. Masters, *The Reciprocity Treaty of 1854: Its History, Its Relation to British and Foreign Policy and to the Development of Canadian Fiscal Autonomy* (Toronto: McClelland & Stewart, 1963).
36 NA, RG 17, vol. 2314, 37. The UC Board reported in May 1853 that, lacking instructions, it had dropped the subject. RG 17, vol. 1, 113; vol. 2314, 191. Holwell's 'Supplementary Report to the Hon. John Rolph, Minister of Agriculture' is in *JLAPC* 13 (1854), appendix I.I.
37 Toronto *Globe*, Sept. 1853, passim; NA, RG 17, vol. 2314, 184.
38 Provincial Archives of Newfoundland, GN1/1/3, Letterbooks of Despatches to the Colonial Office, James Crowley to M. Sedgwick, 27 Dec. 1852; *Newfoundlander*, 20 Dec. 1852; *Official Catalogue of the New-York Exhibition of the Industry of All Nations* (New York: G. Putnam, 1853); *Newfoundlander*, 26 May 1853; Mrs M.S. Peace, *The Convict Ship and Other Poems* (Greenock: Robert A. Baird, 1850), 164.
39 *Islander*, 12 Aug. and 7 Oct. 1853.
40 McDougall's report is also in *JLAPC* 13 (1854), appendix I.I.

41 Commission impériale, *Rapport sur l'Exposition universelle de 1855 présenté à l'Empereur* (Paris: Imprimerie impériale, 1857).
42 Careless, *Brown of the 'Globe'*, 1: 154.
43 Peter Baskerville, 'Sir Allan Napier MacNab,' *DCB*, vol. 9, 525–6.
44 Montreal *Gazette*, 18 Sept. 1855.
45 Ibid., 14 Sept. 1855.
46 Legislative Council, 20 Nov. 1854, in the *Scrapbook Debates*, Legislative Assembly of the Province of Canada; see Antony Rasporich, 'Imperial Sentiment in the Province of Canada during the Crimean War, 1854–1856,' in W.L. Morton, ed., *The Shield of Achilles: Aspects of Canada in the Victorian Age* (Toronto: McClelland & Stewart, 1968), 145.
47 J.-C. Taché, *Canada at the Universal Exhibition* (Montreal: John Lovell, 1856); NA, RG 17, vol. 2315, Minutes of Proceedings relating to the Paris Exhibition; ANQ, Fonds Taché, G113, Paris Exhibition Commission Minutes, 2 Feb. 1855.
48 See the article by Suzanne Zeller in the *DCB*, vol. 12, 539–42.
49 NA, RG 17, vol. 2315, Minutes of Proceedings.
50 McGill University Archives, Logan Papers, A. Ramsay to Logan, 5 Dec. 1854; *JLAPC* 1854, appendix L.
51 *Exposition Universelle de 1855: Rapports du jury mixte international* (Paris: Imprimerie Impériale, 1856), I: 3–4, 55.
52 In the Montreal *Gazette*, 9 June and 13 Sept. 1855.
53 Montreal *Gazette*, 3 Oct. 1855.
54 Taché, *Canada at the Universal Exhibition*, 279.
55 *The Times*, 5 Sept. 1855 and 8 Oct. 1867.
56 Toronto *Leader*, 1 and 2 May, 1856, in MTRL, BR, Logan Scrapbook, 103. See Morris Zaslow, *Reading the Rocks: The Story of the Geological Survey of Canada, 1842–1972* (Toronto: Macmillan, 1974); Suzanne Zeller, *Inventing Canada: Early Victorian Science and the Idea of a Transcontinental Nation* (Toronto: University of Toronto Press, 1987); also Robert A. Stafford, *Scientist of Empire: Sir Roderick Murchison, Scientific Exploration and Victorian Imperialism* (Cambridge: Cambridge University Press, 1989), 216.
57 Toronto *Globe*, in the Logan Scrapbook, 98–9; Zeller, *Inventing Canada*.
58 Nancy Christie, 'Sir William Logan's Geological Empire and the "Humbug" of Economic Utility,' *Canadian Historical Review* 75 (1994): 161–204.
59 This argument is largely a gloss on the remark by Michael Bliss that Canada's natural wealth depends on the state of the market for these goods: 'Drop the price low enough and the black gold and the bountiful forest and the glittering ore turned into worthless goo and rocks and trees, the northern cornucopia became a northern wilderness again, *cap de nada*.' *Northern*

Enterprise: Five Centuries of Canadian Business (Toronto: McClelland & Stewart, 1987), 561–2.

60 Robert A. Stafford, 'Geological Surveys, Mineral Discoveries and British Expansion, 1835–1871,' *Journal of Imperial Commonwealth History* 12 (1984): 5–32.

61 *JLAPC* 1852, appendix O.

62 Logan Papers, Logan to Hincks, 11 May 1855.

63 *Rapport sur l'Exposition Universelle* (Paris: Imprimerie Impériale, 1857), I: 169, 275.

64 Toronto *Globe*, 4 Aug. 1855.

65 Taché, *Canada at the Universal Exhibition*, 202–3.

66 Montreal *Gazette*, 22 Nov. 1855.

67 John Clarke, 'Account of the application of steam power to the cultivation of the land,' *Journal of the Royal Agricultural Society of England* 19 (1858): 174–228.

68 At the 1861 trials of the Royal Agricultural Society, 'the work done, though but little, was certainly the best in the field, the soil being finely pulverized to the depth of 7 inches; but the expenditure of coal, oil, and water was something fearful. This cost, coupled with the wages of the men, the wear and tear of the implement, and the interest of capital invested in its purchase, would very far exceed the value of the work.' *JRASE* 22 (1861): 464.

69 Hope Morritt, *Rivers of Oil: The Founding of North America's Petroleum Industry* (Kingston: Quarry Press, 1993), 24; Edward Phelps, 'Foundation of the Canadian Oil Industry,' in Nick and Helma Mica, eds., *The Shaping of Ontario from Exploration to Confederation* (Belleville: Mica, 1985), 189. Records of sales are in the Logan Papers, passim.

70 William Logan and Thomas Sterry Hunt, *A Sketch of the Geology of Canada* (Paris: Hector Bossange, 1855); and Taché, *Descriptive Catalogue of the Productions of Canada Exhibited in Paris in 1855* (Paris; G.A. Pinard, 1855). Both are reprinted in Taché, *Canada at the Universal Exposition.*

71 Taché, *Canada at the Universal Exhibition*, 15.

72 J.M.S. Careless, *The Union of the Canadas: The Growth of Canadian Institutions, 1841–1857* (Toronto: McClelland & Stewart, 1967).

73 Bowmanville *Canadian Statesman*, 6 Sept. 1855; Logan scrapbook, 57.

74 ANQ, Fonds Taché, G119; *Journal du Québec*, 14 Aug. 1855; Walter Benjamin, *Charles Baudelaire: A Lyric Poet in the Era of High Capitalism* (London: NCB, 1973), 54; Dana Brand, *The Spectator and the City in Nineteenth-Century American Literature* (Cambridge: Cambridge University Press, 1991).

75 Montreal *Pays*, 19 July 1855; Toronto *Globe*, 24 Aug., 1 and 9 Sept. 1855, quoting the Quebec *Chronicle* and the *British Whig.*

76 Macdonald to Brown Chamberlin, 7 Aug. 1855. J.K. Johnson, ed., *The Letters of Sir John A. Macdonald, 1836–1857* (Ottawa: PAC, 1968), 292.
77 Logan scrapbook, 102; 'J.-C. Taché,' by Jean-Guy Nadeau, *DCB*, vol. 12, 1013.
78 J.-C. Taché, *The Lunatic Asylums of the Province of Quebec and Their Defamers*, trans. J.-P. Tardival (Quebec: Léger Brousseau, 1885); Bossé, *Joseph-Charles Taché*, 43.
79 J. Sheridan Hogan, *Canada: An Essay* (Montreal: John Lovell, 1855); Alexander Morris, *Canada and Her Resources* (Montreal: Dawson, 1855); J.-C. Taché, 'Sketch of Canada, Its Industrial Condition and Resources' (Paris: Hector Bossange and Sons, 1855), printed in J.-C. Taché, *Canada at the Universal Exhibition*; H.-L. Langevin, *Le Canada: Ses institutions, resources, produits, manufactures* (Quebec: Lovell & Lameroux, 1855); William Hutton, *Canada: A Brief Outline of her Geographical Position, Productions, Climate, Capabilities, Educational and Municipal Institutions, Fisheries, Railroads* (Toronto, 1857) (This book was published by authority, but the 1855 essay served as its basis); Adam Lillie, *Canada: Physical, Economic, and Social* (Toronto: Maclean & Co, 1855).
80 Hogan, *Canada*, 109.
81 Quoted in Taché, *Canada at the Universal Exhibition*; Montreal *Minerve*, 3 Nov. 1855.
82 The article continued: 'Has the philosophic mind not reason to feel confidence in the future destinies of human industry when it traces in the rising States of the earth manufacturing energies so forward and vigorous, directed to objects so strictly in accordance with the material necessities of the position they occupy?' *The Times*, 7 Sept. 1855.
83 Montreal *Gazette*, 12 Nov. 1855; Laurence S. Fallis Jr, 'The Idea of Progress in the Province of Canada: A Study in the History of Ideas,' in Morton, *Shield of Achilles*, 175.
84 Imperial College Archives, Playfair Papers, 458, Logan to Playfair, 20 Nov. 1855.
85 Montreal *Gazette*, 28 Feb. and 3 Oct. 1855; Montreal *Herald*, 28 July 1855.
86 Montreal *Pays*, 25 Aug. 1855. See Jean-Paul Bernard, *Les Rouges: libéralisme, nationalisme et anticléricalisme au milieu du XIXe siècle* (Montreal: Les Presses de l'Université, Québec, 1971).
87 Nadeau, 'Taché,' *DCB*, vol. 12, 1014; L.F.S. Upton, "The Idea of Confederation, 1754–1858," in W.L. Morton, *Shield of Achilles*, 184–207; Arthur Silver, *The French-Canadian Idea of Confederation, 1864–1900* (Toronto: University of Toronto Press, 1982), 40–1; J.-C. Taché, *Des provinces de l'Amérique du Nord et d'une Union fédérale* (Québec: J.-T. Brousseau, 1858). It first appeared as a series of articles in 1857.

88 Taché, *Des provinces*, 181.
89 Alexander Morris, *Nova Britannia; or, British North America, Its Extent and Future* (Montreal: John Lovell, 1858), 27; Monet, *Last Cannon Shot*, 384–5; Andrée Désilets, *Hector-Louis Langevin: un père de la Conféderation canadienne (1826–1906)* (Quebec: Les presses de l'Université Laval, 1969).
90 *Canadian Agriculturist* 13 (1861), 194.
91 *Principal Speeches and Addresses*, 113.
92 On the leviathan as unity, see Hobbes, *Leviathan*, 112.
93 NA, MG 24 D30, Brown Chamberlin Papers (CP), II, 123.
94 *Saturday Review* 13 (1862): 684–5.
95 Peterborough *Review*, 7 Nov. 1862.
96 The minutes of the collecting committee are in the Chamberlin Papers, vol. 6.
97 Montreal *Herald*, 22 March 1862.
98 Montreal *Witness*, 6 Aug. 1862.
99 Quoted in the Charlottetown *Islander*, 11 July 1862.
100 *Nova Scotian*, 19 Aug. 1861 and 27 Jan. 1862.
101 Ibid., 2 Sept. 1861.
102 Charlottetown *Islander*, 25 July, 1862.
103 Quoted in *Newfoundland Express*, 12 July 1862.
104 Victoria *British Colonist*, 12 Feb. 1861.
105 *British Columbian*, 3 Oct. 1861 and 27 March 1862; *British Colonist*, 17 March 1862.
106 Montreal *Herald*, 16 Aug. 1862; Toronto *Globe*, 18 July 1862; Goldwin Smith, *The Empire: A Series of Letters Published in 'The Daily News,' 1862, 1863* (Oxford: John Henry & James Parker, 1863), 73.
107 C.A. Bodelson, *Studies in Mid-Victorian Imperialism* (1924; London: Heinemann, 1960, 37.
108 NA, MG 24 I D15, Tilley Papers, Thomas Daniel to Tilley, 16 May 1862; *The Times*, 2 May 1862.
109 CP, vol. 7, Sandford to Logan, 22 May and 23 June 1862, Chamberlin to Sandford, 10 July 1862, and A.-A. Dorion to Chamberlin, 8 Aug. 1862.
110 Toronto *Globe*, 26 July 1862.
111 Logan Papers, copy 2; Harrington, *Life of Logan*, 347–8.
112 Henry H. Miles, *Canada East at the International Exhibition* (London: G. Norman, 1862); NA, CP, vol. 2, 130, Dawson to Chamberlin, 2 Oct. 1862.
113 Logan Papers, Chamberlin to Logan, 4 Oct. 1862.
114 Johnson, *Letters of Sir John A. Macdonald*, 2: 371; T. McIlroy, *JBAMUC* 4 (1864): 72.
115 NA, CP, vol. 7, Chamberlin to A.-A. Dorion, 26 July and 7 Aug. 1862.

116 Order in Council, 11 Nov. 1862.

117 *Dublin International Exhibition of Arts and Manufactures, 1865, Official Catalogue* (Dublin: John Falconer, 1865).

118 *Report of the Nova Scotia Department of the Dublin International Exhibition, 1865* (Halifax: A. Grant, 1866), 13.

119 *Nova Scotian*, 9 Oct. 1865, 21 May 1866.

120 NA, RG 17, vol. 5, 309, Cuff, 7 June 1865.

121 Ibid., vol. 6, 416, Cuff, 7 Oct. 1865; *JBAMUC* 5 (1865): 320.

122 Josephine Phelan, *The Ardent Exile: The Life and Times of Thomas D'Arcy McGee* (Toronto: Macmillan, 1951), 241–2. A copy is in the MTRL, BR. Robert Rumilly claims that this cost him many Irish votes in Montreal. *Histoire de la Province de Québec*, I: 46.

123 NA, RG 17, vol. 5, 348; Cuff, 27 July 1865.

124 *JBAMUC* 1 (1861): 309–12.

125 NA, RG 17, vol. 9, 635 and 667; vol. 18, 1565 (dated January 1868).

126 Ibid., vol. 6, 406 and 430, 7 Sept. 1865, has a full account.

127 Ibid., 416, and vol. 8, 573.

128 J.S. Ingram, *The Centennial Exposition Described and Illustrated* (Philadelphia: Hubbard Bros., 1876), 38.

129 'Appendice sur l'avenir des expositions,' Commission impériale, *Rapport sur l'exposition universelle de 1867, à Paris* (Paris: Imprimerie impériale, 1869–72), 265–328.

130 Scadding, 'On Museums and Other Classified Collections, Temporary or Permanent, As Instruments of Education in Natural Sciences,' *Canadian Journal*, ns 13 (1871): 11; NA, MG 24 I9, Hill Papers, 3570–7.

131 ANQM, Board of Arts and Manufactures for Lower Canada, 4 Jan. 1866; ANQ, Fonds Chapais, Collection Langevin, vol. 23, Taché to Langevin, 27 Oct. 1865.

132 NA, RG 17, vol. 17, 1486; vol. 22, 1963.

133 Raymond Vézina, *Napoléon Bourassa, 1827–1916: Introduction à l'étude de son art* (Montréal: Elysée, 1976), 85; Fonds Siméon LeSage, vol. 4, 2: 279–80.

134 Donald Creighton, *John A. Macdonald: The Young Politician* (Toronto: Macmillan, 1952), and Phelan, *Ardent Exile*, 271–2.

135 Montreal *Minerve*, 13 April 1867.

136 Ibid., 1 and 15 June 1867.

137 *Nova Scotian*, 13 May and 10 June 1867.

138 Montreal *Gazette*, 20 July 1867.

139 MTRL, BR, Logan Scrapbook; Zeller, *Inventing Canada*.

140 Careless, *Brown of the Globe*, 2: 80.

141 One Eastern Townships member cited his district's display as proof of its rich resources. In *Parliamentary Debates on the Subject of the Confederation of the British North American Provinces* (Quebec: Hunter, Rose & Co., 1865).

142 Ian Hacking, *Representing and Intervening: Introductory Topics in the Philosophy of Natural Science* (Cambridge: Cambridge University Press, 1983), 136. He continues: 'It will be protested that reality, or the world, was there before any representation or human language. Of course. But conceptualizing it as reality is secondary. First there is this human thing, the making of representations. Then there was the judging of representations as real or unreal, true of false, faithful or unfaithful. Finally comes the world, not first but second, third or fourth.'

143 *Parliamentary Debates on the Subject of the Confederation*, 60.

Chapter 7: Exhibitions in Europe after Confederation

1 *Sessional Papers* 7, 40 (1874): vi. This number did not include inland arrivals. Full statistics are provided in Reg Whitaker, *Canadian Immigration Policy since Confederation* (Ottawa: Canadian Historical Association Booklet, 1991), 2.

2 NA, RG 17, Department of Agriculture, vol. 24, 2117 (Toronto *Globe*, 1868); Montreal *Gazette*, 10 July 1871.

3 NA, RG 17, vol. 38, 3511, Taché to William Dixon, May–June 1870; ANQ, P134, Collection Chapais, fonds Langevin, vol. 23, Taché to Langevin, 19 Oct. 1868. On the Department of Agriculture, see Garth Stevenson, *Ex Uno Plures: Federal-Provincial Relations in Canada, 1867–1896* (Montreal: McGill-Queen's University Press, 1993).

4 *Announcement of the Forthcoming Series of Annual International Exhibitions of Selected Works of Fine and Industrial Art and Scientific Inventions* (London: Her Majesty's Commissioners, 1870). NA, RG 17, vol. 37, 3440; Findling, *Historical Dictionary*, 45.

5 NA, RG 17, files 6957, 7200, 9297, 9327, 9519; see Hugh Cunningham, *Leisure in the Industrial Revolution* (London: Croon Helm, 1980); J.M. Golby and A.W. Purdue, *The Civilisation of the Crowd: Popular Culture in England, 1750–1900* (London: Batsford, 1984); also Bailey, *Leisure and Class in Victorian England.*

6 Sir Henry Cole, *Fifty Years of Public Work of Sir Henry Cole* (London: G. Bell, 1884), 1: 269.

7 NA, RG 17, 8598, Childers, 29 Nov. 1872 (commas added).

8 Ibid., Rose to Macdonald, 25 Nov. 1872.

9 Ibid., Macdonald to Pope, 17 March 1873.

10 Findling, *Historical Dictionary of World's Fairs and Expositions*, 48–54.

11 NA, RG 17, vol. 82, 7955, 7995.
12 NA, MG 26 D13, Macdonald Papers, D. McCulloch to Macdonald, 5 June 1873, 65104–5; Palmer, *A Culture in Conflict*, 148.
13 Macdonald Papers, 645106–8, Witton to Macdonald, 6 June 1873.
14 NA, MG 29 E18, Lowe Papers, vol. 3, 492–3, John Lowe to Cartier, 27 June 1872.
15 Toronto *Globe*, 5 May 1874; NA, RG 17, vol. 109, 10637.
16 Simon, Introduction, *Rapports du jury international*, 243–55; Raymond Isay, 'L'Éxposition de 1878,' *Revue des deux mondes* 38 (1937): 896–923.
17 T.C. Keefer, *Paris Universal Exhibition: Report of the Canadian Commissioner* (Ottawa: Maclean, Roger & Co., 1881), 35.
18 Keefer is quoted in Hodgetts, *Pioneer Public Service*, 58; NA, RG 72, Canadian Exhibition Commission, vol. 26, 24839, Paul DeCazes to Begg, 28 Jan. 1879. On Keefer, see See H.V. Nelles, Introduction to T.C. Keefer, *Philosophy of Railroads and Other Essays*, ed., H.V. Nelles (Toronto: University of Toronto Press, 1972), ix–lxiii.
19 ANQ, Fonds Joly, vol. 10, 728–31, Perrault to Joly, 12. Aug. 1878.
20 NA, RG 72, vol. 25, Keefer, 19 Jan. 1878.
21 T.C. Keefer, *Paris Universal Exhibition 1878: Handbook and Official Catalogue of the Canadian Section* (London: Eyre & Spottiswoode, 1878).
22 NA, Watkins Papers, MG 29 C75, Thomas C. Watkins, 'Paris Exhibition.'
23 Harriet A. Boomer, *Notes from Our Log in South Africa and On Foot through the Colonies at the Paris Exhibition* (London, Ont.: Free Press, 1880) (CIHM 08833), 86.
24 *Farmer's Advocate* 13 (1878): 273.
25 NA, RG 72, vol. 25, Keefer memorandum, 20 Oct. 1877.
26 Ibid., 23144; vol. 28, 19624.
27 NA, MG 28 III44, Montreal Board of Trade, vol. 6, 240, 18 Aug. 1877.
28 Ibid., vol. 4, 13 Nov. 1877.
29 NA, RG 72, vol. 25, 22175.
30 Boomer, *Notes from Our Log*, 87; *Farmer's Advocate* 13 (1878): 273.
31 NA, RG 72, vol. 192, Accounts; vol. 27, 27663.
32 *Gazette des campagnes* 6 (1866–7): 173.
33 NA, RG 72, vol. 34, passim.
34 Ibid., vol. 25, 23806, Keefer, 13 March 1878.
35 Ibid., 24763; also 40434 (Keefer letter dated 12 July 1883); and see Perrault's testy letter of 25 March 1879 in vol. 26, 25210.
36 Ibid., 24638.
37 Report of Ignatius Kormann, Ontario, *Sessional Papers* 11, 40 (1879): 2.
38 Drolet, *Zouviana*, 332.

39 *A Monthly Review Devoted to Canadian Emancipation and Commercial Union with the United States* (Montreal, 1880). See especially February 1880, passim.
40 Drolet, *Zouviana*, 342.
41 Yvan Lamonde, *Louis-Antoine Dessaulles: Un seigneur libéral et anticlérical* (Quebec: Fides, 1994), 265–7.
42 Magella Quinn, 'Les capitaux français et le Québec, 1855–1900,' *Revue d'histoire de l'Amérique française* 24 (1971): 527–66.
43 Frédéric Gerbie, *Le Canada et l'emigration française*, 13th ed. (Quebec: G. Darveau, 1884), 355–441. Copious letters from Perrault are in ANQ, Correspondence of the Quebec Commissioner of Agriculture.
44 Sylain Simard, *Mythe et reflet de la France, L'image du Canada en France, 1850–1914* (Ottawa: University of Ottawa Press, 1987).
45 *Colonies and India* Supplement, 30 Nov. 1878.
46 Jan Morris, *The Spectacle of Empire: Style, Effect, and the Pax Britannia* (London: Faber & Faber, 1982), 10; John MacKenzie, *Propaganda and Empire: The Manipulation of British Public Opinion, 1880–1960* (Manchester: University of Manchester Press, 1985); Anne McClintock, *Imperial Leather: Race, Gender and Sexuality in the Colonial Context* (New York: Routledge, 1995): 56. See also Thomas Richards, *The Imperial Archive: Knowledge and the Fantasy of Empire* (London: Verso, 1993).
47 NA, MG 26 E1 A, Mackenzie Bowell Papers, vol. 3, 856–63. Samuel Wilmot, *Great International Fisheries Exhibition, London, 1883* (Ottawa: A.S. Woodburn, 1884); W.F. Whicher, 'Practical Results of Fish Culture in the Dominion of Canada,' *Forest and Stream* 20 (1883): 408; see the article on 'Samuel Wilmot,' by A.B. McCullough, *DCB*, vol. 12, 1106–7. Related publications are L.Z. Joncas, *The Fisheries of Canada* (New York: Erastus Wiman, 1884), and his *Fisheries of Canada* (London: William Clowes, 1883).
48 Sandford Fleming, *A Summer Tour between Old and New Westminster* (Montreal: Dawson Brothers, 1884), 72–81; Geoff Meggs, *Salmon: The Decline of the British Columbia Fishery* (Vancouver: Douglas & McIntyre, 1991), 20.
49 Provincial Archives of Newfoundland, GN 2/1/A, vol. 57, 433–4.
50 Recent surveys on imperialism during this period are Bernard Semmel, *The Liberal Ideal and the Demons of Empire: Theories of Imperialism from Adam Smith to Lenin* (Baltimore: Johns Hopkins University Press, 1993); Bernard Porter, *The Lion's Share: A Short History of British Imperialism, 1850–1995*, 3rd ed. (London: Longman, 1996); C.C. Eldridge, ed., *British Imperialism in the Nineteenth Century* (London: Macmillan, 1984).
51 Bernard Semmel, *The Rise of Free Trade Imperialism: Classical Political Economy, the Empire of Free Trade, and Imperialism, 1750–1850* (Cambridge: Cambridge University Press, 1970); but see B.A. Knox, 'Reconsidering Mid-Victorian

Imperialism,' *Journal of Imperial and Commonwealth History* 1 (1973): 155–72.
52 *Imperial Federation* 1 (1886): 42–3, 54, 58–9, 256.
53 Sir Charles Tupper, *Recollections of Sixty Years in Canada* (London: Cassell & Co., 1914), 282; NA, Macdonald Papers, MG 26 D13, vol. 282, 128980–3, passim.
54 R. Craig Brown argues that the Conservatives turned to other markets in the 1890s (*Canada's National Policy, 1883–1900: A Study in Canadian-American Relations*, Princeton: Princeton University Press, 1964, 195–228): but the Colonial Exhibition was an early step in that direction.
55 Lance Davis and Robert A Huttenbach, 'The Export of British Finance, 1865–1914,' *Journal of Imperial and Commonwealth History* 13 (1985): 28–76.
56 Saint John *Daily Sun*, 4 July 1885, 1 Sept. 1886.
57 NA, RG 17, vol. 447, 48787.
58 Ibid., vol. 238, 24202.
59 *The Empire*, 181.
60 NA, RG 17, vol. 1629, Report by A.T. Galt, 20 May 1880.
61 Donald Creighton, *John A. Macdonald: The Old Chieftain* (Toronto: Macmillan, 1955, 1966): 368.
62 *Imperial Federation* 1 (1886): 28.
63 Newcastle *Chronicle*, quoted in *Colonial and Indian Exhibition of 1886: A Revelation of Canada's Progress and Resources. Extracts from British and Colonial Journals* (Ottawa: Department of Agriculture, 1888), 74.
64 NA, RG 17, vol. 35, 3244.
65 *Manitoba Free Press*, 14 July 1886.
66 Ibid., 9 Sept. 1876.
67 Napanee *Standard*, 3 Oct. 1861.
68 *Canadian Agriculturist* 12 (1860): 657.
69 *Canada in 1880*, *Sessional Papers*, 12 (1880–1), appendix.
70 Andrew Wernick, *Promotional Culture: Advertising, Ideology, and Symbolic Expression* (London: Sage, 1991).
71 NA, RG 17, vol. 98, 9532.
72 From a lecture to the Saint John Mechanics' Institute, in his *Recollections of Sixty Years in Canada* (London: Cassell, 1914), 28–9.
73 *Confederation Debates*, 86; in *Brown of the Globe*, 2: 183.
74 Norman Macdonald, *Canada Immigration and Colonization, 1841–1903* (Aberdeen: Aberdeen University Press, 1966), 31.
75 Saint John *Daily Sun*, 23 Sept. 1885 (emphasis added).
76 In W. Hamish Fraser, *The Coming of the Mass Market, 1850–1914* (London: Macmillan, 1981), 134.
77 As propounded by C.S. Pierce, William James, and John Dewey, pragmatism

suggested that, in the absence of a correspondence theory of truth, 'truth is whatever is in the end delivered to the community of inquirers who pursue a certain end in a certain way.' C.S. Pierce in Ian Hacking, *Representing and Intervening*, 59.Timothy Mitchell provides one example whereby an exhibit reshaped the reality it represented in its own image: A 'Cairo Street,' built for the Paris Exposition of 1878, later became a model for imperial agents who redesigned Cairo to make it orderly and more visible. *Colonizing Egypt* (New York: Cambridge University Press, 1988).

78 *Nova Scotian*, 12 Oct. 1868.

79 Canada, Parliament, *Hansard*, 5 May 1886.

80 Ibid., 21 May 1888; and see NA, RG 72, vol. 52, 3618, Lowe to Tupper.

81 NA, Tupper Papers, MG 26 F, 3279–80, Macdonald to Tupper, 10 March 1886.

82 NA, RG 72, vol. 38, 270, and 393; vol. 39, 510.

83 Ibid., vol. 64, 273–4; *Hansard*, 10 April 1885.

84 In *Colonial and Indian Exhibition of 1886: A Revelation*, 55.

85 *Dominion Annual Register*, 1885; Sir Charles Hibbert Tupper, ed., *The Life and Letters of the Rt. Hon. Sir Charles Tupper*, Supplement (Toronto: Ryerson Press, 1926), 106.

86 NA, RG 72, vol. 65, 665.

87 Ibid., 573–4.

88 Ibid., vol. 38, 204; Michael Bliss, *Plague: A Story of Smallpox in Montreal* (Toronto: HarperCollins, 1991).

89 NA, RG 72, vol. 36, 5158, and vol. 38, 343; Charlottetown *Islander*, 3 July 1886; Victoria *British Colonist*, 8 Aug. 1886.

90 NA, RG 72, vol. 42, 1178.

91 This term is from William Leiss et al., *Social Communication in Advertising: Persons, Products and Images of Well-Being*, 2nd ed. (Scarborough: Routledge/ Nelson Canada, 1990), 106.

92 *Colonial and Indian Exhibition, London 1886: Official Catalogue of the Canadian Section, Published under the Authority of the Hon. Sir Charles Tupper* (London: McCorquedale & Co., 1886).

93 NA, RG 72, vol. 57, 2: 152–4, Tupper to Carling, 11 Feb. 1886; vol. 65, 502, Carling to Tupper, 12 Dec. 1886.

94 Ibid., 39, 583.

95 Macdonald, *Canada Immigration and Colonization, 1841–1903*, 43.

96 Alexander Somerville, *The Autobiography of a Working Man* (1848; London: Turnstile Press, 1951), Introduction by John Carswell. See NA, RG 17, 1599, 1837, 9553, 18135, and passim. He was kept on because Sir John A. Macdonald pitied him: ANQ, Chapais Papers, vol. 18, 29 May.

97 NA, RG 72, vol. 44, 1660.
98 Ibid., vol. 46, 1936.
99 Ibid., vol. 39, 536; the files contain dozens of letters from Begg.
100 Doug Owram, *Promise of Eden: The Canadian Expansionist Movement and the Idea of the West, 1856–1900* (Toronto: University of Toronto Press, 1980).
101 NA, RG 72, vol. 39, 529. See W.A. Waiser *The Field Naturalist: John Macoun, the Geological Survey, and Natural Science* (Toronto: University of Toronto Press, 1989).
102 Ontario, *Sessional Papers* 22, 24 (1890): 181; 19, 50 (1887): 90.
103 Canada, *Sessional Papers* 27, 8H (1894): 24; J.A. Ruddick, 'The Development of the Dairy Industry in Canada,' in H.A. Innis, ed., *The Dairy Industry in Canada* (Toronto: Ryerson Press, 1937), 13–123.
104 NA, RG 72, vol. 50, 3113.
105 Ibid., vol. 48, 2524, and vol. 49, 2836.
106 Quoted in *Colonial and Indian Exhibition of 1886: A Revelation*, 73, 84.
107 *Report of Sir Charles Tupper*, 8.
108 NA, RG 72, vol. 49, 2999; vol. 51, 3847; on the government's encouragement of hunting, see Keith Wamsley, 'Cultural Significance and National Ideologies: Rifle Shooting in Late Nineteenth-Century Canada,' *Social History* 20 (1995): 63–72.
109 NA, RG 72, vol. 51, 3741; vol. 43, 1337.
110 Ibid., vol. 42, 1129; vol. 46 2109; vol. 47, 2307.
111 Ibid., vol. 37 101.
112 Quoted in *Colonial and Indian Exhibition of 1886: A Revelation*, 121.
113 Denison, *Harvest Triumphant*, 96–7.
114 NA, RG 72, vol. 61, Harkom report, 9 Sept. 1886; vol. 67, 167–8.
115 Ibid., vol. 56, 4637; 'John Watson of Ayr,' *Waterloo Historical Society Report* 17 (1929): 144.
116 Stanstead *Journal*, 7 Oct. 1886.
117 NA, RG 72, vol. 52, 3462.
118 Ibid., 3549; vol. 51, 3846. See also Eskano Heikkonen, *Reaping the Bounty: McCormick Harvesting Machine Company Turns Abroad, 1878–1902* (Helsinki: Finnish Historical Society, 1995).
119 Ibid., vol. 37, 196.
120 Ibid., vol. 41, 892.
121 Ibid., vol. 47, 2257; vol. 48, 2593; vol. 50, 3026 and 3053.
122 Ibid., vol. 45, 1841; vol. 54, 4095; vol. 56, 4637; vol. 54, 4177; AO, MS571, Heinztman and Co., file 9. Gayle M. Comeau suggests their export sales did later increase: 'Theodor August Heintzman,' *DCB*, vol. 12, 421–4.

123 NA, RG 17, vol. 460, 50189; Tupper Papers, October 1885; 3064 ff. RG 72, vol. 37, 50.

124 See his letter to Tupper, Tupper Papers, 3064. The Royal Canadian Academy also demanded public money: NA, MG 28 I126, passim.

125 NA, RG 72, vol. 2, 325.

126 NA, MG 27 IB6, Lansdowne Papers, 313–5, 15 and 23 Oct. 1886.

127 The report is appended to Tupper's *Report*, 61–8.

128 Quoted in *A Revelation*, 132–6.

129 NA, RG 72, vol. 52, 3527.

130 *Imperial Federation* 1 (1886), Supplement.

131 *Colonial Conference, London 1887: Proceedings and Papers* (London: H.M. Stationary's Office, 1887). A useful condensation, boiled down to Canadian matters, is in Maurice Oliver, ed., *The Colonial and Imperial Conferences from 1887 to 1937*, vol I: *Colonial Conferences* (Ottawa: Edmond Cloutier, 1954).

132 NA, Lansdowne Papers, 53, Lansdowne to Macdonald, 29 Jan. 1887.

133 Jeffrey L. Lant, *Insubstantial Pageant: Ceremony and Confusion at Queen Victoria's Court* (London: Hamish Hamilton, 1979); MacKenzie, *Propaganda.*

134 NA, Tupper Papers, 3427–8, Macdonald to Tupper, 20 Dec. 1886.

135 Ibid., 393–4, Macdonald to Tupper, 13 Oct. 1886.

136 NA, Macdonald Papers, 129012, Tupper to Macdonald, 15 Nov. 1886.

137 NA, Tupper Papers, 3427–8, Macdonald to Tupper, 20 Dec. 1886.

138 Ibid., 3396–9, 3401–9, and 3443–4.

139 NA, Lansdowne Papers vol. 4, 57–8 and 89–92.

140 Manitoba *Free Press*, 12 May 1886.

141 *Imperial Federation* 1 (1886): 188–9; also *England and Her Colonies: The Five Best Essays on Imperial Federation, Submitted to the London Chamber of Commerce for their Prize Competition* (London: Swan Sonnenschein, 1887).

142 In *Imperial Federation* 1 (1886): 212.

143 Hacking, 'Historical Epistemology,' paper presented at the conference on Historical Epistemology, University of Toronto, November 1993, 12.

144 Eric Hobsbawm, 'Mass-Producing Traditions: Europe, 1870–1914,' in Eric Hobsbawm and Terence Ranger, eds., *The Invention of Tradition* (Cambridge: Cambridge University Press, 1983), 268–9.

145 The 1880s is usually given as the period of transition to the new, popular imperialism. James Sturgis, 'Britain and the New Imperialism,' in Eldridge, ed., *British Imperialism in the Nineteenth Century*, 85–105; on the old imperialism, see C.A. Bodelson, *Studies in Mid-Victorian Imperialism* (1924; London: Heinemann, 1960), and C.C. Eldridge, *Disraeli and the Rise of a New Imperialism* (Cardiff: Wales University Press, 1996); and on the new, see Robert H.

MacDonald, *The Language of Empire: Myths and Metaphors of Popular Imperialism, 1880–1918* (Manchester: Manchester University Press, 1994).

146 Quoted in Lant, *Insubstantial Pageant*, 126.

147 London *Times*, 13 Jan. 1887; see Lant, *Insubstantial Pageant*, 132–4; MacKenzie, *Propaganda and Empire*; Michael Worboys, 'Science and British Colonial Imperialism, 1895–1940' (PhD thesis, School of Mathematical and Physical Sciences, 1979), 143–90.

148 G.T. Denison, *The Struggle for Imperial Unity: Recollections and Experiences* (London: Macmillan, 1909), 68; Donald F. Warner, *The Idea of Continental Union: Agitation for the Annexation of Canada to the United States, 1849–1893* (Kentucky: University of Kentucky Press, 1960).

149 NA, MG 30 D44, George Parkin Papers, 722–3, 739–42; Carl Berger, *The Sense of Power: Studies in the Ideas of Canadian Imperialism* (Toronto: University of Toronto Press, 1970); Terry Cook, 'George R. Parkin and the Concept of Britannic Idealism,' *Journal of Canadian Studies* 10 (1975): 15–31.

150 Goldwin Smith, *Canada and the Canadian Question* (1891; Toronto: University of Toronto Press, 1971).

151 Brigitte Schroeder-Gudehus, 'Les grandes puissances devant l'Exposition universelle de 1889,' *Le mouvement social* 149 (Oct.–Dec. 1989): 15–24.

152 Quoted in Lamonde, *Louis-Antoine Dessaulles*, 279.

153 Kenneth J. Munro, *The Political Career of Sir Adolphe Chapleau, Premier of Quebec, 1789–1882* (Lewiston: Edwin Mellen, 1992), 156; Jules Helbronner, *Rapport sur la section d'economie sociale de l'Exposition universelle internationale de 1889 à Paris* (Ottawa: B. Chamberlin, 1890).

154 Denison, *Harvest Triumphant*, 106–10; *Massey's Illustrated* 7 (1889): 158.

155 Montreal *Gazette*, 19 Dec. 1889.

156 NA, RG 17, vol. 634, 71963, 72012; vol. 635, 72027.

157 Ibid., vol. 625, 70851.

158 Ibid., vol. 639, 72555.

159 NA, RG 72, vol. 75 passim; RG 17, vol. 623, 70694.

160 NA, RG 72, vol. 117, 94996.

161 Ibid., vol. 644, 73188.

162 Thomas August, *The Selling of the Empire: British and French Imperial Propaganda, 1890–1940* (Westport: Greenwood Press, 1985), 72.

163 NA, RG 72, vol. 74, Preston to Fisher, 1 Nov. 1905.

164 Richard D. Mandell, *Paris 1900: The Great World's Fair* (Toronto: University of Toronto Press, 1967), 112.

165 Henry Adams, *The Education of Henry Adams: An Autobiography* (Boston: Houghton Miflin, 1918), ch. 25; and see John Patrick Diggins, *The Promise of*

Pragmatism: Modernism and the Crisis of Knowledge and Authority (Chicago: University of Chicago Press 1994).

166 Mandell, *Paris 1900*, 94–5.

167 Montreal *Patrie*, 14 April 1900.

168 NA, RG 72, vol. 196, Strathcona to Sidney Fisher, 1 April 1898.

169 George Collins Levey, *Report of the Secretary of the Colonial Committee of the Royal Commission for the Paris Exhibition 1900* (London: Charles Jones, 1901), 13.

170 NA, RG 72, vol. 196, Strathcona to Fisher, 31 March 1898.

171 Ibid., vol. 196, Fisher to Strathcona, 31 May 1898.

172 Ibid., J.C. Colmer to G.C. Levey, 10 Dec. 1901.

173 Ibid., Report of the Canadian Commission at the Paris Exhibition, 3½. The report was never published. The author originally wrote 'large inducements,' but the first word was crossed out by an editor.

174 Auguste Dupuis, *The Province of Quebec at the Paris Universal Exposition of 1900* (Quebec: Mercury, 1901), 81.

175 William D. Scott, *Official Catalogue of the Canadian Section*, and George Johnson, *Canada: Its History, Productions, and Natural Resources* (Ottawa: Department of Agriculture, 1900).

176 NA, MG 28 I126, Royal Canadian Academy of Arts, vol. 17, 80.

177 Emile Falardeau, *Marc-Aurèle-alias-Suzor-Côté: Peintre et sculpteur 1869–1937* (Laval: Galerie des anciens, 1969); Sylvaine Allaire, 'Les canadiens au salon officiel de Paris entre 1870 et 1910: sections peinture et dessin,' *Journal of Canadian Art History* 4 (1977–8): 141–54.

178 The *Patrie* printed columns full of names of Canadians at Paris.

179 Montreal *Patrie*, 30 Aug. 1900.

180 Ibid., 2 June 1893.

181 Ibid., 23 June 1900.

182 Ibid., passim; NA, MG 27 III, B3, Dandurand Papers, correspondence between Raoul Dandurand and Sir Wilfrid Laurier, Nov. 1899.

183 NA, RG 72, vol. 74, E. Girardot to Hutchison, 7 Aug. 1905.

184 Ibid., Hutchinson to Fisher, 30 July 1907.

185 Burton Benedict, *The Anthropology of World's Fairs: San Francisco's Panorama Pacific International Exhibition of 1915* (London: Scholar Press, 1983), 46.

186 August, *The Selling of the Empire*, 135.

187 NA, RG 72, vol. 192, Wembley Scrapbook.

188 Erving Goffman, *The Presentation of the Self in Everyday Life* (New York: Doubleday, 1959); Kenneth Burke, *A Grammar of Motive* (New York: Prentice Hall, 1945).

189 *Patrie*, 11 Oct. 1900.

Chapter 8: Exhibitions in America after Confederation

1 James D. McCabe, *The Illustrated History of the Centennial Exhibition* (Philadelphia: National Publishing, 1975); John Maass, *The Glorious Enterprise: The Centennial Exhibition of 1876 and H.J. Schwarzmann, Architect in Chief* (New York: American Life, 1973); Robert Rydell, *All the World's a Fair: Visions of Empire at American International Expositions, 1876–1916* (Chicago: University of Chicago Press, 1984), 9–37; Thomas J. Schlereth, 'The Material Universe of American World Expositions,' in his *Cultural History and Material Culture: Everyday Life, Landscapes, Museums* (Charlottesville: University of Virginia Press, 1992).
2 J.D. Morrell, quoted in McCabe, *Illustrated History*, 286; David E. Nye, *American Technological Sublime* (Cambridge, Mass.: MIT Press, 1994), 120–2.
3 Mabel Burkholder, *The Story of Hamilton* (Hamilton: Davis-Lisson, 1938).
4 Quoted in Maass, *The Glorious Enterprise*, 105.
5 United States Centennial Commission, in NA, RG 72, vol. 2, 202; and see McCabe, *Illustrated History*, 298.
6 United States Centennial Commission, *International Exhibition, 1876: Reports and Awards* (Washington: Government Printing Office, 1880).
7 Rydell, *All the World's a Fair*, 21–6; Raymond Corbey, 'Ethnographic Showcases, 1870–1930,' *Cultural Anthropology* 8 (1993): 338–69.
8 Michael Adas, *Machines as the Measure of Men: Science, Technology, and Ideologies of Western Dominance* (Ithaca: Cornell University Press, 1989), 3.
9 Leo Marx, *The Machine in the Garden: Technology and the Pastoral Ideal in America* (London: Oxford University Press, 1964), 192.
10 Montreal *Gazette*, 22 July 1876; James Gilbert, *Perfect Cities: Chicago's Utopia of 1893* (Chicago: University of Chicago Press, 1991), 108–9.
11 S.P. May, 'Report on the Ontario Educational Exhibit at Philadelphia, 1876,' in 'Report of the Commissioner of Agriculture on the Products, Manufactures, &c., of Ontario, Exhibited at the International Exhibition, Philadelphia, 1876,' Ontario, *Sessional Papers* 9, 33 (1877): 185.
12 Forster, *A Conjunction of Interests*.
13 NA, RG 72, vol. 4, F.W. Glen, 1 Dec. 1874.
14 Ronald D. Tallman, 'Reciprocity 1874: The Failure of Liberal Diplomacy,' *Ontario History* 65 (1873): 87–106.
15 *Monetary Times*, 20 Aug. 1875; NA, MG 24 D16, vol. 111, 72955; RG 72, vol. 1, 84.
16 *Report of the Canadian Commission at the Philadelphia Exhibition* (Ottawa: By Authority, 1877), 19, 30–2.
17 NA, RG 72, vol. 5, 451 and 469.

18 P.B. Casgrain, *Letellier de Saint-Just et son temps* (Quebec: C. Darveau, 1885), 190.

19 Parti conservateur (Quebec), *Dépenses de la Commission canadienne sous le gouvernement des Libéraux à l'Exposition de Philadelphie.* For example, on 17 June were purchased three dozen bottles of Bass ale, half a dozen of champagne, a bottle of Hennessy brandy, and one of wine; five days later it was two dozen of Clavel claret, a case of sparkling wine, four dozen of Bass again, and a bottle and a gallon of sherry; on 13 July, three cases of Clavel Medoc, three of sparkling wine, and half a dozen bottles of Hennessy brandy; and so on. A dinner for thirty-two at the Lafayette Restaurant cost $176.

20 Toronto *Mail*, 5 May 1879.

21 NA, RG 72, vol. 3, 243; Montreal *Gazette*, 1, 3, and 5 June 1876.

22 McGill Archives, Dawson Papers, MG 1022, vol. 61, B.J. Harrington to his wife, 26 July 1876.

23 Montreal *Gazette*, 14 Aug. 1876. Letters on 6 Oct. and 6 Sept. railed against sending 'an expensive political parasite.'

24 NA, RG 72, vol. 2, 131.

25 Ibid., vol. 1, 110; 'Report of Dr S.P. May,' 219–20; United States Centennial Commission, *Reports and Awards*, vol. 7, 5; vol. 6, 277.

26 J. Coleman, 'Report on the Agricultural Implements at the Philadelphia Centennial Exhibition,' *Journal of the Royal Agricultural Society of England* 13 (1877): 1–91.

27 Saint John *Daily News*, 29 Sept. 1875; advertisement in L.W. Bailey and Edward Jack, *The Woods and Minerals of New Brunswick* (Fredericton: Daily Telegraph, 1876).

28 Robert Bell: NA, RG 72, vol. 37, 155.

29 *Reports and Awards* vol. 6, 705. The jury situated the province in British America rather than Canada.

30 ANQ, E25, vol. 41, 31278, and vol. 42, 31741; ANQ, P149, Fonds Siméon LeSage, vol. 4, 2: 308, LeSage to Auguste Dupuis, 10 Jan. 1876.

31 NA, RG 72, vol. 1, 39½; vol. 2, 220; ANQ, E25, vol. 46, 33739.

32 *Reports and Awards*, vol. 4, 677.

33 H.A. Innis and A.R.M. Lower, *Select Documents in Canadian Economic History*, 2: 505.

34 Canada, *Sessional Papers* 27, 8H (1894): 21; 'Report of J. Ballantyne, Esq. M.P.P., et al.,' in 'Report of the Commissioner of Agriculture,' 142–8.

35 *Reports and Awards*, vol. 4, 430.

36 This was the verdict of American experts as well as the Canadians. See the report on the 'Centennial Exhibition,' by the Ontario Fruit Growers' Association, in the annual report of the Ontario Commissioner of Agriculture, Ontario, *Sessional Papers* 9, 4 (1877): 295–326.

37 Ingram, *The Centennial Exposition*, 676.

38 Grant MacEwan, *Highlights of Shorthorn History* (Calgary: Comprint, 1982), 67.

39 *Illustrated History*, 256; Peterborough *Times*, 9 Sept. 1876.

40 *The Awards and Claims of Exhibitors at the International Exhibition 1876* (Boston: National Association of Wool Manufactures, 1877), 309, 451; Richard Reid, 'The Rosamond Woollen Company of Almonte: Industrial Development in a Rural Setting,' *Ontario History* 75 (1985): 275–6.

41 Canada, *Sessional Papers*, Reports on Trade and Commerce.

42 Bruce Kidd, 'Edward Hanlan,' *DCB*, vol. 13, 437–40.

43 Halifax *Citizen*, Sept. 1876 passim.

44 Maass, *The Glorious Enterprise*, 75; *Le naturaliste canadien* 8 (1876): 371–84.

45 Daniel Fowler, 'Autobiography or Recollections of an Artist,' in Frances K. Smith, *Daniel Fowler of Amherst Island, 1810–1894* (Kingston: Agnes Etherington Art Centre, 1979), 176.

46 *The Centennial Exposition*, 448.

47 J. Russell Harper, *Painting in Canada: A History*, 2nd ed. (Toronto: University of Toronto Press, 1988), 124; Reid, *A Concise History*, 89 (who insists that 'he painted them with feeling').

48 NA, RG 72, vols. 2–3, 194, 209, 213, 269; J.M. Gregory, 'Leading Features of the Educational Exhibit,' in the *Reports and Awards*, vol. 8, 267–8.

49 *Historical and Other Papers*, 317.

50 Robert M. Stamp, 'Ontario at Philadelphia: The Centennial Exposition of 1876,' in Neil McDonald and Alf Chaiton, eds., *Egerton Ryerson and His Times* (Toronto: Macmillan, 1978), 314–17; see his *The Schools of Ontario, 1876–1976* (Toronto: OHSS/University of Toronto Press, 1982).

51 NA, RG 72, vol. 2, 183.

52 ANQ, P149, Fonds LeSage, vol. 7, 6, Lesage to Lamarche, 21 Jan. 1876.; vol. 4, 2: 327–8, Lesage to O'Neill, 21 Feb. 1876.

53 Gabriel Dussault, *Le Curé Labelle, messianisme, utopie et colonisation au Québec, 1850–1900* (Montreal: Hurtubise, 1983), ch. 4.

54 ANQ, E13, Education, General Correspondence, 1876, vol. 342, 2088.

55 Quoted in Hodgins, *Special Report*, 159.

56 *Journal of Education for the Province of Quebec* 29 (1879): 17–20; NA, MG 29, D73, Archambault Papers; ANQ, E13, vol. 554, 1953, Archambault to Ouimet, 13 July 1877.

57 ANQ, E13, year 1880 file 168; also André Labarrère-Paulé, *Les instituteurs laiques au Canada français, 1836–1900* (Quebec: Les presses de l'Université Laval, 1965), 336.

58 ANQ, E13, 1880 file 168, Archambault to Ouimet, 28 Sept. 1880. The award was cancelled by the Council of Arts, which competed with the Christian

Teaching Brothers' drawing schools and in 1878 tried to impose an American method on them, even after the Brothers pointed out that their system had won prizes in Paris and Vienna. ANQM, Council of Arts, vol. 2, Minutes, 6 Feb. and 14 May 1878. See also *La Minerve*'s complaint that the council decided in advance not to let another French-Canadian educator, Abbé Chabert, win a prize for his school of design: 18 Sept. 1882; Bernard Molaire, 'Joseph Chabert,' *DCB*, vol. 12, 170–1.

59 Labarrère-Paulé, *Les instituteurs laiques*, 336–7.

60 J. Larke, *Report of the Executive Commissioner for Canada to the World's Columbian Exposition Chicago, 1893* (Ottawa: S.E. Dawson, 1894), 42; NA, RG 72, vol. 87, W.D. Dimock to Larke, 16 March 1893; John McIntosh, *Report of the Commissioner from the Province of Quebec World's Columbian Exposition* (Montreal: Gazette Office, 1893), 24–5; ANQ, E13, vol. 731, 1891 file 91, has the Chicago correspondence. *La Vérité*, 29 July and 14 Oct. 1893; see also *Le Manitoba*, 20 Aug. 1893.

61 *Journal de l'instruction publique* 29 (1879): 9; Hodgins, *Special Report*, 24.

62 Jules Simon, 'Introduction,' *Rapports du jury international*, 158.

63 Both passages are from Hodgins, *Historical and Other Papers*, 317 and 321.

64 B.G. Northrup, Connecticut educator, in Hodgins, *Special Report*, 217–18.

65 Bennett, *The Birth of the Museum*, 197; see also the discussion of 'panoptical time' in McClintock, *Imperial Leather*, 36.

66 'Report of Dr. S.P. May,' 47.

67 Rousseau, *Emile*, 168.

68 The phrase is taken from Bryan D. Palmer, *Descent into Discourse: The Reification of Language and the Writing of Social History* (Philadelphia: Temple University Press, 1990).

69 Montreal *Gazette*, 24 Aug. 1876.

70 NA, RG 17 A II5, vol. 2317, 22–4, Lowe to A. Morris, 30 Dec. 1876.

71 Roderick William Cameron, 'Report of the Canadian Commission at the Melbourne International Exhibition, 1880–81,' Canada, *Sessional Papers* 15, 11 (1882), appendix 46, vii; and NA, RG 17, vol. 246, 25388.

72 NA, MG 11, CO 448, vol. 18 (1856A); RG 17, vols. 245–54, 24585, 25217, 25426, 25440; vol. 287, 29694; 'Reports of the Australian International Exhibitions,' appendix 46 to the 'Report of the Minister of Agriculture for the Year 1881,' *Sessional Papers* 11 (1882).

73 Brown, *Canada's National Policy*, 225.

74 Montreal *Gazette*, 9 Sept. 1876.

75 London *Advertiser*, 12 Jan. 1885; *Farmer's Advocate* 20 (1885): 66.

76 Toronto *Globe*, 3 Sept. 1884.

77 *Mineral Exhibit of Ontario, Descriptive Catalogue* (Cincinatti: Robert Clarke, 1888).

78 Karen Booth, 'When Jamaica Welcomed the World: The Great Exhibition of 1891,' *Jamaica Journal* 18, 3 (1988): 39–52; *Report of the Honorary Commissioner (Mr. Adam Brown) Representing Canada at the Jamaica Exhibition, Held at Kingston, Jamaica, 1891* (Ottawa: Queen's Printer, 1891).
79 John E. Findling, *Chicago's Great World's Fairs* (Manchester: Manchester University Press, 1994); Robert Mucciguosso, *Celebrating the New World: Chicago's Columbian Exposition of 1893* (Chicago: Ivan R. Dee, 1993); David F. Burg, *Chicago's White City of 1893* (Lexington: University of Kentucky Press, 1976); Reid Badger, *The Great American Fair: The World's Columbian Exposition and American Culture* (Chicago: Nelson Hall, 1979).
80 Toronto *Globe*, 2 May 1893; W.E. Cameron, *The World's Fair, Being a Pictorial History of the Columbian Exposition* (Philadelphia: Home Library, 1893), 854.
81 Toronto *Globe*, 11 Nov. 1893.
82 Toronto *Empire*, 8 July 1893; *Globe*, 26 Aug. 1893.
83 Quebec City Archives, Expositions, Conseils et comités, vol. 53, Correspondence October 30, 1889; Ontario, *Sessional Papers* 22, 8 (1890): 215.
84 Schroeder-Gudehus, *Les fastes du progrès*, 121.
85 *Hansard*, 5 April 1892.
86 NA, RG 72, vol. 105, 61–5.
87 Brown, *Canada's National Policy*, 222–3.
88 Louis Aubrey Wood, *A History of Farmers' Movements in Canada: The Origins and Development of Agrarian Protest, 1872–1924* (Toronto: University of Toronto Press, 1975); Vittorio M.G. de Vecchi, 'Science and Scientists in Government, 1878–1896,' *Scientia Canadensis* 8 (1984): 112–42, and 9 (1988): 97–113.
89 Michael Bliss, *A Living Profit: Studies in the Social History of Canadian Business, 1883–1911* (Toronto: McClelland & Stewart, 1974), 36.
90 *Hansard*, 28 March and 26 April 1892.
91 NA, RG 72, vol. 90, Huot, 9 Sept. 1892; vol. 106, 473–6. (For Chicago, outgoing correspondence appears in paginated letterbooks, and incoming correspondence is unnumbered but sorted alphabetically by author or company.)
92 Ibid., vol. 106, 957–8.
93 Ibid., vol. 86, Coombs; vol. 107, 987–8.
94 Ibid., vol. 109, 385, 362, 886–7.
95 Ibid., 151–2.
96 Ibid., 708–9.
97 Ibid., 181–2.
98 Ibid., vol. 102, 154–5; vol. 106, 768–9; vol. 86, Consumer Cordage; vol. 94, David Morrice; vol. 113, 116–17.

99 Ibid., vol. 89, F. Fitzgerald for the Hamilton Conservative Association, 18 April 1893.

100 Ibid., Angers to Larke, 8, 12 and 26 June 1893.

101 Ibid., Anderson.

102 Ibid., vol. 86, J.J. Cassidy (secretary).

103 *Canadian Manufacturer* 23 (1892), 106.

104 The phrase is from Michael Bliss, *A Living Profit.*

105 NA, RG 72, vol. 83, P. Beneteau; vol. 100, H. Walker,

106 Larke, *Report of the Executive Commissioner,* 21.

107 Hubert Howe Bancroft, *The Book of the Fair: An Historical and Descriptive Presentation of the World's Science, Art and Industry* (Chicago: Bancroft, 1893), 2: 185.

108 Toronto *Globe,* 1 July 1893.

109 NA, RG 72, vol. 87, W.D. Dimock passim.

110 Bliss, *Northern Enterprise,* 307; Denison, *Harvest Triumphant,* ch. 7.

111 *Massey's Illustrated,* ns 5, 5 (May 1893): 3. Holdings at the Agricultural Museum, Milton, Ont.

112 Ibid., and *Empire,* 6 April 1893.

113 NA, RG 72, vol. 95, Lyman Jones, 18 Nov. 1893.

114 Ibid., vol. 112, 915–20, 2 Nov. 1893.

115 Canada, *Sessional Papers* 27, 8E (1894), *Special Report of the Executive Commissioner on Awards on Agricultural Implements at Chicago, 1893.*

116 NA, RG 72, vol. 84, Cockshutt, 3 Jan. 1894.

117 Ibid., vol. 92, Hugh Johnston, 3 Jan. 1894.

118 Ibid., vol. 112, 36–7.

119 Archives Ste-Anne de la Pocatière, Fonds J.-C. Chapais, vol. 801, 226.

120 His report is in Canada, *Sessional Papers* 27, 8B (1894): 173–94.

121 Toronto *Globe,* 27 May 1893.

122 *Manitoba Free Press,* 1 Nov. 1893.

123 Toronto *Globe,* 29 July 1893.

124 Ibid., 11 Nov. 1893.

125 Ibid., 22 April 1893.

126 Ibid., 6 May 1893; Canada, *Sessional Papers* 27, 8B (1894): 193.

127 NA, RG 72, vol. 97, Robertson to Dimock, 14 March 1893.

128 *Farmer's Advocate* 12 (1877): 197.

129 Larke, *Report of the Executive Commissioner,* 31.

130 Ibid., 21.

131 NA, RG 72, vol. 112, 622–3 and 531–2.

132 Quoted in Awrey, *Report of the Ontario Commissioner.*

133 NA, RG 72, vol. 98, 'Selwyn' (letter by Ernest Thompson to Selwyn, 4 March

1893); vol. 100, 'Thompson,' passim; A.M. Machar, 'Side-lights on the Columbian Exposition: The Aesthetic Aspect,' *The Week* 10 (1893): 1162. The episode is discussed in John Henry Wadland, *Ernest Thompson Seton: Man in Nature and the Progressive Era, 1880–1915* (New York: Arno Press, 1978), 122–42.

134 British Columbia, Department of the Provincial Secretary, *Report of the Executive Commissioner for British Columbia at the World's Columbian Exposition, Chicago* (Victoria: Richard Wolfenden, 1894).

135 Ontario had 1832 such factories, and Quebec, 958: Canada, *Sessional Papers* 27, 8D (1894): 6.

136 Larke, *Report of the Executive Commissioner*, 17.

137 Naylor, *History of Canadian Business*, 2: 237 and 85.

138 Toronto *Empire*, 15 and 29 July 1893.

139 Quoted in Robert Rydell and Nancy Gwinn, 'Introduction,' in Rydell and Gwinn, eds., *Fair Representations: World's Fairs and the Modern World* (Amsterdam: Vu University Press, 1994), 8, and see also Robert Rydell, *World of Fairs: The Century of Progress Expositions* (Chicago: University of Chicago Press, 1993).

140 T.C. Keefer, *Canadian Water Ways from the Great Lakes to the Atlantic* (Boston: Damrell & Upham, [1893]) (CIHM 07801); B.E. Walker, *Banking in Canada* [np], (CIHM 25418); *Globe*, 24 June 1893.

141 Bliss, *A Living Profit*, 51.

142 In Rev. John Henry Barrows, ed., *The World's Parliament of Religions* (Chicago: Parliament Publishing, 1893), I: 21–2. See also Richard Hughes Seeger, *The World's Parliament of Religions: The East-West Encounter, Chicago, 1893* (Bloomington: University of Indiana Press, 1995).

143 Dennis B. Downey, 'Tradition and Acceptance: American Catholics and the Columbian Exposition,' *Mid-America* 63 (1981): 79–92.

144 *La Vérité*, 30 Sept. 1893; *Manitoba Free Press*, 27 Sept. 1893.

145 Toronto *Mail*, 14 and 17 March 1893.

146 In *World's Parliament of Religions*, I: 103–6; 2: 942–6 and 1502–4. Grant's inaugural speech is also in *Queen's Quarterly* 1 (1893–4): 158–61. The *Nova Scotian* had objected to Grant's excessive tolerance in 1867 (18 Feb.).

147 See Ramsay Cook, *The Regenerators: Social Criticism in Late Victorian English Canada* (Toronto: University of Toronto Press, 1985), 48–58.

148 Allen Pringle, *True Religion versus Creeds and Dogmas: A Discussion between Two Clergymen, a Layman, and Allen Pringle* (Toronto: C.M. Ellis, 1894) (CIHM 12170), 13.

149 Toronto *Globe*, 16 Oct. 1893.

150 *World's Parliament of Religions*, 2: 1112 (later misquoted by Pringle).

151 Toronto *Empire*, 8 May 1893.
152 Fidelis, 'Impressions of the Parliament of Religion,' *The Week* 10 (1893): 1114–17.
153 G.M. Grant, *The Religions of the World*, rev. ed. (London: A.C. Black, 1895). See D.B. Mack, 'George Monroe Grant,' *DCB*, vol. 13, 403–9.
154 Owen Chadwick, *The Secularization of the European Mind in the Nineteenth Century* (Cambridge: Cambridge University Press, 1975).
155 Cook, *The Regenerators*, and Downey, 'Tradition and Acceptance.'
156 Armstrong and Nelles, *The Revenge of the Methodist Bicycle Company*.
157 NA, RG 72, Columbian Exposition correspondence, passim, contains dozens of letters. Others were sent directly to Chicago, but appear in newspapers.
158 Toronto *Globe* and *Mail*, 3 June 1893.
159 Armstrong and Nelles, *The Revenge of the Mehtodist Bicycle Company*, 51; Charleton, *Speeches and Addresses Political, Literary and Religious* (Toronto: Morang, 1905), 253–89.
160 Toronto *Mail*, 17 Aug. 1893. On Faith Fenton, see Jill Downie, *A Passionate Pen: The Life and Times of Faith Fenton* (Toronto: HarperCollins, 1996).
161 Toronto *Empire*, 17 June 1893.
162 Rasmussen, 'State Fair,' 345.
163 Daniel Francis, *The Imaginary Indian: The Image of the Indian in Canadian Culture* (Vancouver: Arsenal Pulp Press, 1992): 'Performing Indians.'
164 Toronto *Mail*, 2 Sept. 1893.
165 Ibid., 9 Sept. 1893.
166 Douglas Cole, *Captured Heritage: The Scramble for Northwest Coast Artifacts* (Vancouver: Douglas & McIntyre, 1985), 128.
167 Bothwell *Times*, 5 Oct. 1893.
168 Toronto *Empire*, 4 Sept. 1893.
169 AO, MU2823, David Smith Papers, file 9.
170 NA, RG 72, vol. 96, Oronhyatekhon, 3 March 1894.
171 Toronto *Empire*, 6 July 1893.
172 British Columbia Archives and Records Service, Add. Mss. 2657, World's Columbian Exposition, British Columbia Visitor's Books, passim.
173 NA, RG 72, vol. 108, 658–9; *Manitoba Free Press*, 10 Feb. 1893.
174 *Manitoba Free Press*, 8 Sept. 1893.
175 *Farmer's Advocate* (Manitoba) 4 (1893): 142; *Free Press*, 7 March 1893.
176 Montreal *Minerve*, 20 Sept. 1893.
177 *Manitoba Free Press*, 19 and 22 Aug. 1893.
178 NA, RG 72, vol. 106, 574–8.
179 Ibid., vol. 82, Angers, 21 July 1893.

180 Donald Swainson, 'Ontario on the Rise,' *Horizon Canada* 66 (1986): 1561–7.
181 NA, RG 72, vol. 76, 94202.
182 Ibid., vol. 88, Dimock, 3 July 1894.
183 Antiquarian and Numismatic Society, *Shall We Have a World's Fair in Montreal in 1892, to Celebrate the 250th Anniversary of the Founding of Ville Marie?* (Np, Oct., 1888) (CIHM 64110); NA, RG 17, file 66308.
184 See Hervé Gagnon, 'Divertissement et patriotisme: la genèse des musées d'histoire à Montréal au XIXe siècle,' *Revue d'histoire de l'Amérique française* 48 (1995): 317–50; correspondence is in the Baby collection at the University of Montreal.

Chapter 9: Women and the Political Economy of Exhibitions

1 Joan B. Landes, *Women and the Public Sphere in the Age of the French Revolution* (Ithaca: Cornell University Press, 1988); Mary P. Ryan, *Women in Public: Between Banners and Ballots, 1825–1880* (Baltimore: Johns Hopkins University Press, 1990). Some Canadian studies of women in public are Bonnie Huskins, 'The Ceremonial Space of Women: Public Processions in Victorian Saint John and Halifax,' in Janet Guildford and Suzanne Morton, eds., *Separate Spheres: Women's Worlds in the Nineteenth-Century Maritimes* (Fredericton: Acadiensis Press, 1994), 145–60; David C. Jones, '"From Babies to Buttonholes": Women's Work at Agricultural Fairs,' *Alberta History* 29, 4 (1981): 26–32; and Cecilia Morgan, *Public Men and Virtuous Women: The Gendered Languages of Religion and Politics in Upper Canada, 1791–1850* (Toronto: University of Toronto Press, 1996). Political aspects of female participation are also discussed in Walden, *Becoming Modern in Toronto.*
2 Habermas, *The Structural Transformation of the Public Sphere*; see the essays in Calhoun, *Habermas and the Public Sphere*, and Johanna Meehan, ed., *Feminists Read Habermas: Gendering the Subject of Discourse* (New York: Routledge, 1995).
3 Pierre Manent, *An Intellectual History of Liberalism*, trans. Rebecca Balinski (Princeton: Princeton University Press, 1994), ch. 4.
4 John Stuart Mill was one outstanding exception: a liberal who extended to women the rights of men. However, even Mill believed that married women would prefer the domestic hearth to waged labour.
5 Michael Oakeshott, *Rationalism in Politics and Other Essays* (Indianapolis: Liberty Press, 1991), chs 1 and 2; 57.
6 Montreal *Gazette*, 24 May 1820; Rousseau, *Politics and the Arts*, 48; Linda M.G. Zerilli, *Signifying Women: Culture and Chaos in Rousseau, Burke, and Mill* (Ithaca: Cornell University Press, 1994), 28. Allan Bloom remarks: 'It is essential to underline that Rousseau was more potent for thoughtful and

sensitive men and women in the nineteenth century than either Marx or Freud is for us.' See his *Love and Friendship* (New York: Simon & Schuster, 1993), 158.

7 Greer, *The Patriots and the People*, 201–2.

8 Carolyn Strange, *Toronto's Girl Problem: The Perils and Pleasures of the City, 1880–1930* (Toronto: University of Toronto Press, 1995), 89 and passim, on the 'relentless sexualization of working girls' behaviour'; Backhouse, *Petticoats and Prejudice*, 263–4.

9 Montreal *Gazette*, 31 Dec. 1833. Rousseau had a low opinion of female novelists as well, thinking them unwomanly creatures.

10 Toronto *Leader*, 12 Sept. 1853.

11 In Ramsay Cook and Wendy Mitchinson, eds., *The Proper Sphere: Woman's Place in Canadian Society* (Toronto: Oxford University Press, 1976), 302; see Glenna Matthews, *The Rise of Public Woman: Woman's Power and Woman's Place in the United States, 1630–1970* (New York: Oxford University Press, 1992), and Judith R. Walkowitz, *City of Dreadful Delight: Narratives of Sexual Danger in Late-Victorian London* (Chicago: University of Chicago Press, 1992), 67.

12 Rousseau, *Emile, or On Education*, 364–5.

13 *Canadian Agriculturist* 7 (1855): 217–18.

14 Montreal *Gazette*, 2 June 1860, citing the *Atlantic Monthly*; *Canadian Agriculturist* 7 (1855): 217–18.

15 *Canada Education Monthly* (1879), quoted in Cook and Mitchinson, *The Proper Sphere*, 137.

16 Montreal *Gazette*, 3 April 1822; *Canada Farmer*, ns 2 (1870): 467; London *Advertiser*, 15 May 1876.

17 Her Majesty's Commissioners, *First Report of the Commissioners for the Exhibition of 1851* (London: Spicer Brothers, 1852), liv; Day, *Address Delivered at the Provincial Industrial Exhibition*; Nelles, 'Thoughts on the Great Exhibition'; Toronto *Globe*, 9 Oct. 1851.

18 Bliss Carman, quoted in Cook and Mitchinson, *The Proper Sphere*, 79.

19 Burke, *A Grammar of Motives*, 332.

20 Quoted in Victoria *Daily Colonist*, 23 Oct. 1892.

21 John Ruskin, *The Two Paths* (1859; Orpington: George Allan, 1888), 19.

22 William Leach, *True Love and Perfect Union: The Feminist Reform of Sex and Society* (1894; New York: Basic, 1980), 159; Mary Kavanaugh Oldham Eagle, *The Congress of Women* 1894; (New York: Arno Press, 1974), 172.

23 *JLALC* 1820, appendix F; *Missiskoi Standard*, 7 Nov. 1837.

24 For example, Charlottetown *Islander*, 19 April 1866.

25 Montreal *Gazette*, 8 Sept. 1892. The figure is not exact because the number of Guernsey prizes is not given. The women won 17 of 161 prizes listed.

26 Quebec, *Sessional Papers* 22, 2 (1889); NA, Philippe Landry Papers, Montmagny Agricultural Society, vol. 19.
27 Talman, 'Agricultural Societies,' 551; Kingston *Chronicle*, 11 Oct. 1822, 17 Oct. 1823
28 Sidney Smith, *The Settler's New Home* (London: John Kendrick, 1849), 62. See Margerie Griffin Cohen, *Women's Work: Markets and Economic Development in Nineteenth-Century Ontario* (Toronto: University of Toronto Press, 1988), and Rosemary R. Ball, '"A Perfect Farmer's Wife": Women in 19th Century Rural Ontario,' *Canada: An Historical Magazine* 3, 2 (1975): 3–21; Terry Crowley, 'Rural Labour,' in Paul Craven, ed., *Labouring Lives: Work and Workers in Nineteenth-Century Ontario* (Toronto: University of Toronto Press, 1995), 13–104; Martine Tremblay, 'La division sexuelle du travail et la modernisation de l'agriculture à travers la presse agricole, 1840–1900,' *Revue d'histoire de l'Amérique française* 42 (1993): 221–42.
29 *Canadian Agricultural Reader* (Niagara: John Simpson, 1845), 94.
30 *Gazette des campagnes* 8 (1869–70): 237–8, and 22 (1884–5): 65–70.
31 Montreal *Gazette*, 10 Feb. 1831.
32 *The Novascotian*, 23 Oct. 1854; and see Marling, *Blue Ribbon*, 96.
33 Ontario, *Sessional Papers* 13, 3 (1881): appendix A, 28; 6, 1 (1874): appendix A, 72; 2, 5 (1869): appendix A.
34 Toronto *Mail*, 5 Sept. 1881, 6 Sept. 1882, 15 Sept. 1880; Toronto *Globe*, 14 Sept. 1883. On the arts and crafts movement, see Elisabeth Ashin, *The Aesthetic Movement: Prelude to Art Nouveau* (London: Elek, 1969); Peter Stansky, *Redesigning the World: William Morris, the 1880s, and the Arts and Craft Movement* (Princeton: Princeton University Press, 1985); also Maria Tippett, *By a Lady: Celebrating Three Centuries of Art by Canadian Women* (Toronto: Viking, 1992).
35 Niagara *Mail*, 31 Oct. 1855.
36 *Farmer's Advocate* 9 (1874): 161–2, and 6 (1871): 161–2.
37 London *Advertiser*, 7 Oct. 1880.
38 Ibid., 7 Oct. 1879.
39 *Gazette des campagnes* 22 (1884–5): 65–70. John A. Macdonald's bill to give spinsters and widows the vote failed not long after.
40 Ontario, *Sessional Papers* 11, 3 (1879).
41 Walkowitz, *City of Dreadful Delight*, 46–7; Jackson Lears, *Fables of Abundance*, 209.
42 *Farmer's Advocate* 20 (1885): 291.
43 *Le Bas Canada*, 1 Sept. 1856.
44 Quoted in Marling, *Blue Ribbon*, 22; *Saturday Night*, 17 Sept. 1892, 1; on the association between performing women and prostitution, see Tracy C. Davis, *Actresses as Working Women: Their Social Identity in Victorian Culture* (London: Routledge, 1991).

45 *Report of the Seventh Convention of the Dominion W.C.T.U.* (Toronto: J.J. Crabbe, 1894), 88; *Report of the Sixth Convention of the Dominion W.C.T.U.* (Toronto: Timms and Co, 1893), 83.

46 Montreal *Gazette* 9 July and 25 Sept. 1891; 23 Sept. 1881.

47 *Canadian Agriculturist* 5 (1852): 295.

48 In England, in 1851, there were half a million more women than men, and one million unmarried women; by 1911, the number of unmarried women rose to one and a half million. Joan Perkin, *Victorian Women* (London: John Murray, 1993), 153. The problem was less acute in Canada, but in Montreal there were 113 marriageable women for every 100 men. With the depression of the 1870s, fewer women married, and in 1881, 20 per cent of women aged forty-one to sixty-one were single: Bettina Bradbury, *Working Families*, 52. In Toronto, in 1851, there were 102.7 women for every 100 men, and, by 1901, the number had risen to 112.5 to 100. Among single adults, there were 120.9 women for every 100 men in 1901: Strange, *Toronto's Girl Problem*, 24.

49 Elizabeth Bloomfield and G.T. Bloomfield, *Canadian Women in Workshops, Mills, and Factories: The Evidence of the 1871 Census* (Guelph: Guelph University, 1991), 35; D. Suzanne Cross, 'The Neglected Majority: The Changing Role of Women in Nineteenth-Century Montreal,' in Susan Mann Trofimenkoff and Alison Prentice, eds., *The Neglected Majority: Essays in Canadian Women's History* (Toronto: McClelland & Stewart, 1977), 66–86.

50 Ontario, *Sessional Papers* 22, 8 (1890).

51 Toronto *Globe*, 23 Sept. 1893; *Saturday Night*, 10 Sept. 1898, 9.

52 Short, 'Canada Exhibited 1851–1867'; Montreal *Gazette*, 27 March 1850.

53 Charlottetown *Islander*, 2 Feb. and 7 March 1851.

54 Halifax *Nova Scotian*, 14 and 28 Oct. 1850, 13 Jan. 1851.

55 A. Holwell, 'Supplementary Report to the Hon. John Rolph, Minister of Agriculture,' *JLAPC* 13 (1854): appendix 2; *Official Catalogue of the New-York Exhibition*; Mrs Peace, *The Convict Ship*.

56 Toronto *Leader*, 12 Sept. 1853.

57 Taché, *Canada at the Universal Exhibition*.

58 *The International Exhibition of 1862, the Illustrated Catalogue* (London: Her Majesty's Commissioners, 1862), 3: 10–15; *Nova Scotia in 1862: Papers Relating to the Two Great Exhibitions in London That Year* (Halifax: T. Chamberlain, 1864), 25–6.

59 *JBAMUC* 5 (1865): 315; 7 (1867): 202–4; *Report of the Nova Scotia Department of the Dublin International Exhibition, 1865*, 16.

60 Mary Frances Cordato, 'Representing the Expansion of Woman's Sphere: Women's Work and Culture at the World's Fairs of 1876, 1893, and 1904' (PhD dissertation, New York University, 1989); Virginia Grant Darney,

'Women and World's Fairs: American International Expositions, 1876–1904' (PhD dissertation, Emory University, 1982).

61 NA, RG 72, vol. 2, 130, 203; vol. 1, 112.

62 ANQ, Fonds LeSage, vol. 7, 6; NA, RG 72, vol. 2, 183; Hodgins, *Special Report to the Honourable the Minister of Education, on the Ontario Educational Exhibit*; Stamp, 'Ontario at Philadelphia'; *New Century for Women* (1876): 100–1.

63 In Cordato, 'Representing the Expansion,' 60.

64 Brampton *Conservator*, 13 Oct. 1876.

65 Saint John *Daily News*, 23 May 1876.

66 *New Century for Women*, 107, 62.

67 Saint John *Daily Sun*, 1 June 1876; Maass, *The Glorious Enterprise*, 124.

68 *New Century for Women*, 26.

69 Catalogue and prize list are in the *Report of the Canadian Commission at the International Exhibition of Philadelphia.*

70 NA, RG 17 A5 II, vol. 2319, 84.

71 Keefer, *Paris Universal Exhibition 1878: Handbook.*

72 Rydell, *All the World's a Fair*; Raymond Corbey, 'Ethnographic Showcases, 1870–1930,' *Cultural Anthropology* 8 (1993): 338–69; Marian Fowler, *Redney: A Life of Sara Jeannette Duncan* (Toronto: Anansi, 1983).

73 NA, RG 72, vol. 37, 26; vol. 40, 704.

74 Ibid., vol. 3, 267; Toronto *Mail*, 28 Nov. 1878.

75 Patrick Kay Bidelman, *Pariahs Stand Up! The Founding of the Liberal Feminist Movement in France, 1858–1889* (Westport: Greenwood Press, 1982), and in *Mil neuf cent: Cahiers Georges Sorel* 7 (1989): Anne Rasmussen, 'Les Congrès internationaux liés aux Expositions universelles de Paris (1867–1900),' 23–44, and Laurence Klejman, 'Les Congrès féministes internationaux,' 71–86.

76 *Report of the Honorary Commissioner (Mr. Adam Brown) Representing Canada at the Jamaica Exhibition*; Booth, 'When Jamaica Welcomed the World.'

77 See Andrew H. Miller, *Novels behind Glass: Commodity Culture and Victorian Narrative* (Cambridge: Cambridge University Press, 1995), ch. 2.

78 LeSage Papers, vol. 2, Fabre to Lesage, 9 June 1885 and 10 June 1886.

79 NA, RG 72, vol. 38, 372.

80 World's Columbian Exposition, Board of Lady Managers, *Addresses and Reports of Mrs. Potter Palmer* (Chicago: Rand McNally, 1894), 119; see Cordato, 'Representing the Expansion'; Darney, 'Women and World's Fairs'; and Jeanne Madeleine Weimann, *The Fair Women: The Story of the Woman's Building, World's Columbian Exposition, Chicago 1893* (Chicago: Academy, 1981).

81 Toronto *Globe*, 11 May 1893; *Empire*, 7 Aug. 1893; *Mail*, 8 Aug. 1893.

82 NA, RG 72, vol. 91, Mary Holden to the Commission, 23 June 1892, and

Hoodless, in the same bundle, 29 June 1892; on Hoodless, see Cheryl MacDonald, *Adelaide Hoodless: Domestic Crusader* (Toronto: Dundurn Press, 1986).

83 NA, RG 72, vol. 106, 380–2; vol. 107, 31–3; vol. 91, Holden, 28 July 1892; vol. 109, 253–8; vol. 108, 788–9. The suggestion came from the WCTU.

84 Ibid., vol. 95, Angus MacKay, 31 Oct. 1892 and 15 March 1893.

85 Ibid., vols. 95–6, various letters from Lillian Phelps; vol. 82, J. Archibald, 3 Jan. 1893; *Woman's Journal*, March 1893, 3; Sept. 1893, 7; Feb. 1899, 7.

86 NA, RG 72, vol. 108, 565; Toronto *Globe*, 12 Aug. 1893; Toronto *Empire*, 29 July 1893.

87 See Joy Parr, *The Gender of Breadwinners: Women, Men, and Change in Two Industrial Towns, 1880–1950* (Toronto: University of Toronto Press, 1990).

88 NA, RG 72, vol. 84, C.S. Clark to J. Larke, 13 June 1893; vol. 92, Eliza Jones, 5 Aug. 1893; also vol. 112, 319, 363–2, 412.

89 Darney, 'Women and World's Fairs,' 748, quoting an 1898 source.

90 *Le coin du feu* 1 (1893): 67–8.

91 Toronto *Globe*, 17 May 1893; Mary Beacock Fryer, *Emily Stowe: Doctor and Suffragist* (Toronto: AMS/Hannah Institute for the History of Medicine, 1990), 110.

92 *World's Congress of Representative Women*, and Toronto *Globe*, 18 May 1893.

93 Toronto *Globe*, 27 May 1893; *Saturday Night*, 8 Jan. 1898, 8. See Veronica Strong-Boag, 'Setting the Stage: National Organization and the Woman's Movement in the Late Nineteenth Century,' in Linda Kealey, ed., *A Not Unreasonable Claim: Women and Reform in Canada, 1880s–1920s* (Toronto: Women's Press, 1979), 97.

94 Toronto *Globe*, 3 June 1893.

95 London *Advertiser*, 26 June 1876; Noel, *Canada Dry*, 100–1; Letitia Youmans, *Campaign Echoes: The Autobiography of Mrs. Letitia Youmans* (Toronto: William Briggs, 1893); Sharon Anne Cook, *'Through Sunshine and Shadow': The Woman's Christian Temperance Union, Evangelicalism, and Reform in Ontario, 1874–1930* (Montreal: McGill-Queen's University Press, 1995). Youmans's autobiography describes her fears of and other women's repugnance for public speaking in 1875, although she had been proud to win first prize for butter at a Picton exhibition a quarter century earlier: *Campaign Echoes*, 78, 105–8.

96 *Addresses and Reports of Mrs Potter Palmer*, 116, 131–40; Weimann, *Fair Women*, 251.

97 Cook and Mitchinson, *The Proper Sphere*, 44–5; May Wright Sewell, ed., *The World's Congress of Representative Women* (Chicago: Rand McNally, 1894), 440; Eagle, *The Congress of Women*, 85.

98 Landes, *Women and the Public Sphere,* 179–80; Kit Coleman in Toronto *Mail,* 15 July 1893; Alice Freeman in *Empire,* 7 Oct. 1893. Coleman attended the exposition, but not the Congress. The mock parliament is described in *Manitoba Free Press,* 10 Feb. 1892.

99 Darney, 'Women,' 121; Cordato supports the same argument.

100 National Woman Suffrage Association, *Report of the International Council of Women* (Washington: Rufus H. Darby, 1888).

101 Eleanor Gordon, *Women and the Labor Movement in Scotland, 1850–1914* (Oxford: Clarendon Press, 1991), 216–7.

102 NA, MG 28 I 25, National Council of Women, vol. 1, 1. See *'We Twa': Reminiscences of Lord and Lady Aberdeen* (London: W. Collins Sons, 1925), 2: 98; Veronica Strong-Boag, *The Parliament of Women: The National Council of Women of Canada, 1893–1929* (Ottawa: National Museum of Man, 1976); and N.E.S. Griffiths, *The Splendid Vision: Centennial History of the National Council of Women* (Ottawa: Carleton University Press, 1993); also the first annual report of the National Council of Women, *Women Workers of Canada* (Ottawa: Thoburn, 1894), and subsequent volumes.

103 Mary Quayle Innis, *Unfold the Years: A History of the Young Women's Christian Association in Canada* (Toronto: McClelland & Stewart, 1949), 38–9; Diana Pederson, 'The Young Woman's Christian Association in Canada, 1870–1920: A Movement to Meet a Spiritual, Civic, and Natural Need,' (PhD dissertation, Carleton University, 1987), 80–2.

104 Toronto *Empire,* 28 Oct. and 4 Nov. 1893.

105 In Backhouse, *Petticoats and Prejudice.*

106 *Structural Transformation of the Public Sphere,* 51–2.

107 Toronto *Empire,* 27 May 1893.

108 NA, MG 28 I 25, vol. 4, meeting of 9 May 1899 and vol. 5, passim.

109 *Women of Canada,* 254 and 30.

110 NA, MG 27 III B 3, Dandurand Papers, vol. 8A, Josephine Dandurand's diary, entry for 21 Oct. 1898. I am grateful to Madame Bonar for permission to cite this source.

111 Nadia Fahmy-Eid, 'La presse féminine au Québec (1890–1920): Une pratique culturelle et politique ambivalente,' in Yolande Cohen, ed., *Femmes et politique* (Montreal: Le Jour, 1981), 101–15.

112 See note 111; also Yolande Pinard, 'Les débuts du mouvement des femmes à Montréal, 1893–1902,' in Marie Lavigne and Yolande Pinard, eds., *Travailleuses et féministes: Les femmes dans la société québécoise* (Montreal: Boréal Express, 1983), 177–98.

113 *Deuxième congrès international des oeuvres et institutions féminines* (Paris: Charles Blot, 1902), 2: 595–618.

114 Quebec *La Patrie*, 10 July 1900.
115 *Deuxième congrès*, 4: 349.
116 Raymond Williams, *Culture and Society, 1780–1950* (London: Chatto and Windus, 1958).
117 See the discussion of Hegelian allegory, according to which 'nature, history, and art ... move dialectically and harmoniously through better and better toward a goal which will be their absolute spiritualization, their vanishing or disembodiment in the fulfilment of a total meaning.' J. Hillis Miller, 'The Two Allegories,' in Morton W. Bloomfield, ed., *Allegory, Myth, and Symbol* (Harvard: Harvard University Press, 1981), 355–70.

Chapter 10: Making a Spectacle: Exhibitions of the First Nations

1 Bahktin, *Rabelais and His World*; DeBord, *Society of the Spectacle.*
2 'What Knowledge Is of Most Worth,' *Westminster and Foreign Quarterly Review* 72 (1859): 1–4; *Islander* 18 Jan. 1861.
3 Donald B. Smith, *Sacred Feathers: The Reverend Peter Jones (Kahkewaquonaby) and the Mississauga Indians* (Toronto: University of Toronto Press, 1987), 201; J.C.H. King, 'A Century of Indian Shows: Canadian and United States Exhibitions in London, 1825–1925,' *European Journal of Native American Studies* 5 (1991): 35–42; see the collection edited by Christian F. Feest, *Indians and Europe: An Interdisciplinary Collection of Essays* (Aachen: Herodot, 1987), especially Dale Idiens, 'Eskimos in Scotland, c. 1682–1924,' 161–74; J.C.H. King, 'Family of Botocudos Exhibited on Bond Street, in 1822,' 243–52; Christopher Mulvey, 'Among the Sag-a-noshes: Ojibwa and Iowa Indians with George Catlin in Europe, 1843–1848,' 253–76. See also the discussion on otherness as spectacle by Edward Ball, 'Constructing Ethnicity,' in Lynne Cooke and Peter Wollen, eds., *Visual Display: Culture beyond Appearances* (Seattle: Bay Press, 1995), 142–53.
4 Olive Patricia Dickason, *Canada's First Nations: A History of Founding Peoples from Earliest Times* (Toronto: McClelland & Stewart, 1992), 232. On civilizing by agriculture, see David Rich Lewis, *Neither Wolf nor Dog: American Indians, Environment, and Agrarian Change* (New York: Oxford University Press, 1994).
5 J. Douglas Leighton, 'The Development of Federal Indian Policy in Canada, 1840–1890' (PhD dissertation, University of Western Ontario, 1975), 237. See also J. Donald Wilson, '"No Blanket to Be Worn in School": The Education of Indians in Nineteenth-Century Ontario,' in Jean Barman, Yvonne Hébert, and Don McCaskill, eds., *Indian Education in Canada: The Legacy* (Vancouver: UBC Press, 1986), 64–87.
6 J.R. Miller, *Skyscrapers Hide the Heavens: A History of Indian–White Relations in*

Canada, rev. ed. (Toronto: University of Toronto Press, 1991), 100–2; Edward S. Rogers, 'The Algonquian Farmers of Southern Ontario, 1830–1945,' in Edward S. Rogers and Donald B. Smith, eds., *Aboriginal Ontario: Historical Perspectives on the First Nations* (Toronto: Dundurn, 1994), 122–66; see also the survey in J.R. Miller, *Shingwauk's Vision: A History of Native Residential Schools* (Toronto: University of Toronto Press, 1996); Hodgins, *Historical and Other Papers*, 3, 102; Dickason, *Canada's First Nations*, 236; David A. Nock, *A Victorian Missionary and Canadian Indian Policy: Cultural Synthesis vs. Cultural Replacement* (Waterloo: Wilfred Laurier University Press, 1988).

7 NA, RG5 A1, Upper Canada Sundries, 58018–30; Toronto *Mail*, 12 July 1878.

8 Graham Burchell, 'Liberal Government and the Techniques of the Self,' in Barry et al., *Foucault and Political Reason*, 19–36.

9 John L. Tobias, 'Protection, Civilization, Assimilation: An Outline History of Canada's Indian Policy,' in R. Douglas Francis and Howard Palmer, eds., *The Prairie West: Historical Readings*, rev. ed. (Edmonton: Pica Pica Press, 1992), 211.

10 Leighton, 'Development of Federal Indian Policy,' 525.

11 Sarah Carter, *Lost Harvests: Prairie Indian Reserve Farmers and Government Policy* (Montreal: McGill-Queen's University Press, 1990), 175–6.

12 Canada, *Sessional Papers* 20, 6 (1887): lx and 92–3.

13 Ibid. 6, 22 (1873): 31; on the potlatch, see Douglas Cole and Ira Chaikin, *An Iron Hand upon the People: The Law against the Potlatch on the Northwest Coast* (Vancouver: Douglas & McIntyre, 1990).

14 Thomas M. Charland, *Histoire des Abénakis d'Odonak (1875–1937)* (Montreal: Lévier, 1964), 331–4; Gaby Pelletier, *Abenaki Basketry* (Ottawa: National Museum of Man, 1982), 5–6; Canada, *Sessional Papers* 20, 6 (1887): xxvii; L.F.S. Upton, *Micmacs and Colonists: Indian–White Relations in the Maritimes, 1713–1867* (Vancouver: UBC Press, 1979), 128–9; Joseph Doutre, 'Les sauvages du Canada en 1852,' in J.L Fontaine, éd., *Institut Canadien en 1855* (Montreal: Senecal & Daniel, 1855), 216. Ruth Phillips has studied these artifacts extensively: See her 'Why Not Tourist Art: Significant Silences in Native American Art Museum Representations,' in Gyan Prakash, ed., *After Colonialism* (Princeton: Princeton University Press, 1995), 98–125; and *Trading Identities: Souvenir Arts in Northeastern North America, 1700–1900* (Washington: University of Washington Press, 1997).

15 Canada, *Sessional Papers* 14, 14 (1881): 45; ibid., 15, 6 (1882): xlvii.

16 Ibid., 13, 4 (1880): 5, and 17, 4 (1884): xii.

17 Sally M. Weaver, 'The Iroquois: The Grand River Reserve in the Late Nineteenth and Early Twentieth Centuries, 1875–1945,' in Rogers and Smith, *Aboriginal Ontario*, 213–57; Brantford *Expositor*, 30 Oct. 1885.

18 Brantford *Expositor,* 4 Nov. 1870; Smith, *Sacred Feathers,* 244.
19 Canada, *Sessional Papers* 17, 4 (1884): 8.
20 NA, RG 10, Department of Indian Affairs, vol. 2071, 10626; *Farmer's Advocate* 20 (1885): 355.
21 Bothwell *Times,* 8 Nov. 1883.
22 NA, RG 10, vol. 7555, 1007–1.
23 Ibid., vol. 7556, 1029–1. The last example is from 1909.
24 Ibid., vol. 7555, 1025–2.
25 Ibid., vol. 7556, 1007–2, J.J. Wilson, 12 Feb. 1894.
26 Ibid., 1032, has the exhibition correspondence for the Six Nations, 1880–3.
27 Ibid., 1032-A, covers the years 1886–1917; see vol. 2534, 110639, on Dingman.
28 Alan D. McMillan, *Native Peoples and Cultures of Canada,* rev. ed. (Vancouver: Douglas & McIntyre, 1995), 91–2; NA, RG 10, vol. 7556, 1032-B, 1032-C.
29 NA, RG 10, vol. 7556, 1019–4.
30 Ibid., vol. 2234, 45307, for the years 1883–9.
31 Ibid., vol. 7555, 1005.
32 Annie E. Coombes, *Reinventing Africa: Museums, Material Culture, and Popular Imagination in Late Victorian and Edwardian England* (New Haven: Yale University Press, 1994), ch. 8.
33 St Boniface *Le Manitoba,* 19 Oct., 2 Nov., and 27 July 1892.
34 See *The Work of a Few Years among the Indians of Manitoba and the North-West Territories* (Chicago: Rupert's Land Indian Industrial School, 1893), 15.
35 Canada, *Sessional Papers* 18, 3 (1885): 101.
36 Bennett, *Birth of the Museum,* 68.
37 Canada, *Sessional Papers* 17, 4 (1884): 20.
38 Brantford *Expositor,* 4 Oct. 1883.
39 *Manitoba Free Press,* 15 Sept. 1886.
40 Report from the Peigan agency, in Canada, *Sessional Papers* 24, 18 (1891): 167.
41 Canada, *Sessional Papers* 25, 14 (1892): 94.
42 Ibid., 25, 14 (1892): xx.
43 T.J. Jackson Lears, 'The Concept of Cultural Hegemony: Problems and Possibilities,' *American Historical Review* 90 (1985): 567–93; David Harris, *From Class Struggle to the Politics of Pleasure: The Effects of Gramscianism on Cultural Studies* (London: Routledge, 1992), 154.
44 *Minerve,* 10 Nov. 1831; Smith, *Sacred Feathers,* 216, erroneously remarks that he won £15.
45 *Canadian Agriculturist* 11 (1859): 279.
46 *Nova Scotian,* 3 July 1854.

47 London *Advertiser*, 2 Oct. 1878; Toronto *Leader*, 6 Oct. 1864.

48 Thomas Vennum Jr, *American Indian Lacrosse: Little Brother of War* (Washington: Smithsonian Press, 1994), 261; *Gazette*, 28 Aug. 1860.

49 Montreal *Gazette*, 22 June 1876; NA, MG 29 C17, Christopher Massiah Papers; Vennum, *American Indian Lacrosse*, 265.

50 See John F. Sears, 'Beirstad, Buffalo Bill, and the Wild West in Europe,' in R. Kroes, R.W. Rydell, and B.F.J. Bossher, eds., *Cultural Transmissions and Receptions: American Mass Culture in Europe* (Amsterdam: Vu University Press, 1993), 15–23.

51 *Canadian Live-Stock and Farm Journal* 5 (1888): 233.

52 Francis, *Imaginary Indian*; Keith Regular, 'On Public Display,' *Alberta History* 34 (1986): 1–12; NA, RG 10, vol. 4010, 253430.

53 *Art Journal*, ns 3 (1851): 181; *A Few Words upon Canada*, 2.

54 *Extraits du rapport sur l'Exposition de Paris relativement aux produits du Canada* (Toronto: S. Derbishire and G. Desbarats, 1857), 25.

55 Taché, *Canada at the Universal Exhibition of 1855*, 199–200, 250.

56 *The Crystal Palace and Its Contents* (London: W.M. Clark, 1852), 42.

57 Day, *Address Delivered at the Provincial Industrial Exhibition.*

58 Whewell, 'The General Bearing of the Great Exhibition,' in Society of Arts, *Lectures on the Results of the Great Exhibition of 1851*, 3–34.

59 Upton, *Micmacs and Colonists*, 140.

60 *The International Exhibition of 1862: The Illustrated Catalogue of the Industrial Department* (London: Her Majesty's Commissioners, 1862); Toronto *Globe*, 6 June 1862.

61 Victoria *British Colonist*, 14 and 19 Feb. 1862; Jean Barman, *The West beyond the West: A History of British Columbia* (Toronto: University of Toronto Press, 1993), 153.

62 ANQM, Board of Arts and Manufactures of Lower Canada, 4 Jan. 1866. Two collections of native works were entered: *Catalogue of the Canadian Contributions to the Dublin Exhibition* (1865) (CIHM 47185).

63 *Report of the Nova Scotia Department of the Dublin International Exhibition, 1865*, 16.

64 *Nova Scotian*, 19 Feb. 1862.

65 ANQM, Board of Arts Minutes, 1: 201.

66 E.J. Devine, *Historic Caughnawaga* (Montreal: Messenger 1922), 404.

67 Quoted in the *Newfoundlander*, 14 May 1867.

68 NA, RG 17, files 10388, 10363, 10575.

69 In Joan M. Schwartz, 'The Past in Focus: Photography and British Columbia, 1858–1914,' *BC Studies* 52 (winter 1981–2): 9.

70 S.P. May, 'Report of the Ontario Educational Exhibit at Philadelphia,' Ontario, *Sessional Papers* 9, 33 (1877): 185.

71 *Handbook and Official Catalogue of the Canadian Section*, 53.

72 NA, RG 72, vol. 39, 147; vol. 38, 372, 289, 352; vol. 37, 90.

73 Ibid., vol. 42, 1215.

74 *Winnipeg Free Press*, 28 April 1886; *British Colonist*, 13 Dec. 1885.

75 Ian Ross Robertson, 'The 1850s: Maturity and Reform,' in Philip A. Buckner and John G. Reid, eds., *The Atlantic Region to Confederation: A History* (Toronto and Fredericton: University of Toronto Press/Acadiensis Press, 1994), 333–59.

76 NA, RG 72, vol. 67, 30.

77 Toronto *Mail*, quoted in the Saint John *Daily Sun*, 29 May 1886.

78 Saint John *Daily Sun*, 13 April, 9 June, 15 July, 22 Oct. 1886.

79 St John's *Times and General Commercial Gazette*, 29 April 1893; Victoria *British Colonist*, 22 April 1893.

80 Victoria *British Colonist*, 20 April 1893.

81 Cole, *Captured Heritage*, 129.

82 *British Colonist*, 25 Aug. and 10 Oct. 1893.

83 NA, RG 10, vol. 3865, 85529, Reed, 28 Dec. 1891; T.J. Morgan to Reed, 12 Sept. 1892.

84 Ibid., A. Vowell, 19 Oct. 1892, and A. Cameron, 28 Dec. 1892.

85 *Canadian Indian*, 15 Sept. 1893.

86 *The Work of a Few Years among the Indians*, 14.

87 Toronto *Empire*, 7 Oct. 1893.

88 See its attempts to secure some mention for the school: NA, RG 72, vol. 99, Duncan Campbell Scott, 9 March 1894; also vol. 103, 341–2, 349–51, 369.

89 NA, MG 29 E106, Hayter Reed Papers, vol. 20, 5 Aug. 1892, 27 Feb., 16 June, and 9 Nov. 1893; and see Miller, *Shingwauk's Vision*.

90 NA, MG 29 E106, vol. 20, 1722, R.H. McColl, 9 Dec. 1892.

91 Toronto *Globe*, 26 Sept. 1893.

92 Robert A. Trennert, 'Selling Indian Education at World's Fairs and Expositions, 1893–1904,' *American Indian Quarterly* 11 (1987): 203–20.

93 The report on their population was published in 1877 by the representative of the Department of the Interior to the Philadelphia Exhibition: John Eaton, *Are the Indians Dying Out? Preliminary Observations Relating to Indian Civilization and Education* (Washington, November 1877).

94 *British Columbian*, 1 June 1865.

95 Ibid., 25 April and 13 May 1865.

96 *British Colonist*, 18 April 1861.

97 *Winnipeg Free Press*, 11 Nov. 1886.

98 *British Colonist*, 10 May 1885.

99 Saint John *Daily News*, 28 Sept. 1876.

100 Cole, *Captured Heritage.* See also the discussion in Deborah Root, *Cannibal Culture: Art, Appropriation, and the Commodification of Difference* (Boulder: Westview, 1996), 80–1 and passim.

101 Walter Benjamin, *Charles Baudelaire: A Lyric Poet in the Era of High Capitalism* (London: NCB, 1973).

Conclusion

1 Anne Rasmussen, 'Les classifications d'exposition universelle,' in Schroeder-Gudehus and Rasmussen, *Les Fastes du progrès,* 21–39.

2 *Manitoba Free Press,* 21 Oct. 1885; see Audrey Short, 'Canada Exhibited.'

3 Benedict Anderson, *Imagined Communities: Reflections on the Origin and Spread of Nationalism* (London: Verso, 1991).

4 Andrew Wernick, *Promotional Culture: Advertising, Ideology, and Symbolic Expression* (London: Sage, 1991).

5 See the general discussion in Nancy F. Partner, 'Historicity in an Age of Reality-Fictions,' in Frank Ankersmit and Hans Kellner, *A New Philosophy of History* (Chicago: University of Chicago Press, 1995), 21–39.

Note on Sources

This study is based on a variety of published and unpublished documents. The main sources for the chapters on agricultural exhibitions before Confederation are the records of the Provincial Secretary of Upper and Lower Canada and Canada East and West (RG 4 and RG 5) and of the Bureau, later the Department, of Agriculture (RG 17), all of which are stored at the National Archives in Ottawa (NA). I also relied heavily on the published reports of the agricultural societies which appeared among the journals and sessional papers of the Legislative Assembly. After Confederation, the annual reports of agricultural societies were published as sessional papers in Ontario and Quebec; in the latter part of the century, those of Ontario are filed among the unpublished sessional papers at the Archives of Ontario (AO). The administrative records of the Quebec Department of Agriculture are stored at the Archives Nationales du Québec (ANQ) in the papers of the Departments of Agriculture (E9) and Public Works (E25). In Ontario, the records of the Department of Agriculture during this period have not survived. I have drawn extensively on the various monthly agricultural journals published in the two provinces from mid-century on as well as the daily press. Records of individual agricultural societies consulted were the Glengarry Agricultural Society (McGillivray Papers, NA, MG 24 I13), Montmagny Agricultural Society (Philippe Landry Papers, NA, MG 27 IIE4), Missisquoi Agricultural Society (NA, MG 28 I277), Ottawa Agricultural Society (NA, MG 28 I27), Quebec Agricultural Society (NA, MG 24 I37), Darlington Agricultural Society (AO, MU2092), Addington County Agricultural Society (AO MU2086), and Toronto Electoral District Agricultural Society (AO, MU3020). The records of the Quebec Exposition Company were also used (Joseph Landry Papers, NA, MG 30 I17, and holdings at the Archives de la Ville de Québec).

The major source for the chapters on the international exhibitions is the records of the Canadian Exhibition Commission (NA, RG 72) supplemented with the correspondence of the Department of Agriculture (RG 17). Documentation is scanty before Confederation and I have relied heavily on journalistic accounts and published reports for that period as well as the papers of private individuals connected with the exhibits. For the period after Confederation, copious documentation survives, and, again, it has been supplemented with published reports and newspaper accounts.

Chapter 9 is based on the sources used for the earlier chapters, publications relating to the women's congresses, and the papers of the National Council of Women (NA, MG 28 I25). Chapter 10 is based on the manuscript records of the Department of Indian Affairs (NA, RG 10) and its published reports, as well as the Hayter Reed Papers (NA, MG 29 E106) and the press.

Other important manuscript collections consulted include the papers of the Toronto Mechanics' Institute (held at the Metropolitan Toronto Reference Library) and the Montreal Mechanics' Institute (held at Atwater Library in Montreal), the papers of the boards of arts and Manufactures (held at the Archives of Ontario and the Archives Nationales du Québec at Montreal (ANQM). Private collections with substantial relevant records include those of Sir William Logan (at McGill University Archives and the Metropolitan Toronto Reference Library), John Neilson (ANQ P192 and NA, MG 24 B1), Siméon LeSage, Joly de Lotbinè and the Taché Family (all at the ANQ), Brown Chamberlin (NA, MG 24 B19), Sir Charles Tupper (NA, MG 26F) and Sir John A. Macdonald (NA, MG 26 D13).

Illustration Credits

Archives de la Ville de Québec
Quebec Provincial Exhibition.

Bancroft, H.H. ***The Book of the Fair.*** **Chicago: Bancroft, 1893**
Inuit encampment; Kwagiulth encampment.

The British Library
Prize fire-engine, *Illustrated London News*; Colonial Annex, *Illustrated London News*; Canadian agricultural trophy, *Illustrated London News.*

Donald Fyson and E.A. Heaman
Maps: prize money (information on county borders provided by the Ontario Genealogical Society); farmers residing in Quebec (information on county borders provided in Pierre Drouilly, *Atlas des élections au Québec, 1867–1985* [Québec: Bibliothéque de l'Assemblé nationale, 1989]).

Hodgkins, J.G. ***Historical and Other Documents and Papers Illustrative of the Education System of Ontario.*** **Toronto: L.K. Cameron, 1911**
Ontario Education Exhibit.

Metropolitan Toronto Reference Library
Clarence Kirklevington, *Farmer's Advocate,* February 1883; 'Modern Agricultural Exhibition,' *Farmer's Advocate,* October 1885; Oneida exhibition, *Farmer's Advocate,* December 1885; Canadian Court, T14465.

National Archives of Canada
1855 Paris Exposition, C4230; New Brunswick Court, PA 126886.

Seton Memorial Library, Philmont Scout Ranch
'Triumph of the Wolves.'

Index